State-of-the-Art Researches in New Technologies

State-of-the-Art Researches in New Technologies

Edited by **Ed Diego**

NY RESEARCH PRESS

New York

Published by NY Research Press,
23 West, 55th Street, Suite 816,
New York, NY 10019, USA
www.nyresearchpress.com

State-of-the-Art Researches in New Technologies
Edited by Ed Diego

International Standard Book Number: 978-1-63238-427-0 (Hardback)

Printed in China.

Contents

Preface

This book serves as leverage for those researching in the field of new technologies. It presents a compilation of researches in contemporary fields of technology, serving as a significant tool for experts, students and professionals. The book concentrates on numerous aspects of designing and manufacturing, examining composite technological products, and some aspects of the improvement and use of manufacturing and service automation. It also discusses some important subjects like production, development, tools, video methods, textiles, multimedia security, etc. This compilation of data offers information dealing with various facets of technology.

After months of intensive research and writing, this book is the end result of all who devoted their time and efforts in the initiation and progress of this book. It will surely be a source of reference in enhancing the required knowledge of the new developments in the area. During the course of developing this book, certain measures such as accuracy, authenticity and research focused analytical studies were given preference in order to produce a comprehensive book in the area of study.

This book would not have been possible without the efforts of the authors and the publisher. I extend my sincere thanks to them. Secondly, I express my gratitude to my family and well-wishers. And most importantly, I thank my students for constantly expressing their willingness and curiosity in enhancing their knowledge in the field, which encourages me to take up further research projects for the advancement of the area.

Editor

Part 1

Manufacturing Technology

1

CMOS and BiCMOS Regenerative Logic Circuits

Branko L. Dokic
University of Banja Luka, Faculty of Electrical Engineering
Bosnia and Herzegovina

1. Introduction

Schmitt triggers with standard CMOS logic circuits are described, first. Mathematical models for calculating basic parameters and their limits are presented. Most of the chapter is dedicated to different solutions for CMOS and BiCMOS Schmitt logic circuits in monolithic integrated circuits. Two types of inverters with entirely different topologies are described. Also, solutions for Schmitt triggers with voltage-controlled thresholds are described. Beside inverters, NAND and NOR Schmitt logic circuits are analyzed. Basic circuit is inverted Schmitt trigger with three pairs of CMOS transistors. Expansion of the number of inputs is reached in a similar way as in standard CMOS and BiCMOS logic circuits. It is shown that voltage transfer characteristics depend, beside voltage supply and parameters of transistors, on the number of logical circuits' inputs. NAND and NOR Schmitt circuits, in which voltage hysteresis in transfer characteristic is generated only through one input, are also described. Analytic models and SPICE simulations are used for analysis of static and dynamic parameters and conditions for work stability and reliability. Areas of reliability, influence of technology and electrical parameters of transistors and their limits are analyzed.
Concerning the field of application, in literature there are different solutions of Schmitt triggers (Zou et al, 2008, Al-Sarrawi, 2008, Katyal et al, 2008, Lo et al, 2010). In this chapter, solutions with fundamental applications in digital integrated circuits - Schmitt logic circuits are described. The author published most of these solutions (Dokic, 1983, Dokic 1984, Dokic 1996, Dokic, 1988). Today, some of them (Dokic, 1984) are treated as conventional.
The term regenerative is used because every change of state is followed by a regenerative process - positive feedback. Owning to that, transfer characteristic has shape of a hysteresis, like in Schmitt trigger. That is why the term Schmitt logic circuits is most commonly used. Unlike conventional logic circuits, where the output level is uniformly determined for the input voltage value, for Schmitt logic circuits, in certain extent, it is not uniformly determined. In fact, due to hysteresis, in the area of the input voltages between two logic thresholds, logic state at the output depends, beside the input voltage value, also on the previous state. Due to that Schmitt circuits can be used as filters for low frequency interferences. An example of this kind of application is given in Fig.1.
Whenever the value of the input signal passes the value of the threshold voltage V_T of the standard logic circuit, a change of the logic state at the output appears. Therefore, the changes of the input voltage created by noise are transferred to the output as glitches. The change of the logic state at the output of the Schmitt logic circuit can appear only after the noise amplitude of which is greater than the voltage hysteresis.

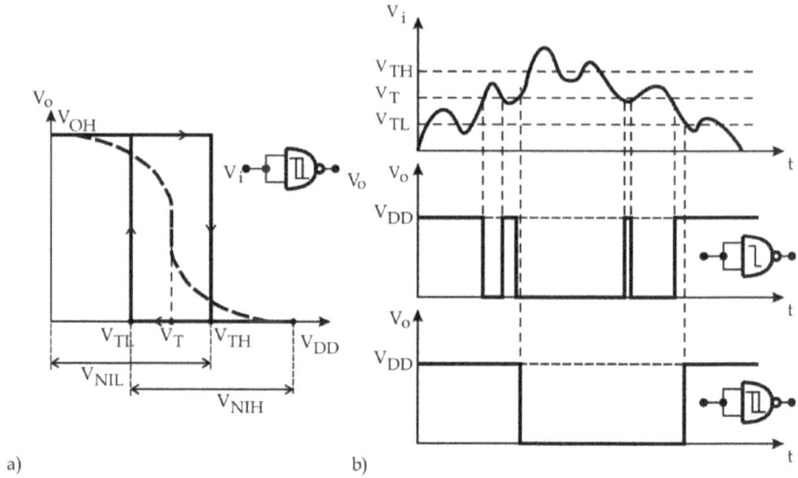

Fig. 1. Transfer characteristic of Schmitt logic circuit (a) and outputs of standard and Schmitt circuit to an input with noise addition (b).

Schmitt trigger is able to hold it's logic state for all changes of the input voltage which are $V_{TL} < V_i < V_{TH}$, where V_{TH} and V_{TL} are the high and the low threshold of the Schmitt trigger. Fig.1 shows the ability of the Schmitt trigger to filter the noise, which, in this case, do not influence the output of the circuit. At the same time at the output of the standard circuit there are two pulse glitches which create system errors.

The hysteresis within the transfer characteristic causes increase of the static noise immunity (Fig.1a). Thus:

$$V_{NIL} = V_{TH}, \tag{1}$$

$$V_{NIH} = V_{DD} - V_{TL} \tag{2}$$

Comparing Schmitt triggers to the standard circuits, the increase of the static noise immunity appears if the threshold voltage V_T of the standard gate lies between the thresholds of the Schmitt circuit, i.e.:

$$V_{TL} < V_T < V_{TH} \tag{3}$$

The transfer characteristic, concerning the noise immunity, is optimum if the thresholds V_{TH} and V_{TL} are symmetric around $V_{DD}/2$. The larger the value of the voltage hysteresis becomes, the noise immunity increases further. The transfer characteristic of the CMOS Schmitt gates is almost perfectly symmetric around $V_{DD}/2$.

There is another advantage of Schmitt circuits compared to the standard ones. Because of the positive feedback, the transfer characteristic is ideal, which means that values of noise margin and noise immunity are equal. Schmitt triggers are used to shape pulses or convert signals that change slowly into pulse signals with short rise and fall times, which is necessary where synchronizing circuits are used. These are the reasons to use Schmitt triggers so often, both as an independent integrated circuit and as a part of a MSI or VLSI circuit. In the second case, Schmitt trigger is almost always used as an input circuit. Also, integrated circuits with mixed signals contain Schmitt triggers (Young, 2010, Chien, 2011, Li, 2009, Hard & Voinigesku, 2009, Wang et al, 2008, Arrabi 2011).

2. Schmitt trigger with logic circuits

The basic Schmitt trigger consists of two CMOS inverters and two resistors (Fig.2).

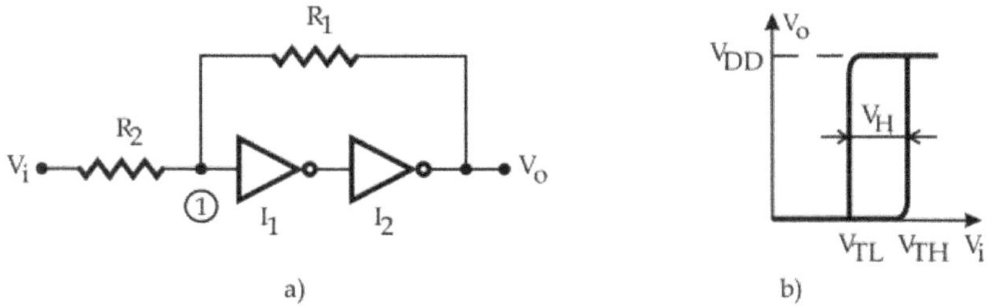

Fig. 2. Schmitt trigger (a) and it's transfer characteristic (b).

The high threshold V_{TH} and the low threshold V_{TL} are defined by the supply voltage V_{DD}, ratio of the resistors R_1/R_2 and the threshold voltage V_T of the inverter I_1. Namely,

$$V_{TH} = V_T(1 + R_1 / R_2) \tag{4}$$

$$V_{TL} = V_T(1 + R_1 / R_2) - V_{DD}R_1 / R_2 \tag{5}$$

and voltage hysteresis is given by:

$$V_H = V_{TH} - V_{TL} = V_{DD}R_1 / R_2 \tag{6}$$

where V_T is given by:

$$V_T = \frac{V_{DD} + V_{tp} + V_{tn}\sqrt{k_n / k_p}}{1 + \sqrt{k_n / k_p}} \tag{7}$$

V_{tn} and V_{tp} are threshold voltages of nMOS and pMOS transistors, respectively, and the constants of transistors are given by:

$$k_n = \frac{\mu_n \varepsilon_{ox}}{2t_{ox}} W_n / L_n, \; k_p = \frac{\mu_p \varepsilon_{ox}}{2t_{ox}} W_p / L_p \tag{8}$$

where μ_n and μ_p are the mobility of the electrons and the holes, ε_{ox} is oxide dielectric constants, t_{ox} the oxide thickness and W and L are width and length of the transistor's channel.

The fact that basic parameters depend on the ratio of the resistance R_1/R_2 yields a wide choice of their absolute values. These values range from several tens of $k\Omega$ ($R_1 + R_2 \gg R_O$, where R_O represents output resistance of the inverter I_2), to several hundreds of $M\Omega$. On the other hand, their ratio can vary within a broad range. Since it is always true: $V_H < V_{DD}$, then $R_1 < R_2$. The other limitations do not exist, it is possible for the ratio to be $R_1/R_2 \ll 1$. This means that the voltage hysteresis can be regulated in a very wide range, from several tens of mV to, approximately, V_{DD}.

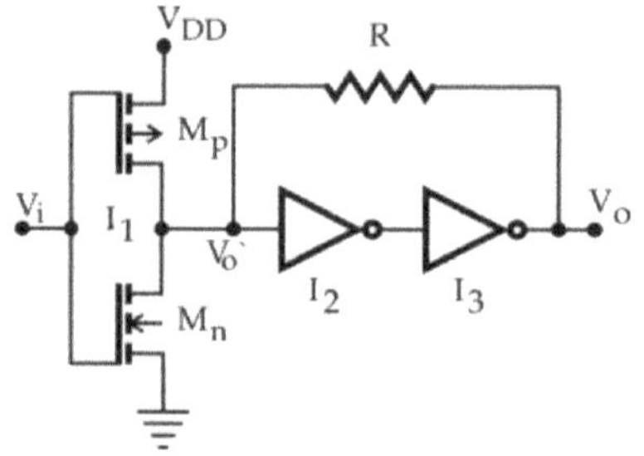

Fig. 3. Schmitt trigger with one resistor (Dokic, 1983).

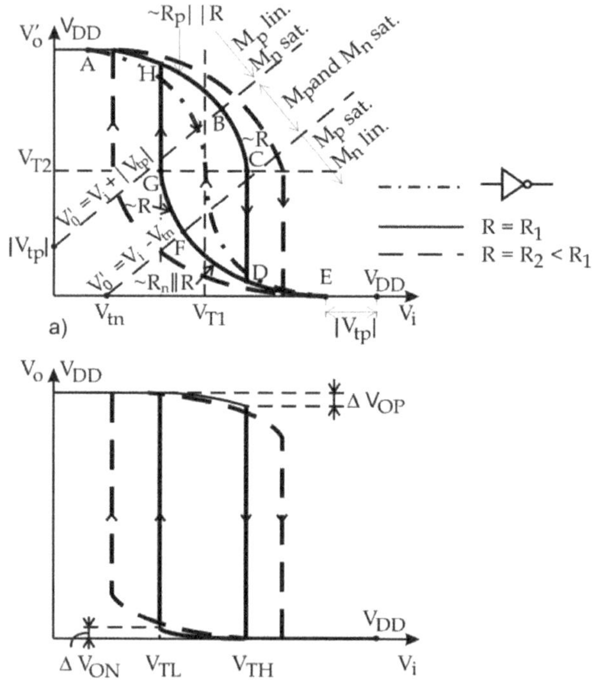

Fig. 4. Characteristics $V_o'(V_i)$ and $V_o(V_i)$ for two values of R.

Another kind of Schmitt trigger which uses discrete CMOS logic circuits (Fig.3) contains three inverters and only one resistor (Dokic, 1983). The advantage of this circuit comparing to the previous one is that it has a CMOS input (very high input resistance). The disadvantage is that the range available for tuning voltage hysteresis is narrower. The resistor R decreases the amplification of the input inverter in the transitional area and moves the characteristic $V_o'(V_i)$ to the right if V_i increases, or to the left if V_i decreases (Fig.4a).

Because of this process, the transfer characteristic is the shape of the hysteresis. In Fig.4a transfer characteristic $V_o' = f(V_i)$ is shown. With line-dot-line, in this same Fig., the transfer characteristic of the inverter I_1 without the resistor R (in other words: $R \to \infty$) is presented. Let V_i increase from 0 to V_{DD}. Before change of the logic state ($V_i < V_{TH}$) the output of the circuit is $V_o = V_{DD}$, thus R and M_p are connected in parallel (Fig.5a). The slope of the function $V_o' = f(V_i)$ in AB area is determined by the resistors connected in parallel, $R_p||R$, where R_p represents resistance of the drain-source of M_p within the linear area of the characteristic. Between the points B and C, where both M_p and M_n are saturated, slope depends only on R. The change of the logic state appears when:

$$V_o' = V_{T2} \tag{9}$$

where V_{T2} is threshold voltage of the inverter I_2. At this moment the change (fall) of the voltage V_o appears, which is, through the resistor R, transferred to V_o', thus decreasing V_o' even more. In this way positive feedback loop is created which leads to step change of V_o' (area between the points C and D), and thusly of V_o.
In the beginning of the process of which the result is change of V_i from V_{DD} to 0, R and the transistor M_n are connected in parallel (Fig.5b). The change of the logic state also appears if the condition (9) is fulfilled.

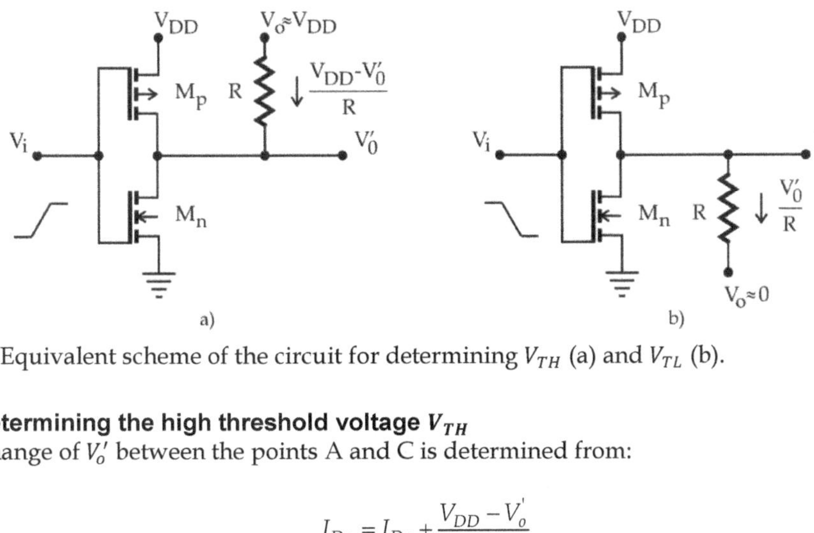

Fig. 5. Equivalent scheme of the circuit for determining V_{TH} (a) and V_{TL} (b).

2.1 Determining the high threshold voltage V_{TH}

The change of V_o' between the points A and C is determined from:

$$I_{Dn} = I_{Dp} + \frac{V_{DD} - V_o'}{R} \tag{10}$$

where I_{Dn} and I_{Dp} represent the drain currents of the transistors M_n and M_p, respectively. During these calculations the working areas of the transistors need to be noted - these are clearly shown in Fig.4. Of special interest is the area BC in which both M_n and M_p are saturated, because that is where the logic state is changed in the circuit. In this area:

$$k_n\left(V_i - V_{tn}\right)^2 = k_p\left(V_{DD} + V_{tp} - V_i\right)^2 + \frac{V_{DD} - V_o'}{R} \tag{11}$$

which leads to:

$$V_o' = V_{DD} - k_p R\left[\frac{k_n}{k_p}(V_i - V_{tn})^2 - \left(V_{DD} + V_{tp} - V_i\right)^2\right] \tag{12}$$

Practically, inverters are symmetric, or almost symmetric, circuits. That is why we, in further text, consider such a case: $k_n = k_p$ and $V_{tn} = |V_{tp}|$ of all transistors. Then (12) can be written as:

$$V_o' = V_{DD}\left[1 + k_p R\left(V_{DD} + 2V_{tp}\right)\right] - 2k_p R\left(V_{DD} + V_{tp} - V_{tn}\right)V_i \tag{13}$$

and the condition of change of the output voltage becomes:

$$V_o' = V_{DD} / 2 \tag{14}$$

When $V_o' = V_{DD}/2$, then $V_i = V_{TH}$, so, taking into account (13) and (14), the high voltage of the Schmitt trigger is given by:

$$V_{TH} = \frac{V_{DD}}{2} + \frac{V_{DD}}{4k_p R\left(V_{DD} + V_{tp} - V_{tn}\right)} \tag{15}$$

2.2 Determining the low threshold voltage V_{TL}

Equivalent circuit used to determine V_{TL} is shown in Fig.5b. Then:

$$I_{Dp} = I_{Dn} + V_o' / R \tag{16}$$

Between points F and G both transistors are saturated, so:

$$k_p\left(V_{DD} + V_{tp} - V_i\right)^2 = k_n\left(V_i - V_{tn}\right)^2 + V_o' / R \tag{17}$$

Presuming that the transistors are symmetric, (17) leads to:

$$V_o' = k_n R\left(V_{DD} + V_{tp} - V_i\right)\left(V_i - V_{tn}\right) \tag{18}$$

Combining (14) and (18) and replacing $V_i = V_{TL}$, low threshold voltage is given by:

$$V_{TL} = \frac{V_{DD}}{2} - \frac{V_{DD}}{4k_n\left(V_{DD} + V_{tp} - V_{tn}\right)R} \tag{19}$$

Voltage hysteresis is:

$$V_H = V_{TH} - V_{TL} = \frac{V_{DD}}{2k_p\left(V_{DD} + V_{tp} - V_{tn}\right)R} \tag{20}$$

From the previous analysis, the following conclusions are derived:

- the thresholds are symmetric relative to $V_{DD}/2$;
- V_{TH}, V_{TL} and V_H are inversely proportional to R.

The dependency of the thresholds V_{TH} and V_{TL} on R and V_{DD} is shown in Fig.6.

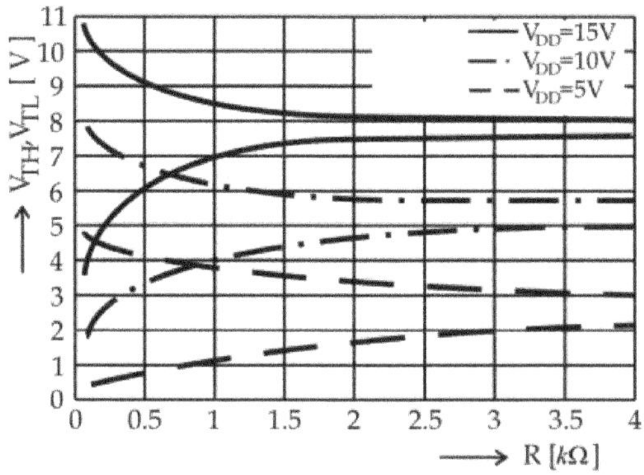

Fig. 6. The high and the low threshold voltages as functions of R and V_{DD}.

The resistance R should be within the range from several hundreds of Ω up to several $k\Omega$. Sensivity of the threshold change decreases if R increases. If $R > 3k\Omega$, this sensitivity is very low.

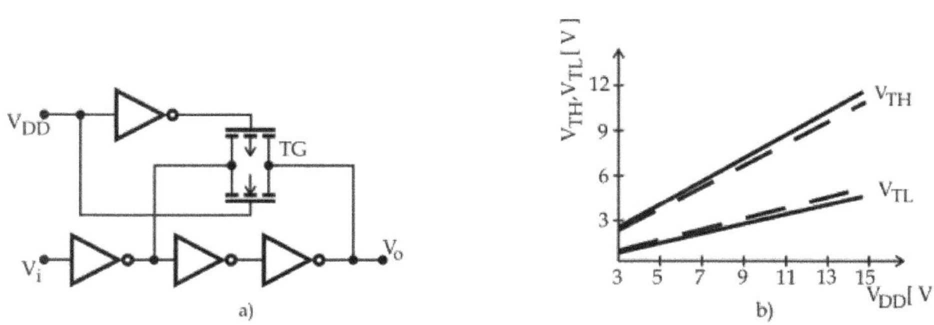

Fig. 7. Schmitt trigger with transmission gate instead of the resistor (a) and the dependency of the high and low threshold voltages on the supply voltage (b).

Instead of resistor R the transmission gate can be used (Fig.7a). The control input of the transmission gate should be in the logic state which keeps it on all the time. In this case, TG acts as a resistor whose resistance depends on the type of TG and the supply voltage. In Fig.7b the dependency of threshold voltages on the supply voltage V_{DD} is shown, if the inverters are CD4069, and TG is CD4066 (full line) or CD4016 (broken line). The resistance of the TG CD4066 is lower, thus the voltage hysteresis of the Schmitt trigger is wider in this case, than when CD4016 is used.

If NAND and NOR logic circuits are used instead of input inverters in Fig.3 and 7a the NAND and NOR Schmitt trigger are obtained.

3. Schmitt trigger – inverter

The basic circuit is the Schmitt trigger - inverter with three pairs of CMOS transistors (Fig.8). This solution has been initially proposed in (Dokic, 1984), which is the most widely cited single-ended Schmitt trigger (Young, 2010). Transistors M_n and M_p form the standard CMOS inverter I (Fig.8b). In wider part of the transfer characteristic transistors M_{n0}, M_{n1} and M_{p0}, M_{p1} are operating as the inverting nMOS and pMOS amplifier, respectively.

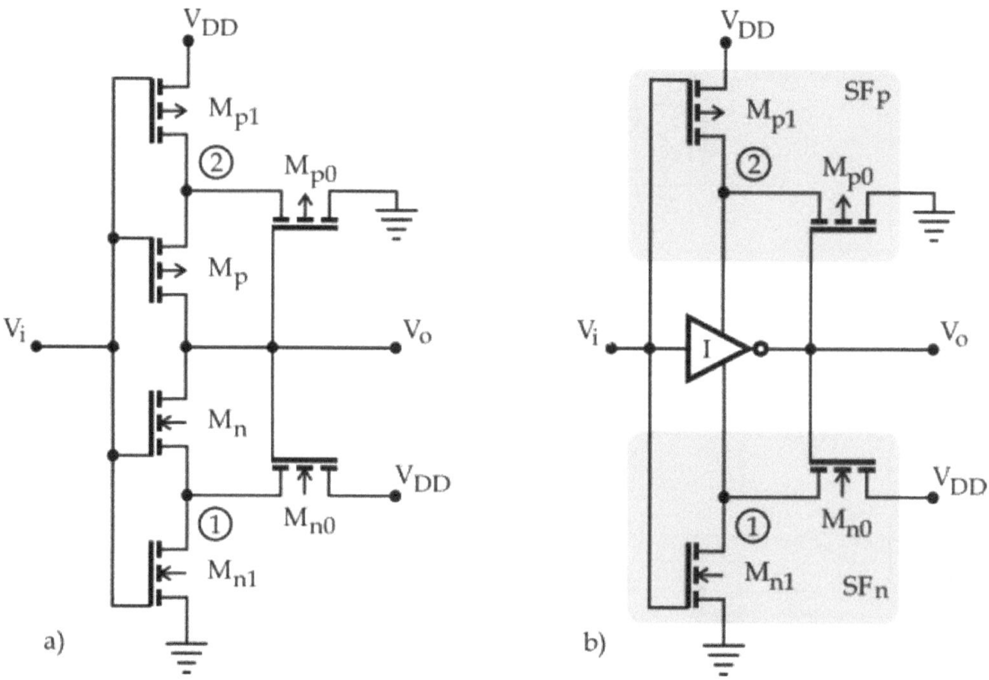

Fig. 8. Inverting Schmitt trigger (a) and it's equivalent (b).

The inverter of nMOS type enlarges the value of input voltage at which M_n turns on, when input voltage increases, and the inverter of pMOS type decreases the value of input voltage at which M_p turns on, when input voltage decreases. Because of this process, the transfer characteristic has the shape of hysteresis. When output voltage V_o changes, nMOS and pMOS inverters have the function of source followers, and through them positive feedback loop is created. They also introduce hysteresis by feeding back the output voltage to points 1 and 2.

To describe the circuit, assume the threshold voltages of all nMOS and pMOS transistors are V_{tn} and V_{tp}, respectively. Constants k of the transistors M_{n0} and M_{p0} are k_{n0} and k_{p0}, and constants k of the other nMOS and pMOS transistors are k_n and k_p, respectively.

For $V_i = 0$, M_n, M_{n1} and M_{p0} are off, and M_p, M_{p1} and M_{n0} are on. Output voltage is V_{DD}. Voltage of point 1 is $V_1 = V_{DD} - V_{tn0}$, because M_{n2} is on and saturated ($V_{GDn0} = 0$). Because of this fact, the high threshold voltage of Schmitt trigger V_{TH} is greater than V_T of standard CMOS inverter (M_n is not on until $V_i = V_{IN} > V_T$). Transistor M_n of standard inverter turns on at $V_i = V_{tn}$.

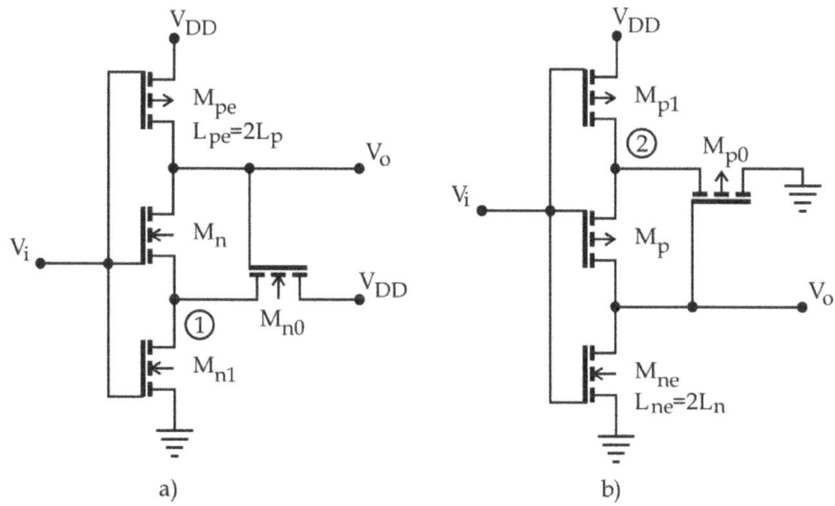

Fig. 9. Equivalent circuits: (a) used to determine V_{TH} and (b) to determine V_{TL}.

Assume V_i increases from 0 up to V_{DD}. Until the output state is changed, M_{po} is off ($V_{GS} = 0$). M_p and M_{p1} are connected in series and can be replaced with one transistor of which the constant is $k_{pe} = k_p/2$ (Dokic, 1988). For further analysis the equivalent circuit shown in Fig.9a is used.

At $V_i \geq V_{tn1}$, M_{n1} turns on, but M_n is off because $V_i - V_1 < V_{tn}$. Now both M_{n1} and M_{no} are saturated, thus forming an inverting amplifier with a voltage gain of about A_n(22). Namely, through equalization $I_{Dn1} = I_{Dno}$ in saturation, we obtain:

$$V_1 = V_{DD} - V_{tn0} - A_n\left(V_i - V_{tn1}\right) \tag{21}$$

where:

$$A_n = \sqrt{\frac{W_{n1} / L_{n1}}{W_{n0} / L_{n0}}} \quad \text{and} \quad V_{tn1} \leq V_i \leq V_1 + V_{tn} \tag{22}$$

Therefore, V_1 decreases as V_i increases (Fig.10a). There is no change of the output state until M_n is off. It turns on when $V_i = V_1 + V_{tn}$, which is equivalent, in regard to (21), to:

$$V_{IN} = V_{tn} + \frac{V_{DD} - V_{tn}}{1 + A_n} \tag{23}$$

where $V_{tn} = V_{tn1} = V_{tno}$. In further analysis we assume $k_n = k_{n1}$, $k_p = k_{p1}$ and that the threshold voltages of pMOS transistors are also equal, i.e. $V_{tp} = V_{tp1} = V_{tpo}$. With standard CMOS inverter the output voltage begins to decrease at $V_i = V_{tn}$, and with Schmitt trigger it does not until $V_i = V_{IN}$. For $k_{n1} = k_{no}$, $V_{IN} = 0.5(V_{DD} + V_{tn})$. Thus, an approximate value of the high threshold voltage V_{TH} can be determined by replacing V_{tn} of eq. (7) with V_{IN}. Out of this claim, we obtain $V_{TH} \approx 0.75V_{DD} - 0.25V_{tn}$, when all transistors are symmetric. More accurate value of V_{TH} is determined in a following way.

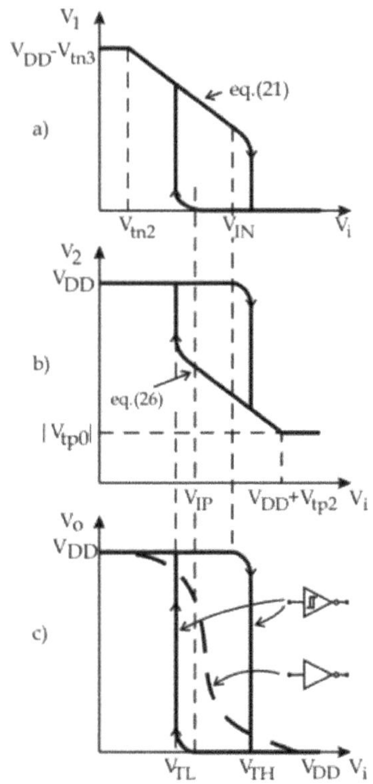

Fig. 10. Voltages in points of interest versus of input voltage.

After M_{n1} is on, V_o starts to decrease slightly. This change, through the gate-source of M_{n0}, is transferred to point 1, which accelerates the process of turning M_n on. When the amplification of the feedback loop achieves the value of (-1), positive feedback leads to step change of the output voltage. During this change, M_{p0} is on, accelerating the process of turning M_p off.

Transistors M_{pe} and M_n (Fig.9a) are in saturation when positive feedback is achieved. Thus, the high threshold voltage V_{TH} can be determined by equalization of drain currents of M_{pe} and M_n in saturation, i.e.:

$$k_n\left(V_i - V_1 - V_{tn1}\right)^2 = k_{pe}\left(V_{DD} + V_{tp} - V_i\right)^2 \tag{24}$$

where $V_1$is determined by (21). Replacing $V_i = V_{TH}$, we obtain:

$$V_{TH} = V_{tn} + \frac{V_{DD} + V_{tp} - V_{tn} + B_1\left(V_{DD} - V_{tn}\right)}{1 + B_1\left(1 + A_n\right)} \tag{25}$$

where:

$$B_1 = \sqrt{k_n / k_{pe}} = \sqrt{2k_n / k_p} \tag{26}$$

When $V_i = V_{DD}$, transistors M_p, M_{p1} and M_{n0} are off, and M_n, M_{n1} and M_{p0} are on, so $V_o = 0$. Then voltage of point 2 is $V_2 = |V_{tp0}|$, because M_{p0} is on and saturated ($V_{GDp} = 0$).
Assume V_i decreases from V_{DD} down to 0. All the time until the process of change of V_0 from 0 up to V_{DD} starts, M_{n0} is off, so the equivalent circuit for determining V_{TL} is shown in Fig.9b. At $V_i = V_{DD} + V_{tp1}$, M_{p1} turns on. Equalizing drain currents of transistors M_{p1} and M_{p0} in saturation, we obtain that the voltage of point 2 increases linearly:

$$V_2 = A_p\left(V_{DD} + V_{tp} - V_i\right) - V_{tp} \tag{27}$$

where:

$$A_p = \sqrt{\frac{W_p / L_p}{W_{p0} / L_{p0}}} \tag{28}$$

Transistor M_p turns on when $V_i = V_2 + V_{tp}$, so, considering (27):

$$V_{IP} = V_{DD} + V_{tp} - \frac{V_{DD} + V_{tp}}{1 + A_p} \tag{29}$$

From (29), for $k_p = k_{p0}$, we obtain: $V_{IP} = 0.5(V_{DD} + V_{tp}) < 0.5V_{DD}$.
Replacing V_{tp} in (7) with V_{IP} ($V_{IP} = -V_{tp}$), we obtain an approximate value of the low threshold voltage of the symmetric Schmitt trigger as follows: $V_{TL} = 0.25V_{DD} - 0.5V_{tp}$. At $V_i = V_{IP}$, V_o starts rising, and this change, through gate-source of M_{p0}, is transferred to point 2 and accelerates the process of turning M_{p1} on. When $|dV_o/dV_i| \geq 1$ positive feedback loop is achieved, so the change of V_o is step. M_p and M_{ne} are saturated, so:

$$k_{ne}\left(V_i - V_{tn}\right)^2 = k_p\left(V_2 + V_{tp} - V_i\right)^2 \tag{30}$$

where V_2 is determined by (27). Replacing $V_i = V_{TL}$, we obtain:

$$V_{TL} = V_{tn} + \frac{A_p\left(V_{DD} + V_{tp} - V_{tn}\right) - V_{tn}}{1 + B_2 + A_p} \tag{31}$$

where:

$$B_2 = \sqrt{k_{ne} / k_p} = \sqrt{k_n / \left(2k_p\right)} \tag{32}$$

When nMOS and pMOS transistors are symmetric, the thresholds V_{TH} and V_{TL} are also symmetric around $V_{DD}/2$, i.e. the transfer characteristic of the Schmitt trigger is optimum. Symmetry in this case is defined with these two conditions:

- symmetric cascode transistors M_n, M_{n1} and M_p, M_{p1};
- symmetric transistors which create positive feedback M_{n0} and M_{p0}.

Therefore, the high V_{TH} and the low V_{TL} threshold voltages, besides supply voltage, depend on the ratio of geometry of cascode transistors and transistors which create positive feedback. This is shown in Fig.11.

Length of the channel is a constant of technology, and most of the transistors within a digital circuit have the same length of the channel. Knowing this, we have:

$$A_n = \sqrt{W_n / W_{n0}}, A_p = \sqrt{W_p / W_{p0}} \tag{33}$$

Fig. 11. shows threshold voltages as functions of squared constants A_n and A_p. Rising widths of channel W_{n0} and W_{p0} of transistors M_{n0} and M_{p0}, the high threshold increases and the low threshold decreases.
Voltage hysteresis $V_H = V_{TH} - V_{TL}$ increases as constants A_n and A_p decrease (Fig.12). When constants k_n and k_p of all transistors are equal, voltage hysteresis is somewhat less than $0.5V_{DD}$.

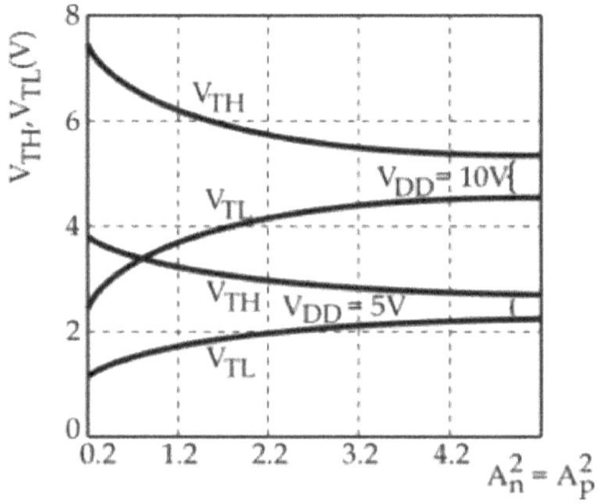

Fig. 11. $V_{TH}$and V_{TL} thresholds versus $A_n^2 = A_p^2$ obtained by SPICE analysis.

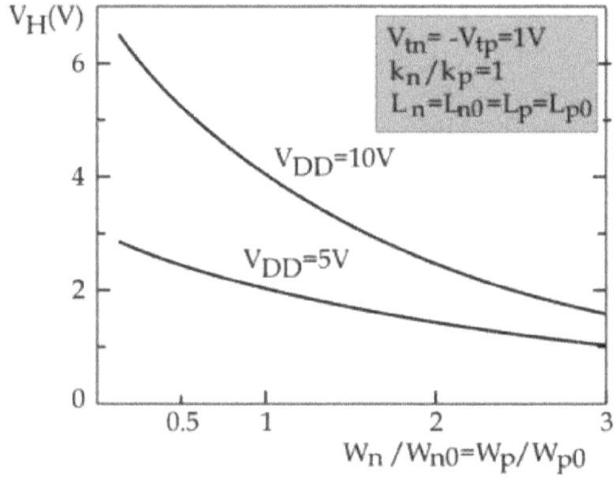

Fig. 12. Voltage hysteresis as function of ratio of transistor channels' widths.

Fig.13. shows the average propagation delay time, obtained by computer simulation using the program SPICE, as a function of the constants $A_n = A_p$ and the capacitive load at $V_{DD} = 5V$. For $A_n = A_p > 1$ the propagation delay time almost does not depend on the M_{n0} and M_{p0} geometry.

Therefore, by controlling the hysteresis through change of A_n and A_p (Fig.12), the propagation delay time is nearly held constant. Thusly, the ratio $W_n/W_{n0} = W_p/W_{p0} \approx 1$ is optimum, when noise immunity and propagation delay time are concerned.
Another, very important, characteristic of this Schmitt trigger is it's absolute stability at a wide number of tolerances of transistor parameters.

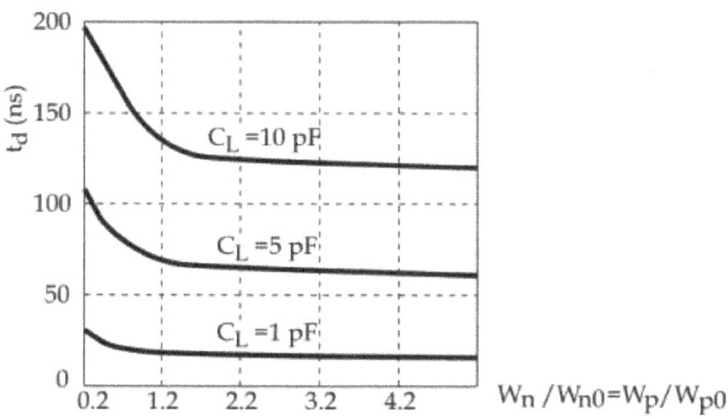

Fig. 13. Average propagation delay time as function of ratio of transistor channels' widths and C_o at $V_{DD} = 5V$.

3.1 Schmitt trigger with four MOS transistors

Schematics of these circuits are shown in Fig.14. and are completely the same as equivalent circuits used to analyze high (Fig.9a) and low (Fig.9b) threshold of Schmitt trigger from Fig.8. That is why, for example, V_{TH} of the circuit from Fig.14a is the same as with the standard Schmitt trigger - only $k_{pe} = k_p/2$ should be replaced with k_p in B_1 (22). Change appears at the low threshold. Namely, while V_i decreases from V_{DD} to V_{TL}, M_{n0} is off. Transfer characteristic, $V_o = f(V_i)$, in that area is determined by CMOS inverter made by M_p and M_n, M_{n1}, so the voltage of the low threshold is determined by (7), where k_n is replaced with $k_n/2$ (M_n and M_{n1} are serially connected), i.e.:

$$V_{TL} = V_{tn} + \frac{V_{DD} + V_{tp} - V_{tn}\sqrt{(k_n/2)/k_p}}{1+\sqrt{(k_n/2)/k_p}} \tag{34}$$

As it has already been said, area of transfer characteristic for which: $V_{TL} < V_i < V_{DD}$, is completely the same as the characteristic of the standard inverter. During the process of state change, as transistor M_n enters saturation, transistor M_{n0} turns on. Through M_{n0} positive feedback loop is formed, thus further changed of V_o are step.
Analogous explanation is given for the circuit shown in Fig.14c. Therefore, V_{TH} is obtained when k_p is replaced with $k_p/2$ in (7) (because M_p and M_{p1} are serially connected), and V_{TL} is determined by (31), where $k_{ne} = k_n/2$ is replaced with k_n.

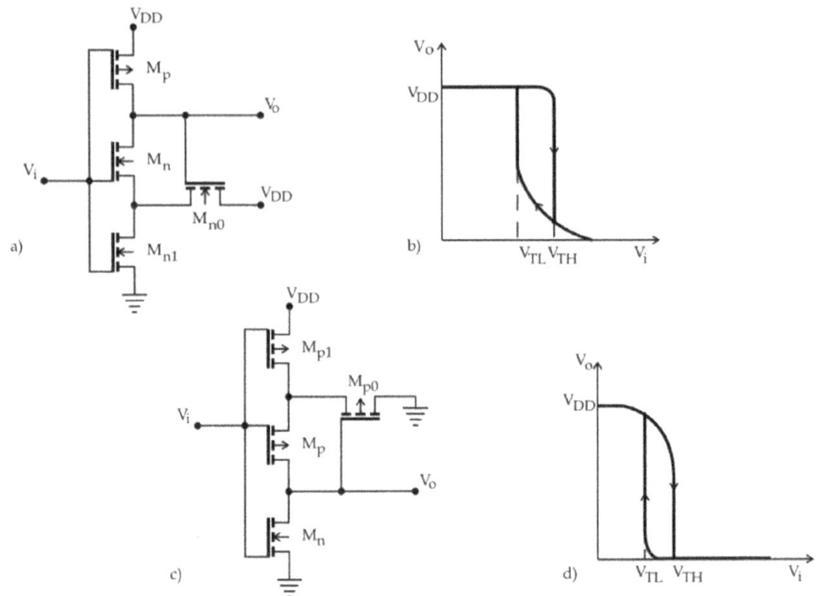

Fig. 14. Schmitt triggers with four transistors and their transfer characteristics.

4. Non-inverting Schmitt trigger

Latch circuit with inverters I_1 and I_2, if applied to output of standard inverter I (Fig.15), can, under certain conditions, work as Schmitt trigger (Bundalo & Dokic, 1985). As it will be shown, transistors of inverter I_1 must be smaller than transistors of the input inverter (smaller W, same L).

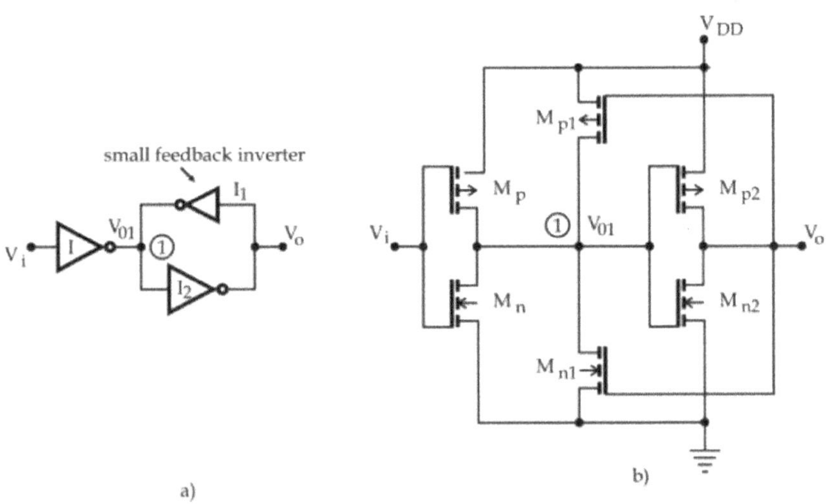

Fig. 15. Logic (a) and expanded schematic of non-inverting Schmitt trigger (b).

Basic idea and characteristics are very similar to the Schmitt trigger in Fig.3. If the input is low, output voltage is: $V_o = 0$. Transistor M_{n1} is off, and M_{p1} on and in linear area of the characteristic. Thus, equivalent circuit, while input voltage increases, is shown in Fig.16a. If M_{p1} is replaced by it's drain-source resistance in linear area, then all equivalent circuits in Fig.s 16a and 5a are completely the same.
While input is high, $V_o = V_{DD}$, M_{p1} is off, and M_{n1} is on. Equivalent circuit (Fig.16b), during the process of input voltage decreasing, is similar to the circuit in Fig.5b, where M_{n1}, which is in linear area, stands instead of resistor R.
Transfer characteristics $V_o(V_i)$ and $V_{o1}(V_i)$ are shown in Fig.17., broken line shows the characteristic of the standard inverter.
Equivalent circuit for determining the high threshold is shown in Fig.16a. Assume that the input voltage increases from 0 up to V_{DD}. For $V_{tn} < V_i < V_{TH}$, all three transistor circuits in Fig.16a are on, so that:

$$I_{Dn} = I_{Dp} + I_{Dp1} \tag{35}$$

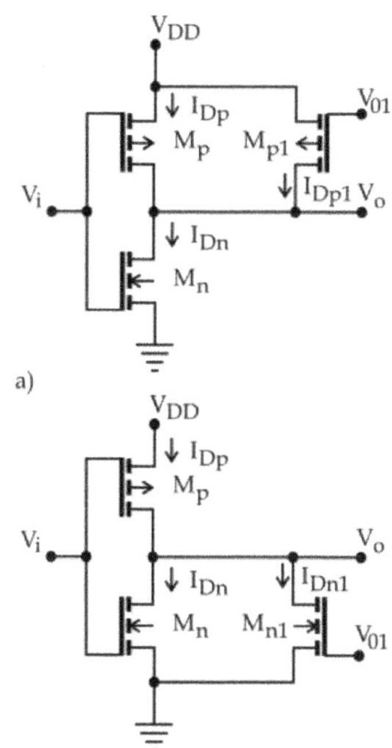

Fig. 16. Equivalent circuit for determi-ning the high (a) and the low (b) threshold voltage.

In the beginning of the process M_{n1} and M_{p1} are in linear, and M_n is in saturation area. When $V_{o1} < V_{DD} - |V_{tp}|$, pMOS transistor of inverter I_2 starts to turn on, so the output voltage V_o increases (to the right from the point C in Fig.17b). This leads to increase of resistance of transistor M_{p1} and of the slope of the characteristic $V_{o1}(V_i)$. Between input and output of latch circuit positive feedback loop is established. The process becomes regenerative and it leads to step changes of V_{o1} and V_o (point A in Fig.17a) when the amplification of the positive feedback loop is:

$$dV_o / dV_i = -1 \tag{36}$$

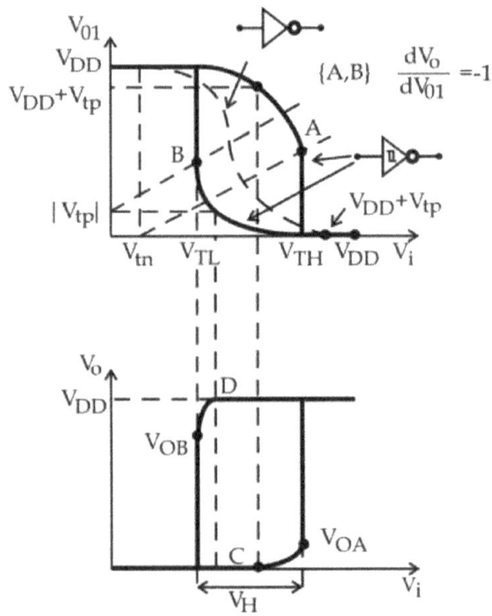

Fig. 17. Transfer characteristics of Schmitt trigger (Fig.15).

Determining the exact value of the high threshold V_{TH}, according to (36), leads to equations of higher degree, making it impossible to find explicit solution for V_{TH}. Because of this approximate method will be used for this purpose. Namely, transistor M_{p1} is in linear area for all values of input voltage: $0 < V_i < V_{TH3}$, so it can be replaced with a resistor of approximate resistance:

$$R_{p1} \approx \frac{1}{2k_{p1}\left(V_{DD} + V_{tp1}\right)} \tag{37}$$

Then the equivalent circuit from Fig.16a is the same as the one in Fig.5a, so the high threshold is obtained by replacing R in (12) with R_{p1}, which leads to:

$$V_{TH} = \frac{V_{DD}}{2} + \frac{k_{p1}}{k_p} \frac{V_{DD} + V_{tp1}}{2\left(V_{DD} + V_{tp} - V_{tn}\right)} V_{DD} \tag{38}$$

As long as the input is high, the output voltage is $V_o = V_{DD}$, which means that M_{p1} is off and M_{n1} is on. The equivalent circuit, when the input voltage starts to decrease, is shown in Fig.16b. Positive feedback is established when $dV_o/dV_i = -1$ (point B in Fig.17). Transistor M_{n1} is in linear area, and can, thus, be replaced with a resistor of approximate resistance:

$$R_{n1} \approx \frac{1}{2k_{n1}\left(V_{DD} - V_{tn1}\right)} \tag{39}$$

Voltage of the low threshold can be determined by replacing R in (16) with R_{n1}, so:

$$V_{TL} = \frac{V_{DD}}{2} - \frac{k_{n1}}{k_n} \frac{V_{DD} - V_{tn1}}{2\left(V_{DD} + V_{tp} - V_{tn}\right)} V_{DD} \tag{40}$$

Transfer characteristic is optimum when inverters are symmetric, because V_{TH} and V_{TL} are symmetric around $V_{DD}/2$. Then voltage hysteresis is given by:

$$V_H = \frac{k_1}{k} \frac{V_{DD} - V_t}{V_{DD} - 2V_t} V_{DD} \tag{41}$$

where: $k_1 = k_{p1} = k_{n1}$, $k = k_p = k_n$ and $V_{tn} = V_{tn1} = |V_{tp}| = |V_{tp1}| = V_t$.
Therefore, basic parameters, V_{TH}, V_{TL} and V_H, besides supply voltage V_{DD} and threshold voltages of the transistors, depend on ratio of geometry of the transistors of feedback inverter I_1 and input inverter I. Range in which ratio can be changed is limited. As it has already been said, changes of state at the output are caused by existence of points in characteristic $V_o(V_{o1})$ with unit amplification, i.e. $dV_o/dV_i = 1$. In the worst case, during static states, voltage V_{o1} has to be, at $V_i = V_{DD}$, less than, and at $V_i = 0$ greater than the threshold of inverter I_2, i.e.:

$$V_o\left(V_{DD}\right) \leq V_{T2} = V_{DD} / 2 \text{ and } V_o\left(0\right) \geq V_{T2} = V_{DD} / 2 \tag{42}$$

The equations show that the changes are conditioned by ratio of geometry of transistors: M_{p1} and M_n (M_p and M_{n1} are off) in first ($V_i = V_{DD}$), and M_p and M_{n1} in second ($V_i = 0$) case.
It can be shown that the Schmitt trigger in Fig.15 will operate reliably for all acceptable supply voltages, if:

$$k_n / k_{p1} > 2 \text{ and } k_p / k_{n1} > 2 \tag{43}$$

4.1 Schmitt triggers with five transistors

The Schmitt trigger shown in Fig.15 has a hysteresis shaped characteristic if one of the transistors of the inverter I_1 is left out. Such simplified circuit is shown in Fig.18. The circuit in Fig.18a will be analyzed. While input voltage increases, this circuit is completely the same as the circuit shown in Fig.15, because during this process M_{n1} was off. Thus, the high threshold voltage is determined in (38).
During negative change of V_i, M_{p1} is off. Since there is no feedback loop, a low threshold voltage is equal to the threshold voltage of input inverter I. During the change of output from high down to low logic level, regenerative process is established. Namely, for $V_o > V_{DD} - |V_{tp}|$, M_{p1} is off and the shape of $V_o(V_i)$ function is determined by cascode inverters I and I_2. At $V_o = V_{DD} - |V_{tp}|$ (point A in the transfer characteristics), M_{p1} turns on and establishes positive feedback loop, thus making change of V_o step.
The whole analysis is analogically applicable for the circuit in Fig.18b, with the difference that the high threshold is equal to the threshold voltage of inverter I, and the low threshold is determined by (40).

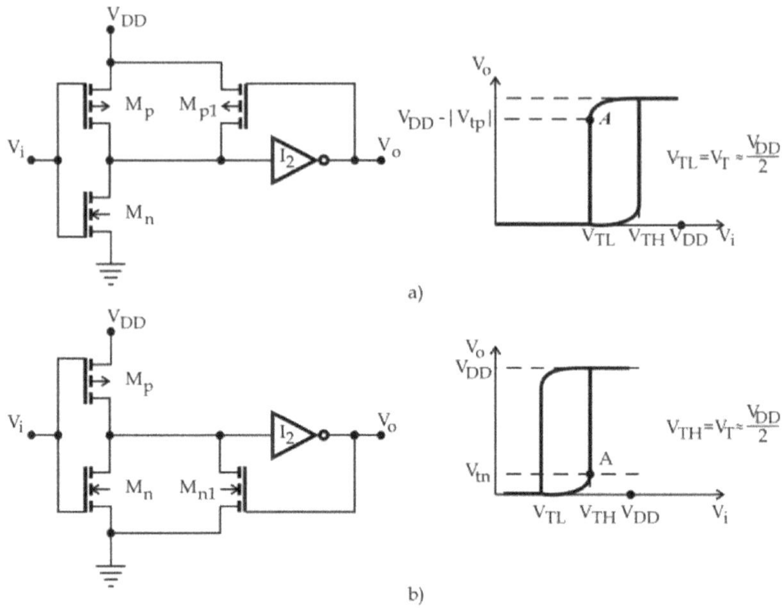

Fig. 18. Schmitt triggers with five transistors and their transfer characteristics.

5. Schmitt trigger with voltage controlled thresholds

All Schmitt circuits analyzed so far have fixed parameters (V_{TH}, V_{TL} and V_H), at defined supply voltage. Often there is a need to control parameters of Schmitt trigger, depending on the field of application. Control of it's parameters from outside of the circuit is demanded. Such possibility exists when the Schmitt trigger shown in Fig.19 is used. Here, in cascode connection with transistors M_{n1} and M_{p1} of the circuit shown in Fig.15, the transistors M_{n0} and M_{p0} are added.

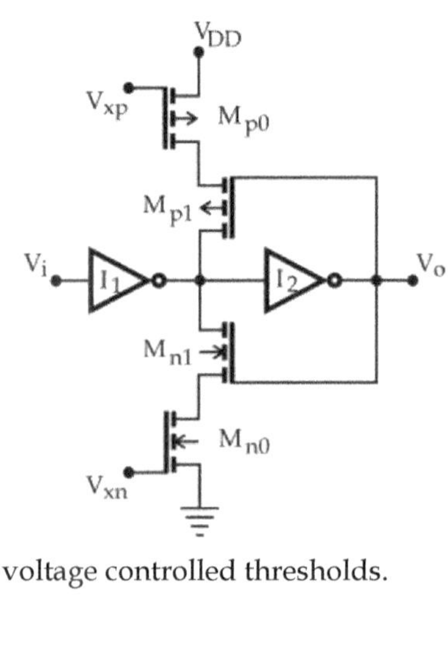

Fig. 19. Schmitt trigger with voltage controlled thresholds.

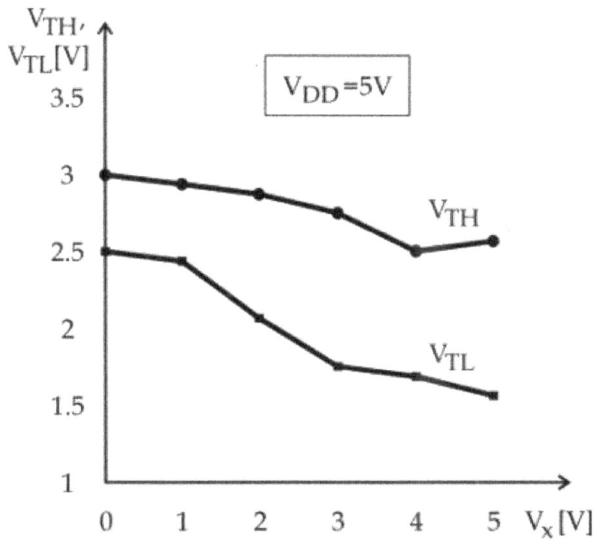

Fig. 20. Threshold voltages of Schmitt trigger in Fig. 19 as functions of common control voltage V_x

Over the gates of transistors M_{n0} and M_{p0}, control voltages V_{xn} and V_{xp} are supplied. These voltages are used to change equivalent resistance of M_{p0} and M_{p1} towards V_{DD}, and M_{n0} and M_{n1} towards ground. While input voltage increases, this resistance is the resistance of transistors M_{p1} and M_{p0} in linear area (M_{n1} is off, so influence of M_{n0} is blocked). Since the resistance of M_{p0} depends on voltage, thus total resistance towards V_{DD} is a function of V_{xp}. Voltage V_{xn} is used to modulate the low threshold of Schmitt trigger by changing the resistance of M_{n0}. Controlling the thresholds is independent for each of them: V_{xp} influences only V_{TH}, and V_{xn} only V_{TL}.

When gates of M_{n0} and M_{p0} are short circuited, control voltage is common. It influences both V_{TH} and V_{TL} in the same amount. In this way voltage hysteresis remains constant (Fig.20).

6. NAND and NOR circuit design

Thus, the Schmitt trigger-inverter (Fig.1) consists of one conventional CMOS inverter (M_n, M_p), one nMOS inverter (M_{n1}, M_{n0}) and one pMOS inverter (M_{p1}, M_{p0}). The same principle is used for design of the NAND and NOR Schmitt circuits shown in Fig.21 (Dokic, 1996). In this case conventional CMOS, nMOS and pMOS NAND and NOR gates are used instead of the corresponding inverters.

Since the transistors M'_{n1}, ..., M'_{nm} and the nMOS transistors of the conventional CMOS NAND gates are connected in series, the output of the circuit in Fig.21a will be low only when all inputs are high, i.e. $\bar{Z} = x_1 x_2 \dots x_m$, so that $Z = \overline{x_1 x_2 \dots x_m}$. Hence this is an m-input NAND gate.

The output of the circuit in Fig.21b will be high only when all inputs are low (M'_{p1}, ..., M'_{pm} and pMOS transistors of the conventional NOR gate are connected in series and must be conducting), i.e. $Z = \bar{x}_1 \bar{x}_2 \dots \bar{x}_m$ or $Z = \overline{x_1 + x_2 + \cdots + x_m}$. Hence this is an m-input NOR gate.

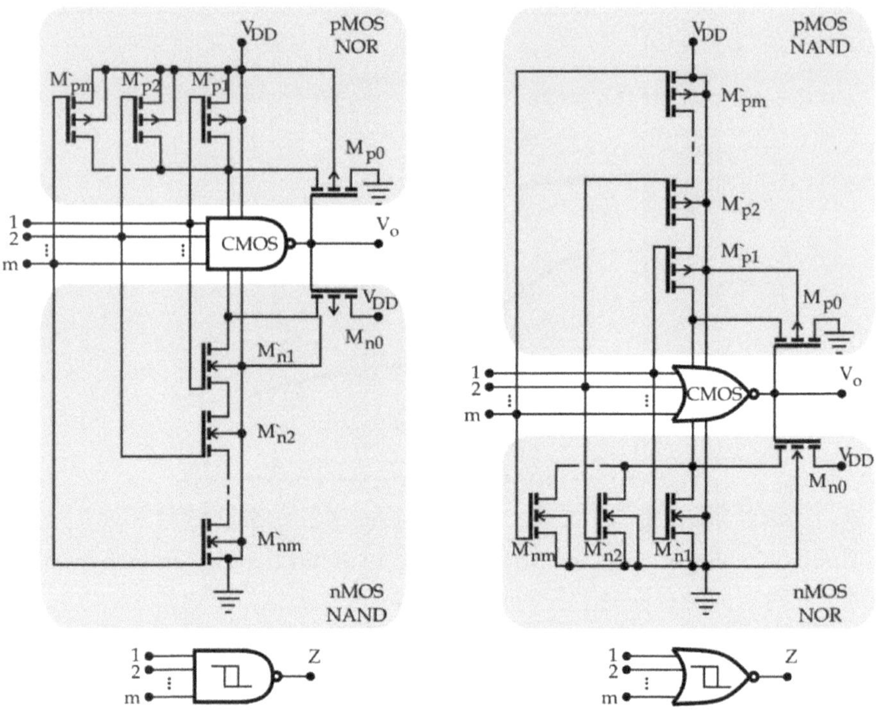

Fig. 21. Principle schematics of m-input NAND and NOR Schmitt circuits (Dokic, 1996).

The transistors M_{n0} and M_{p0} provide feedback to effect rapid change of the output voltage and the transfer characteristic has a shape of the hysteresis curve. Hence the circuits in Figs. 21a and 21b are m-input NAND and NOR Schmitt circuits, respectively.

6.1 NAND circuit analysis

Parallel or series transistors can be replaced by one transistor such as the conventional NAND and NOR circuits (Dokic, 1982). In this way, NAND and NOR Schmitt circuits can be replaced by an equivalent Schmitt trigger-inverter (Fig.9) by dc analysis. It will be shown by analyzing an m-input NAND Schmitt circuit.

Assume that n inputs are active (at V_i), where $1 \leq n \leq m$, and that the other m-n inputs are at V_{DD}. Let the input voltage V_i increase from zero to V_{DD}. For $0 \leq V_i \leq V_{TH}$ the NAND circuits in Fig.21a can be replaced by the equivalent circuit in Fig.9a. M_{p0} and m-n pMOS transistors at $V_i = V_{DD}$ are off. Therefore, the equivalent pMOS transistor M_{pe} consists of n pairs of pMOS transistors M_{pi}, M'_{pi} with active inputs. Hence M_{pe} constant k is given by:

$$k_{pe} = nk_p / 2 \tag{44}$$

Transistor M_{n1} (Fig.9a) needs to be replaced with one equivalent transistor which consists of series nMOS transistors of a conventional NAND gate.

The equivalent constant k of series transistors depends on the number of transistors and position of the first active input (Dokic, 1982). Consequently, for the case of n active inputs the transistors M_{n1}, …, M_{nm} and M'_{n1}, …, M'_{nm} can be replaced by the equivalent transistors M_{ne} and M_{ne1}, respectively, whose constants k are given by:

$$k_{ne} = k_{ne1} = \frac{k_n}{m - p + 1} \tag{45}$$

where p marks the position of the first active input (for example, if $p = 3$, the inputs of the transistors M_{n1}, M'_{n1} and M_{n2}, M'_{n2} are at V_{DD}, and M_{n3}, M'_{n3} are at V_i). Now, eq.(22) and (26) become, respectively:

$$A_{ne} = \sqrt{k_{ne1} / k_{n0}} = A_n (m - p + 1)^{-1/2} \tag{46}$$

$$B_{1e} = \sqrt{k_{ne} / k_{pe}} = B_1 \left[n(m - p + 1) \right]^{-1/2} \tag{47}$$

where A_n and B_1 are given by eqs. (22) and (26), respectively.
From eq. (25), replacing A_n by A_{ne} and B_1 by B_{1e} we obtain the high threshold voltage of the m-inputs NAND Schmitt circuit:

$$V_{TH} = \frac{V_{DD} + V_{tp} + B_1 \left[n(m-p+1) \right]^{-1/2} \left[V_{DD} + A_n (m-p+1)^{-1/2} V_{tn} \right]}{1 + B_1 \left[n(m-p+1) \right]^{-1/2} \left[1 + A_n (m-p+1)^{-1/2} \right]} \tag{48}$$

To calculate V_{TL} the circuit in Fig.21a can be replaced by an equivalent one shown in Fig.9b. Namely, M_{n2} is off. M_{n1}, …, M_{nm}, M'_{n1}, …, M'_{nm} are on and can be replaced by an equivalent one M_{ne} with the constant k:

$$k_{ne} = k_n / (2m) \tag{49}$$

M_{p2} is on. M_{pi} and M_{p1i} $(i = 1, \dots, n)$ with active input can be replaced by equivalent transistors M_{pe} and M_{pe1}, respectively, with the constants k:

$$k_{pe} = k_{pe1} = n k_p \tag{50}$$

The k ratios of M_{pe1} to M_{p0} and M_{ne} to M_{pe1}, respectively, are given by:

$$A_{pe} = \sqrt{k_{pe1} / k_{p2}} = A_p n^{1/2} \tag{51}$$

$$B_{2e} = \sqrt{k_{ne} / k_{pe1}} = B_2 (mn)^{-1/2} \tag{52}$$

From eq. (36), replacing A_p by A_{pe} and B_2 by B_{2e}, we obtain the low threshold voltage of the m-inputs NAND Schmitt circuits:

$$V_{TL} = \frac{A_p n^{1/2} (V_{DD} + V_{tp}) + B_2 (mn)^{-1/2} V_{tn}}{1 + A_p n^{1/2} + B_2 (mn)^{-1/2}} \tag{53}$$

where A_p and B_2 are given by eqs. (27) and (32), respectively.

The threshold voltages V_{TH} and V_{TL} depend on supply voltage V_{DD}, the ratio of the constants k_n/k_p, k_n/k_{n0}, i.e. k_p/k_{p0}, number of inputs m, and number of active inputs n. Besides, V_{TH} depends on the position of the first active input p.

In Fig.22 two-input Schmitt NAND gate and it's transfer characteristics at $V_{DD} = 5V$, for $k_{ni} = k_n = 2k_{pi} = 2k_p$, $i = 0, \ldots, 4$ and $V_{tn} = -V_{tp} = 1V$, are shown. The high threshold depends on the number and the combination of active inputs, and the low only on the number of active inputs. The voltage thresholds, as function of channel widths ratio of cascode transistors and of transistors M_{n0} and M_{p0} in the circuit of positive feedback loop, are shown in Fig.23.

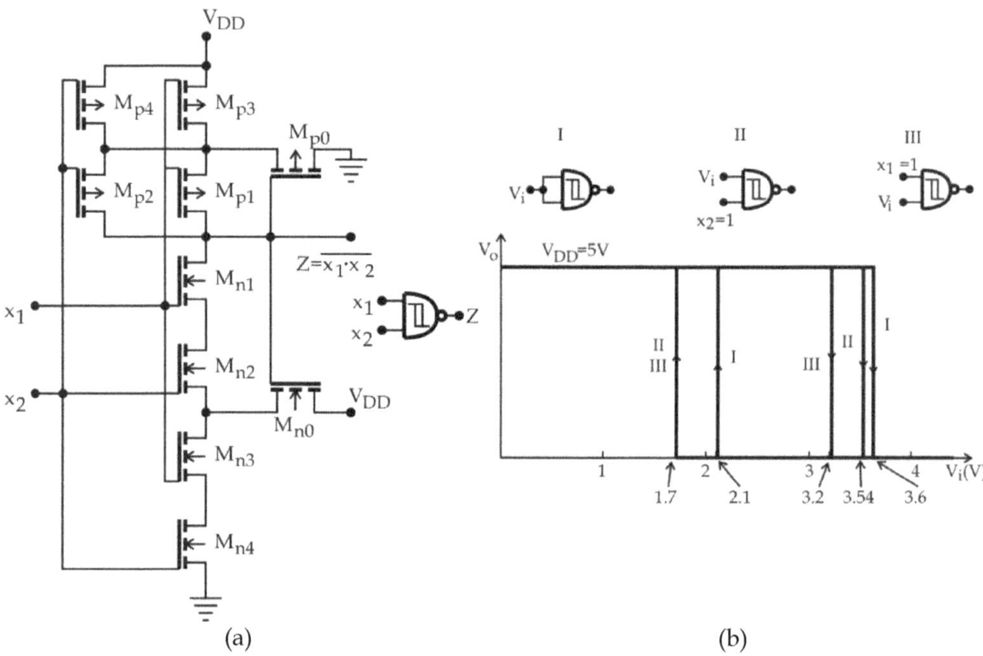

Fig. 22. Two-input NAND Schmitt trigger (a) and it's transfer characteristic (b) at $V_{DD} = 5V$.

6.2 NOR circuit

As the NOR circuit is obtained from the NAND one through the interchange of the p-channel and n-channel transistors and a power supply polarity change, the previous analysis can be applied analogously to this circuit. In this way we obtain:

$$V_{TH} = \frac{V_{DD} + V_{tp} + B_1 (mn)^{1/2} \left(V_{DD} + A_n n^{1/2} V_{tn} \right)}{1 + B_1 (mn)^{1/2} \left(1 + A_n n^{1/2} \right)} \tag{54}$$

$$V_{TL} = \frac{A_p (m-p+1)^{-1/2} \left(V_{DD} + V_{tp} \right) + B_2 \left[n(m-p+1) \right]^{1/2} V_{tn}}{1 + A_p (m-p+1)^{-1/2} + B_2 \left[n(m-p+1) \right]^{1/2}} \tag{55}$$

Therefore, the threshold voltages depend on exactly the same parameters as the thresholds of the NAND circuit except that in the NOR circuit, V_{TH} does not depend on the position of the first active input p.

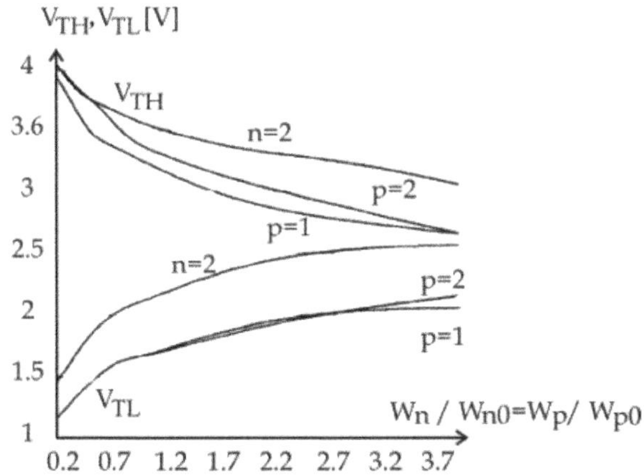

Fig. 23. SPICE values of V_{TH} and V_{TL} for two-input NAND circuit versus $W_n/W_{n0} = W_p/W_{p0}$ for various numbers and combinations of active inputs at $V_{DD} = 5V$ and for optimum geometry ratio of nMOS and pMOS transistors, that is at $k_n/k_p = 2$.

7. CMOS gates with regenerative action at one of inputs

In many applications when NAND and NOR gates are used in an MSI or LSI circuit there is a Schmitt trigger at the external input only. So, for example, a Schmitt trigger action in the clock input of counter provides pulse shaping that allows unlimited clock input pulse rise and fall times. These circuits are made by a conventional NAND or NOR gate and Schmitt trigger on their external input. CMOS NAND or NOR gate and Schmitt trigger action at one input (Fig.24) is a better solution. The Schmitt trigger is an integral part of the gate. The advantages of these circuits, compared with the conventional ones, are: smaller number of transistors, smaller area of the chip and higher switching speed.

7.1 Principle schemes

Fig.24 illustrates the principle schemes of the two input NAND and NOR logic circuits with hysteresis when the input x_1 is active. When the input x_2 is active only, the transfer characteristic is without hysteresis. These circuits consist of the Schmitt trigger on Fig.8 and one pair of CMOS transistors (M_n and M_p). Multiple inputs are made in a conventional way (by adding one pair of CMOS transistors to each input).

Consider the NAND circuit in Fig.24a. When the input x_1 only is active, and $x_2 = 1$, the transistor M_p is off, and M_n is on. Then the transfer characteristic is determined by the Schmitt trigger. If the input x_2 only is active the Schmitt trigger does not operate, so that the transfer characteristic will be the same as that of the CMOS inverter made by the transistors M_n and M_p.

The NOR gate will be described more fully.

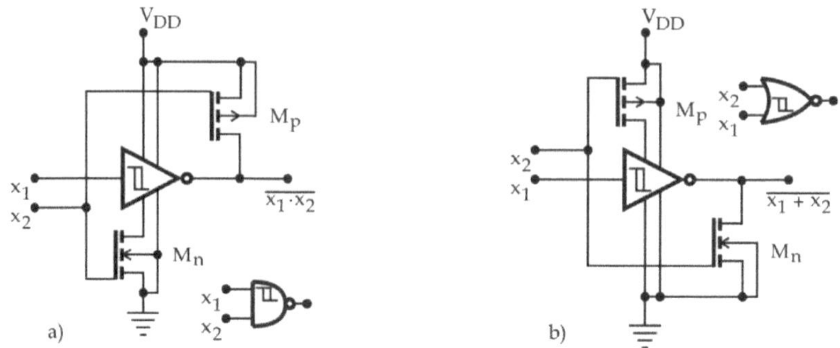

Fig. 24. Two input NAND (a) and NOR (b) gates with the regenerative action only at the input x_1 (Dokic, 1988).

7.2 NOR gate

Fig.25 shows the two-input NOR gate with the regenerative action at the input x_1 only. The transistors M_{ni} and M_{pi} ($i = 0,1,2$) make the Schmitt trigger circuit (Fig.8).

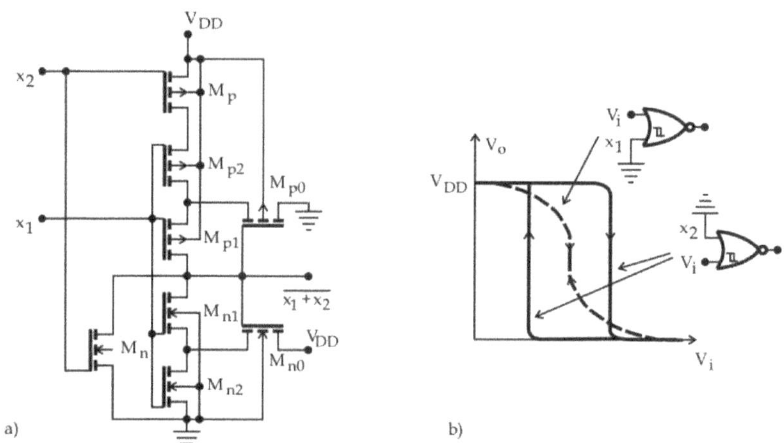

Fig. 25. Scheme of the NOR gate (a) and voltage transfer characteristic (b).

When the input x_2 is active and $x_1 = 0$ the transistors M_{p1} and M_{p2} are on, and M_{n1}, M_{n2} and M_{p0} are off. Hence there is no feedback between the output and the input, the transfer characteristic is without hysteresis (see Fig.25a - broken line) and it is exactly the same as for the conventional two-input gate with the active input.
When the input x_1 is active and $x_2 = 0$ the transistor M_p is on, and M_n is off. Then the circuit in Fig.25 acts as the Schmitt trigger. Namely, the transistors M_p and M_{p2} can be replaced by an equivalent with constant $k_{pe} = k_p/2$, so that the Schmitt trigger on Fig.8 is obtained.

8. BiCMOS Schmitt circuits

Non-inverting BiCMOS Schmitt trigger is shown in Fig.26 (Dokic, 1995). It consists of CMOS Schmitt trigger at the input (Fig.15) and a BinMOS output with the transistors T_1, T_2, M_{n3}

and M_{n4}. Basic static parameters are completely the same as for an analogous CMOS circuit in Fig.15. The output is a standard BiCMOS with the logic amplitude of $\Delta V_o = V_{DD} - 2V_{BE}$. The transistor M_{n2}, at same time, is used to bleed the base charge of T_1.

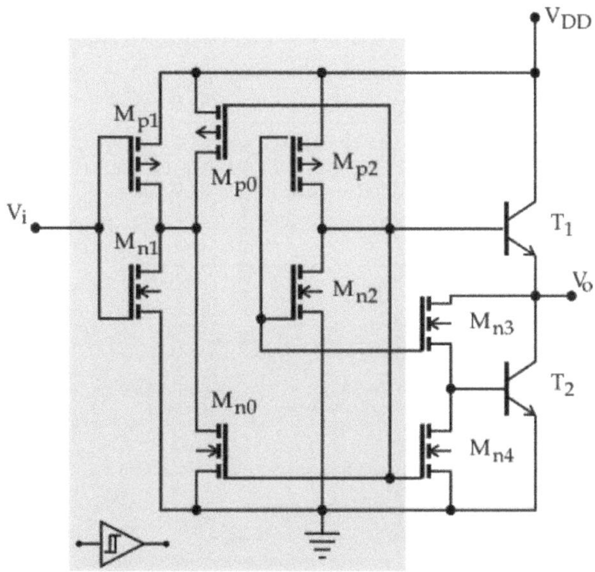

Fig. 26. Non-inverting BiCMOS Schmitt trigger.

Inverting Schmitt trigger is shown in Fig.27. This circuit is very similar to the one shown in Fig.8. Transistors M_{n3} and M_{n2} are MOS bleeding elements. Even the principle of functioning is very similar. Beside increase of speed, bipolar transistor brings certain specifics into static parameters. This increases the amplification with the positive feedback loop, thus leading to a faster change of logic state. Because of this the values of threshold voltages differ from those of the circuit in Fig.8. Practically, as soon as M_n starts conducting at positive, or M_p at negative change of the input voltage, regenerative process is established very fast.

8.1 Short analysis

At $V_i = 0$, transistors M_n, M_{n1}, M_{n3}, M_{po} and T_2 are off, and M_p, M_{p1} and T_1 are conducting. The output voltage is: $V_o = V_{DD} - V_{BE}$. Transistors M_{no} and M_{n2} are on, because of $V_3 = V_{DD}$. Hence, the voltage in point 1 is high and equals: $V_1 = V_{DD} - V_{tn0}$, which postpones the process of turning on the transistor M_n in regard to M_{n1}. During the increase of the input voltage from zero to V_{DD}, transistor M_{n1} is turned on first. Since both M_{n1} and M_{no} are saturated, by equaling their currents, we obtain that voltage V_1 is decreasing, i.e.:

$$V_1 = V_{DD} - V_{tn0} - \sqrt{W_{n1} / W_{n0}} \left(V_i - V_{tn} - V_{BE}\right) \tag{56}$$

where W_{n1} and W_{no} represent channel widths of transistors M_{n1} and M_{no}. We assume that the channel lengths are equal. M_n turns on when:

$$V_{gs} = V_i - V_1 = V_{tn} \tag{57}$$

Afterwards, M_{n3} turns on, which, very quickly, leads to establishment of the regenerative process and change of states at the output. Practically, this means that the high threshold is approximately equal to the input voltage at which the condition (57) is fulfilled. Hence, based on (56) and (57), we obtain:

$$V_{TH} \approx \frac{V_{DD} + \sqrt{W_{n1} / W_{n0}}\,(V_{tn} + V_{BE})}{1 + \sqrt{W_{n1} / W_{n0}}} \tag{58}$$

where the thresholds of all nMOS transistors are equal to V_{tn}.

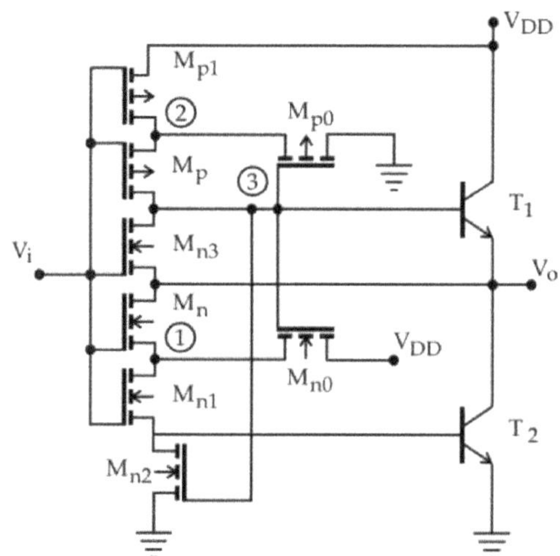

Fig. 27. Inverting BiCMOS Schmitt trigger.

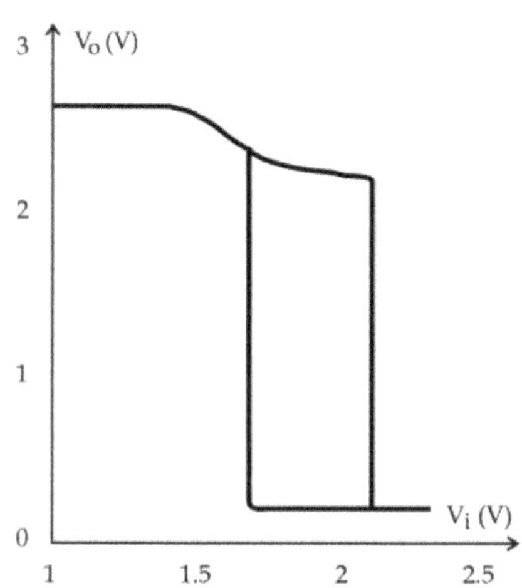

Fig. 28. Static transfer characteristic of inverting Schmitt trigger.

When $V_i = V_{DD}$, transistors M_n, M_{n1}, M_{n3}, M_{p0} and T_2 are on, and M_p, M_{p1}, M_{n0} and T_1 are off. Transistor M_{n3} short-circuits base and emitter of T_1. During the decrease of the input voltage, transistor M_{p1} turns on first. Afterwards, voltage in point 2 increases. Since M_{p1} and M_{p0} are saturated:

$$V_2 = V_{BE2} - V_{tp0} - \sqrt{W_{p1} / W_{p0}}\left(V_{DD} + V_{tp1} + V_i\right) \tag{59}$$

Transistor M_p turns on at:

$$V_i = V_2 + V_{tp} \tag{60}$$

After M_p turns on, the output voltage increases, and a positive feedback loop is established very quickly. Hence V_i in (60) is approximately equal to the low threshold voltage. Based on (59) and (60), we obtain:

$$V_{TL} \approx \frac{\sqrt{W_{p1} / W_{p0}}\left(V_{DD} + V_{tn}\right) + V_{BE}}{1 + \sqrt{W_{p1} / W_{p0}}} \tag{61}$$

Transfer characteristic (Fig.28) is obtained by SPICE analysis at following conditions: supply voltage $V_{DD} = 3V$, $0.8\mu m$ BiCMOS process with minimum channel lengths of all transistors ($L = 0.8\mu m$), transistor threshold voltages $V_{tn} = |V_{tn}| = 0.85V$, channel widths $W_{n1} = W_{n2} = W_{n3} = W_{n4} = 8\mu m$, $W_{p1} = W_{p2} = 24\mu m$, $W_{no} = 2\mu m$ and $W_{po} = 6\mu m$.

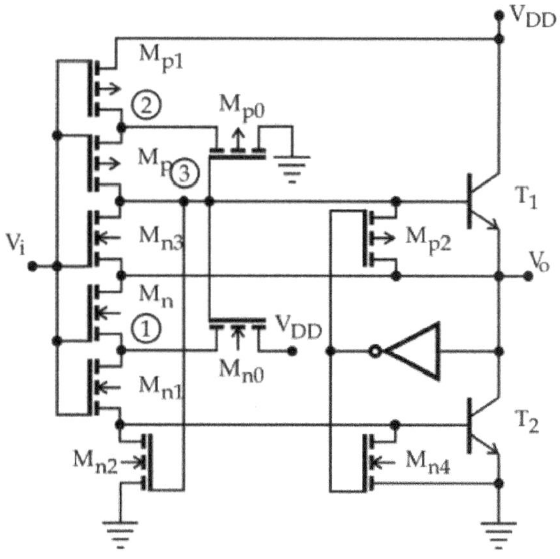

Fig. 29. Low voltage BiCMOS Schmitt trigger.

Calculated values of threshold voltages, by the given parameters including $V_{BE} = 0.7V$, are $V_{TH} = 2.03V$ and $V_{TL} = 1.67V$. Values obtained by simulation are $V_{TH} = 2.12V$ and $V_{TL} = 1.68V$. Errors are very small. These will increase at higher values of supply voltage.

Limitation of application of the circuit in Fig.27 at lower supply voltages is it's decrea-sed logic amplitude of the output voltage $\Delta V_o = V_{DD} - 2V_{BE}$. The low voltage Schmitt trigger is shown in Fig.29. Transistors M_{n4} and M_{p2} through the inverter I hold output voltages at $V_o = V_{DD}$ and $V_o = 0$.
Transistor M_{p4} delivers another improvement. It removes deformation from the transfer characteristic of the standard circuit before the change from high to low level (Fig.28). This deformation is a consequence of the knee-effect of characteristic of bipolar transistor T_1. In Fig.30 SPICE analysis of the output of the low voltage Schmitt trigger designed in $0.8\mu m$ BiCMOS process for equal constants k_n and k_p of all nMOS and pMOS transistors is shown.

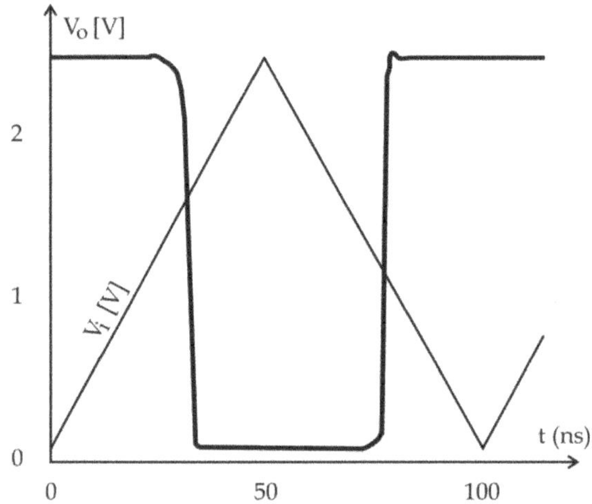

Fig. 30. The output of the low voltage Schmitt trigger to a triangular input at $V_{DD} = 2.5V$.

9. Conclusion

Schmitt logic circuits are produced as independent integrated circuits as well as input circuits of standard MSI/VLSI and ASIC integrated circuits. The author expects that the overview of CMOS and BiCMOS Schmitt triggers will be useful to both engineers who design digital integrated circuits and to those who design digital systems with integrated circuits.
Common characteristics of the described solutions are:

- threshold voltages V_{TH} and V_{TL} are almost symmetric around voltage $V_{DD}/2$, so the transfer characteristics are optimum, concerning noise immunity;
- basic parameters (V_{TH}, V_{TL} and V_H) do not depend on technology.

Overall, the Schmitt inverter (section 3) yields the best characteristic. Its operating stability does not depend on transistor geometries ratio, nor on tolerance of technology. Symmetry of all transistors is optimum, concerning noise immunity and propagation time. While in standard circuits every input is joined by one pair, in Schmitt logic circuits every input is joined by two pairs of CMOS transistors. Taking into account that one pair closes positive feedback loop, it follows that *m*-input NAND and NOR Schmitt circuits are comprised of 2*m*+1 pairs of CMOS transistors. Transfer characteristic of NAND and NOR Schmitt circuits depend on, like those of the standard ones, the number of inputs, number of active inputs and the position of the first active input.

10. References

Al-Sarawi, S.F. Low Power Schmitt Trigger Circuits, *Electronics Letter*, vol. 38, 29th August 2002, pp 1009-1010.

Arrabi, S. A 90nm CMOS Data Flow Processor Using Fine Grained DVS for Energy Efficient Operation from 0.3V to 1.2V, *IEEE SSC Magazine*, Spring 2011, vol. 3, No. 2, pp 52-58.

Bundalo, Z., Dokic, B., Non-inverting Regenerative CMOS Logic Circuits, *Microelectronics Journal*, vol. 16, No. 5, 1985, pp 5-17.

Dokic, B., CMOS Schmitt triggers, *IEE Procedings - Part G*, vol. 131, No. 5, October 1984, pp 197-202. The paper has been copied to Express Information PEAVT, No. 21, 1985, pp 4-14, The Acadamy of Sciences of USSR.

Dokic, B., CMOS Regenerative Logic Circuits, *Microelectronics Journal* vol. 14, No.5, 1983., pp 21-30. The paper has been copied to Express Information PEAVT, No. 29, 1984, pp 4-14, The Acadamy of Sciences of USSR.

Dokic, B., CMOS NAND and NOR Schmitt Circuits, *Microelectronics Journal* 27, No. 8, November, 1996, pp 757-766.

Dokic, B. et al, CMOS Gates with Regenerative action at One of Inputs, *Microelectronics Journal*, vol. 19, No. 3, 1988, pp 17-20.

Dokic, B. and Bundalo, Z., Regenerative Logic Circuits with CMOS Transistors, *Int. J-Electronics*, 1985, vol. 58, No. 6, pp 907-920.

Dokic, B., Influence of Series and Parallel Transistors on DC Characteristics of CMOS Logic Circuits, *Microelectronics Journal*, vol. 13, No. 2, 1982, pp 25-30.

Dokic, B., Three-State BiCMOS Buffers with Positive Feedback, *Proc. 20th Int. Conf. on Microelectronics (MIEL '95)*, vol. 2, 1995, pp 509-511.

Hard, A. and Voinigesku, S. P., A 1GHz Bandwidth Low-Pass $\Delta\Sigma$ ADC with 20-50 GHz Adjustable Sampling Rate, *IEEE Journal of SSC*, vol. 44 No. 5, May 2009, pp 1401-1414.

Chien, H., Switch-controllable Dual-hysteresis Mode Bistable Multivibrator Employing Single Differential Voltage Current Converter, *Microelectronics Journal 42*, 2011, pp 745-753.

Katyal, V. et al, Adjustable Hysteresis CMOS Schmitt Triggers, *Circuits and Systems*, 2008 ISCAS 008. IEEE Int. Symposium on, pp 1938-1941.

Lo, Y. et al, Current-input OTRA Schmitt Trigger with Dual Hysteresis Modes, *Int. J. Circ. Theor. Appl. 2010*, vol. 38, pp 739-746.

Pengfei, L. et al., A Delay-locked Loop Synchronization Scheme for High-Frequency Multiphase Hysteretic DC-DC Converters, *IEEE Journal of SSC* vol. 44, No. 11, November 2009, pp 1401-1414.

Wang, C. et al., A 570kbps ASK Demodulator without External Capacitors for Low Frequency Wireless Bio-implants, *Microelectronics Journal* 39, 2008, pp 130-136.

Young, F., A High-speed Differential CMOS Schmitt Trigger with Regenerative Current Feedback and Adjustable Hysteresis, *Analog Integrated Circuits Process*, 2010, 63: pp 121-127.

Zou, Z., et al., A Novel Schmitt Trigger with Low Temperature Coefficient, *Circuits and Systems*, 2008, APCCAS 2008, IEEE Asia Pacific Conference on, pp 1398-1401.

2

Some Contributions at the Technology of Electrochemical Micromachining with Ultra Short Voltage Pulses

Richard Zemann, Philipp Walter Reiss,
Paul Schörghofer and Friedrich Bleicher
*Vienna University of Technology,
Institute for Production Engineering and Laser Technology,
Austria*

1. Introduction

The tendency to make progressively smaller and increasingly complex products is no longer an exclusive demand of the electronics industry. Many fields such as medicine, biomechanical technology, the automotive, and the aviation industries are searching for tools and methods to realize micro- and nanostructures in various materials. The micro-structuring of very hard materials, like carbides or brittle-hard materials, pose a particularly major challenge for manufacturing technology in the near future. For these reasons the Institute for Production Engineering and Laser Technology (IFT) of the Vienna University of Technology is working in the field of electrochemical micromachining with ultra short voltage pulses (μPECM) in nanosecond duration. With the theoretical resolution of 10 nm, this technology enables high precision manufacturing. [Kock M.]. A question, which can illustrate the motivation to do this research work in this field, is: "Which parameters have to be set at a production machine and which framework conditions have to be managed to reach a desired result?" To answer this question for the materials nickel and steel (1.4301), the IFT has done experimental work.

2. Electrochemical micromachining

Basically, the term machining stands for the removal of material. Furthermore, micromachining is the production of very small scaled shapes and parts in the range of 100 μm - 0,1 μm. DIN 8580 is the classification of all manufacturing processes. Figure 1 illustrates DIN 8590 for ablation, which is a part of DIN 8580.
Ablation is a non-mechanical separation of material. It can be divided into chemical, thermal and electrochemical methods. For example water jet cutting is not yet assigned to either ablation methods or to cutting methods. Electrochemical micromachining (ECM) uses electrochemical reactions to treat a metal work piece. These reactions are for example processes in an electrolyser or a battery. In electrolysers the chemical reaction is driven by an externally applied voltage, whereas in a battery a voltage is created by a chemical reaction. As depicted in figure 1, the group of electrochemical processes are assigned to

ablation, which is a non-cutting technology. Cutting technologies for the realization of microstructures, like high speed cutting, induce mechanical stress, and thermal technologies, like laser ablation, induce thermal stress upon the work piece. Due to the fact that electrochemical technologies have none of these disadvantages, they are of interest to many industrial cases. No stress is induced in the work piece, therefore the structure of the work piece remains unchanged. Another advantage is that there is no machining force necessary and thus it is possible to machine areas which are difficult to reach. Pulsed electrochemical micromachining (PECM) as well as electrochemical micromachining with ultra short pulses (µPECM) belong to the electrochemical micromachining methods. Figure 2 shows the voltage-current curve of metal dissolution. This curve is segmented in active dissolution, passivity and trans-passive dissolution. PECM is positioned in the trans-passive section of the curve (2) whereas µPECM is positioned in the active metal dissolution area (1). Once a voltage of ε_P is reached, the current slopes down rapidly. The current remains low until the end of the passive section. At further increase of the voltage the current rises again to the trans-passive section. Machines, which are working with technologies in the range of active metal dissolution are more precise but obtain lower removal rates as others working in the trans-passive range.

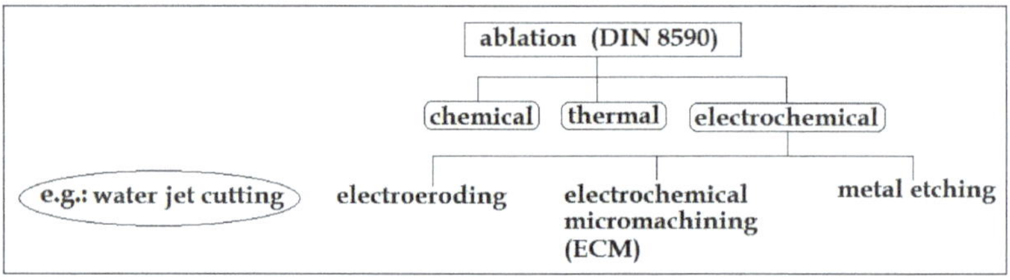

Fig. 1. Classification of ablation (DIN 8590)

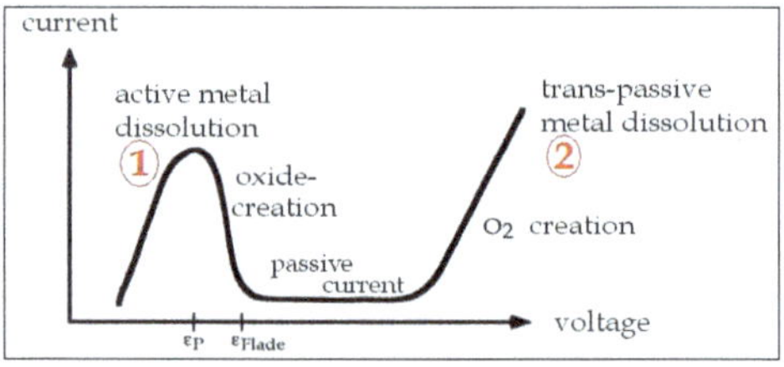

Fig. 2. Schematic illustration of current-voltage curve for metals: The three characteristic sections are: active dissolution, passivity and trans-passive dissolution

Figure 3 shows the main differences of the electrochemical micromachining methods. The conventional ECM uses direct current as energy source. Whereas both PECM and µPECM, use pulsed energy sources, the major difference between these technologies is the pulse width. While the PECM uses pulse widths from milli- to microseconds, the electrochemical micromachining with ultra short pulses uses pulse widths from micro- to picoseconds.

For PECM the removal rate is dependent on the current density distribution. µPECM directly controls the working gap by locally charging and discharging the so called electrochemical double layers. This leads to the advantage of µPECM, that the spatial confinement of electrochemical reactions and the thereby produced resolution is very high.

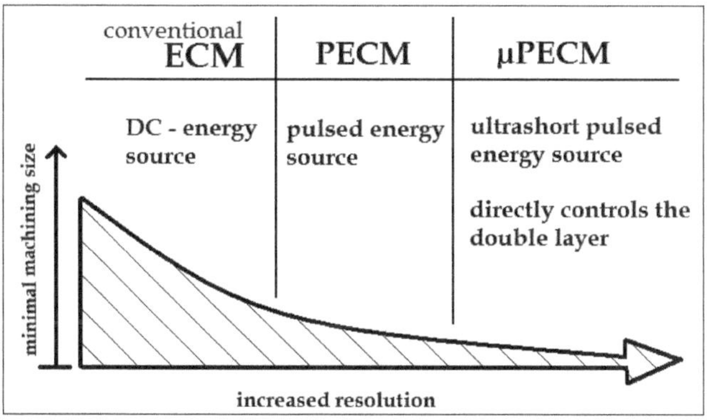

Fig. 3. Comparison of the electrochemical micromachining methods in the field of resolution

3. Electrochemical micromachining with ultra short voltage pulses (µPECM)

3.1 Method and procedure

Electrochemical micromachining with ultra short voltage pulses was developed at the Fritz-Haber-Institute of the Max-Planck-Corporation. Furthermore this innovative method for micromachining was published for the first time in the beginning of 2000. Other universities and companies working on similar topics can be found in Germany, Poland, Korea, and Austria. Since late 2010 the Institute for Production Engineering and Laser Technology (IFT) at the Vienna University of Technology has been working with this method as well. The IFT is striving to deliver machining strategies, new material–electrolyte combinations and production parameters for the industrial applicability. The machining technology of µPECM is based on the already well-established fundamentals of common electrochemical manufacturing technologies. The major advantage of the highest manufacturing precision is derived from the extremely small working gaps that are achievable through ultra short voltage pulses. This describes the main difference to common electrochemical technologies. As previously stated general advantage of electrochemical machining technologies is that the treatment of the work piece takes place without any mechanical forces or thermal influences. Therefore, no abrasive wear of the tool occurs and aspect ratios of >100 are possible which sets the basis for extremely sharp-edged geometries. There is no unintentional rounding of edges and no burring on the part.

These days appropriate electrolytes have already been found for several nonferrous metals such as nickel, tungsten, gold etc., as well as alloys like non-corroding steel 1.4301. Nevertheless, a main research focus for the Institute will be the search for new material-electrolyte combinations to expand the field of application for this technology and to enhance its manufacturing productivity. This needs to be accomplished in order to fulfil the requirements of industrial production because in industries such as the automotive sector the production rate is very important. At the Nano-/Micro-Machining-Center of the IFT, an assortment of high quality measuring devices is available. Based on the technology of µPECM and on the use of high end measuring devices, specimens and parts in the micrometer range are to be manufactured and analyzed in order to investigate material removal rates and the accuracy of resulting work piece geometries.

Due to the multidisciplinary nature of this technology, intensive cooperation with other institutes of the Vienna University of Technology in the fields of electro-technical engineering, high frequency technology and electrochemistry is established. The goal of this research will be to elevate this technology to an appropriate level of possible industrial usage by enhancing the manufacturing accuracy and the process efficiency for current components. Therefore a profound knowledge of material science, electrochemistry, and production technology for extremely small dimensions will be required. The necessary expertise in these fields will be provided by the cooperating institutes and interested companies.
To accomplish these improvements in the technology of electrochemical micromachining with ultra short pulses it will be necessary to merge several research projects which are currently dealing with the topics of piezo-driven nano-positioning devices and the development of high precision machine structures for different types of machines. Table 1 shows all the relevant adjustable parameters for µPECM. In addition to the proper choice of the electrical process parameters like the amplitude of the pulses, the pulse width, the voltages at the tool, and the work piece, the right choice of electrolyte is probably the most important aspect for this process.

Adjustable parameters for the process	abbraviations
amplitude of the pulses	A
pulse width	p
voltage at the tool	T
current through the backing electrode	I
pulse–pause ratio	ppr
diameter of the tool	D
electrolyte solution	E

Table 1. Adjustable parameters which have an influence on the process

In figure 4, the relevant parameters of the applied voltage pulses are illustrated. The duty cycle is the sum of the pulse width and the pause time. A pulse width of 100 ns and a pause time of 800 ns conforms a pulse–pause ratio of 1/8.

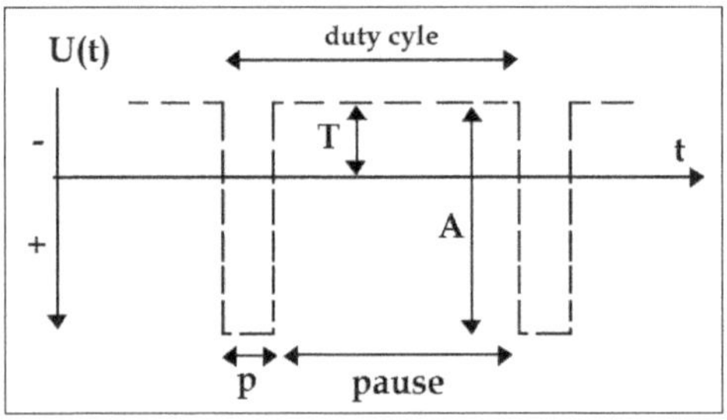

Fig. 4. Pulse-pause ratio of the applied voltage pulse, with pulse width p, length of pause, amplitude A, tool voltage T, applied pulsed voltage signal U(t)

Due to the fact that µPECM is one of the latest elaborated removal technologies, there are no fully developed machines available in the market. All the institutes and companies, which investigate these fields, work with machines in laboratory stage. The machine at the IFT is simple constructed and very easy to maintain, consequently it is adequate for industrial usage. However, a more complex machine structure would give the possibility to reach the highest precision requirement. Figure 5 shows a view inside the IFT´s machine. The whole machining process takes place in a basin filled with an electrolyte solution that has to be adequately adapted to the work piece material used. At the bottom of this electrolyte basin a hole for the connection of work piece and machine can be found. It is important that the basin is well sealed, so that no leakage can occur. The basin is made of Teflon, which has resistance against the electrolytes used in the experiments. Even when filling the basin, caution is required due to the fact that once in contact with the electrolyte, the surface of the material could begin to react. To protect the work piece surface from the influence of the electrolyte-solution, a cathodic protection-current is applied by the backing electrode which is immersed in the electrolyte. At the IFT, a tungsten wire is the preferred tool for the electrochemical micromachining with ultra short voltage pulses. With the basin filled as needed, the process of work piece calibration can be performed.

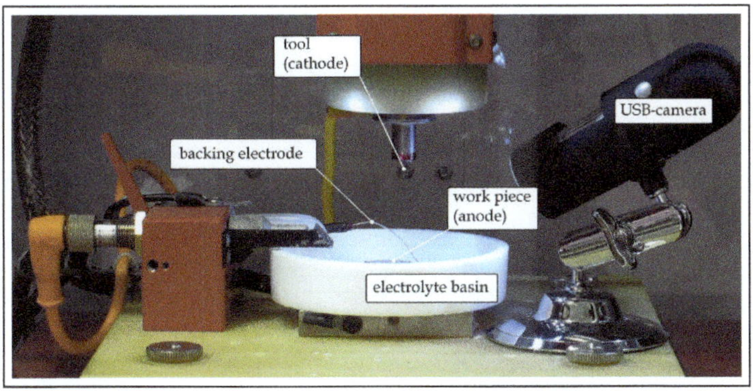

Fig. 5. View inside the electrochemical machine with all important parts for the manufacturing process labelled

The measurement process for finding the work piece surface coordinate is executed automatically by the machine. Therefore a tool potential is necessary to detect the electrical short circuit thru a contact between work piece and tool. Another possible measurement process is to match the local coordinate systems of the work piece with the global coordinate system of the machine structure. With the result of this measurement process and three positioning screws on the plate, whereon the electrolyte basin is mounted, it is now possible to get the necessary congruence between these two coordinate systems. Then the manufacturing program, which conforms to a standard CNC-program, is started. The tool moves along the pre-programmed paths and selectively ablates material due to the principle, that is based on the finite time constant for double layer charging, which varies linearly with the local separation between the electrodes. During nanosecond pulses, the electrochemical reactions are confined to electrode regions in close proximity. [Schuster R.].
To view the manufacturing process and get optical magnification, a USB–camera is used.
Similar to conventional electrochemical manufacturing methods the μPECM process uses an oppositional electric voltage for the work piece and the tool. At the phase boundaries between the tool and the electrolyte and also between the work piece and the electrolyte, an electrochemical double layer is formed. [Schuster R.]

Figure 6 shows the detailed structure of the double layer. The double layer consists of a rigid, outer Helmholtz layer (OHL) and a diffuse area. The inner Helmholtz layer (IHL) is a part of the OHL. In the diffuse area the hydrated metal ions are versatile. The functionality of the OHL can be understood basically as a kind of a plate capacitor, with a plate separation of half of the atom radius. [Hamann C.H.]

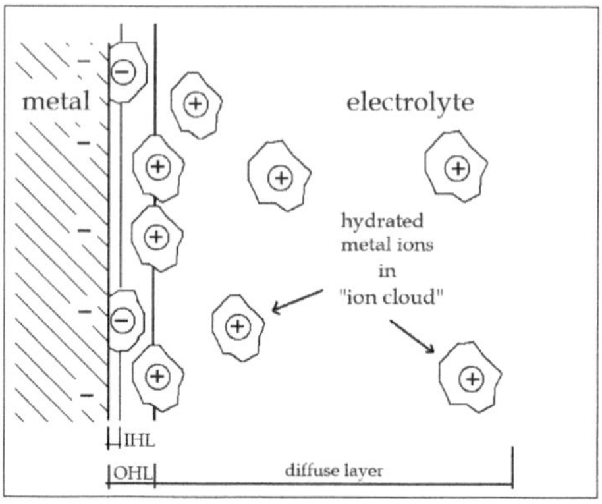

Fig. 6. Simplified Stern-Graham-Model of the electrochemical double layer [Hamann C.H.]

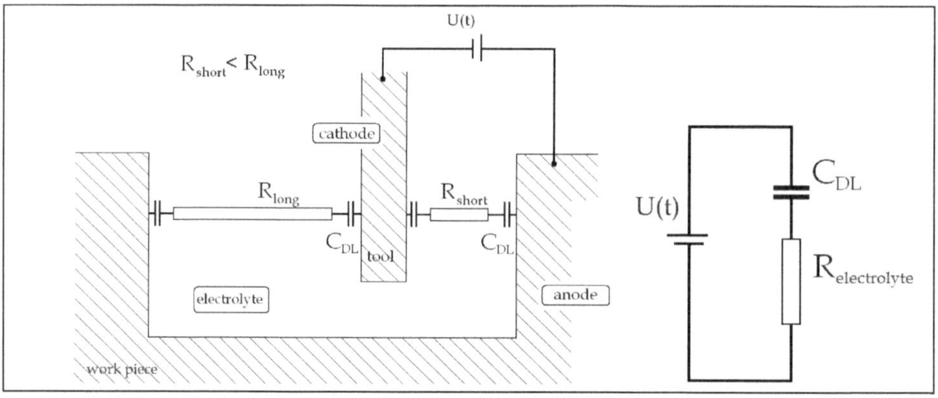

Fig. 7. Schematic illustration of the electrochemical double layers as capacitors and the electrolyte as electrical resistor between tool and work piece (left) and the equivalent circuit diagram (right) with U(t) as energy source, C_{DL} as capacitance of the double layers and $R_{electrolyte}$ as the ohmic resistor of the electrolyte.

The left section of figure 7 shows the schematic illustration of the tool, the work piece in the electrolyte basin, and the electrochemical double layers illustrated as plate capacitors. The electrolyte has comparable characteristics to a linear ohmic resistor with a value that is dependent on the length of the current path. The length of the current path is equal to the distance between the tool and the work piece. The right section of figure 7 shows the equivalent circuit diagram in a simplified version of the left illustration in figure 7. Through charging and discharging the electrochemical double layer, metal ions are solvated out of

the metal surface. If the voltage pulse width is very short, the erosion takes place very closely to the tool (R_{short}), since the ohmic resistance of the electrolyte prevents ablation at areas further away from the tool (R_{long}) due to the double layer capacitor not being able to be sufficiently recharged. [Zemann R.]
The right illustration in figure 8 shows schematically the two different charging curves of the double layers at the work piece for R_{short} and R_{long}. At smaller distances between the tool and the work piece, the charging curve is steeper; this leads to the formulas (1) and (2).

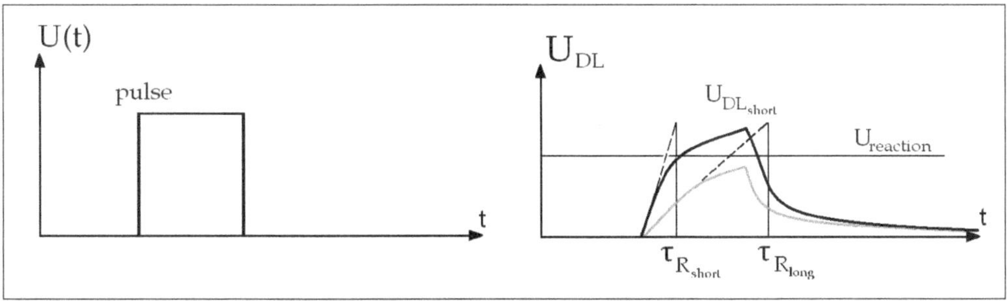

Fig. 8. Applied voltage pulse (left) and time variable voltage curve in the electrochemical double layer (right)

$$\tau = R_{electrolyte} \bullet C_{DL} \tag{1}$$

τ time constant for double capacitor charging
$R_{elektrolyte}$ resistance of the electrolyte
C_{DL} capacitance of the electrochemical double layer

$$U_{DL} = U(t) \bullet \left(1 - e^{(-t/\tau)}\right) \tag{2}$$

U_{DL} charging voltage of the electrochemical double layer
U(t) applied voltage with dependence on time
τ time constant for double capacitor charging

Another important influence on the charge of the double layers has the pulse width and the choice of the electrolyte. Small working gaps between the tool and the work piece of less than 1 μm are produced with pulse widths of less than 100 nanoseconds and lead to a very high resolution of the machined structure. Even more accurate machining can be achieved with pulse widths of less than 1 nanosecond and by separating the processing pulse into a pre-pulse and a main pulse, which is a future research topic for the IFT. In order to elaborate on the research work concerning the technology of using ultra short voltage pulses, the relevant demands of industry, basically increasing the material removal rate, has to be considered as a main goal. Subsequently, an increase in the already high machining accuracy is regarded as a principal target.
Another major advantage of this technology is the possibility to reverse the process electrically. This means that not only the work piece can be machined, but also the tool itself can be defined as the work piece and be machined to its ideal geometry without any further set-up. Regarding all these functionalities, the requirements for precise micromachining are

met. Possible tasks that can be performed with this machining centre include: tooling, milling, turning, sinking, and measuring.

Characteristics of the µPECM process with ultra short voltage pulses:

- High precision (theoretical resolution of 10 nm)
- No thermal load
- No mechanical process forces
- High aspect-ratio >100 (only limited thru the young's modulus of the material)
- No tool wear
- Small working gaps between tool and work piece (< 1 µm)
- Manufacturing of hard materials
- Very small edge-rounding
- No burring
- Adjustable roughness of the work piece surface
- High quality measuring function

Table 2 shows that electrochemical micromachining with ultra short voltage pulses has several advantages compared to other nano- and micromachining technologies. For example the theoretical dissolution range and the aspect ratio are outstanding, whereas in case of the removal rate, µPECM is not competitive against technologies like high speed cutting. For material removal, µPECM is mainly used for post-processing and for producing surfaces with hydrophobic and hydrophilic characteristics at the moment.

	theoretical dissolution range	**aspect ratio**	**treatable materials**	**category**	**removal rate**
µPECM	limit: 10 nm	> 100	electrochem. active materials	electrochem. micro-machining	*
Lithography	>10 nm	~ 1	etch-able, evaporable materials	chemical method	**
LIGA	~ 100 nm	~100	galvanic removable materials	mechanical/ thermal method	**
Laser ablation	~ µm	~ 1	metals and dielectrics	thermal method	**
high speed cutting	~ µm	~1	metals and polymers	cutting method	***
FIB	~ 30 nm	~ 10	conducting materials	thermal method	**
EDM	~ µm	~ 10	metals	thermal method	**

LIGA is the acronym for lithography (LI), electroforming (G) and molding (A)
FIB focussed ion beam milling
EDM electric discharge machining

Table 2. Comparison of nano- and micromachining methods [Kock M.]

3.2 Tooling

The favoured material used for the tool is tungsten. Tungsten can be easily treated with NaOH as electrolyte and has preferable mechanical properties like a Mohs hardness of 7,5 and a Young´s modulus of 410 GPa. For the experimental work wires with a diameter of 75 and 150 μm were used. The first tooling step is, to cut the tungsten wire manually to a length of 15 - 20 mm. The wire is fixed with a collet in the toolholder and should protrude far enough to produce the necessary geometries, mostly that is about 4 - 5 mm. The toolholder has to be protected from the acid to prevent corrosion, which is performed by a layer of Lacomit. It is a dark red fluid, once hardened it isolates the toolholder against the electrolyte. This red fluid functions as a barrier between the electrolyte and the toolholder. Only the top of the upper part of the tungsten wire is free of Lacomit to treat the work piece. Figure 9 shows two toolholders with the different diameters of tool wire.

Fig. 9. Tools ready for manufacturing. The left tool has a diameter of 75 μm and the right tool a diameter of 150 μm, both with Lacomit layer.

As mentioned before the tool/wire is cut off manually. Due to the mechanical characteristics of tungsten it is possible that the cut end splits. If that happens the split section and the usual cut end of the tool (figure 10, left) has to be removed.

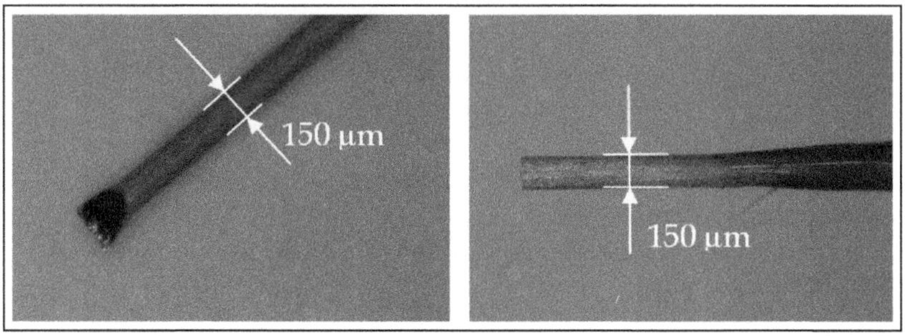

Fig. 10. Tungsten wire with a diameter of 150 μm, untreated with the end after manual cutting (left) and the finished end after electrochemical flattening (right).

The flattening process is performed directly in the μPECM machine. Due to the fact that the spatial resolution and pulse width are linearly related: the higher the pulse width, the higher the spatial resolution [Kock M.], the flattening process is split into two parts to produce a tool with high quality. Another advantage of this sequential machining is that the machining time is reduced. At first a large pulse width (i.e. 400 ns) is used to increase the removal speed of the cut end. Afterwards a smaller pulse width (i.e. 80 ns) is used to create a sharp edged tool with a glossy surface. Only with such tools it is possible to produce geometries with sharp edges on the work piece. Figure 11 illustrates the difference of the radius on the tool´s top for small and large pulse widths.

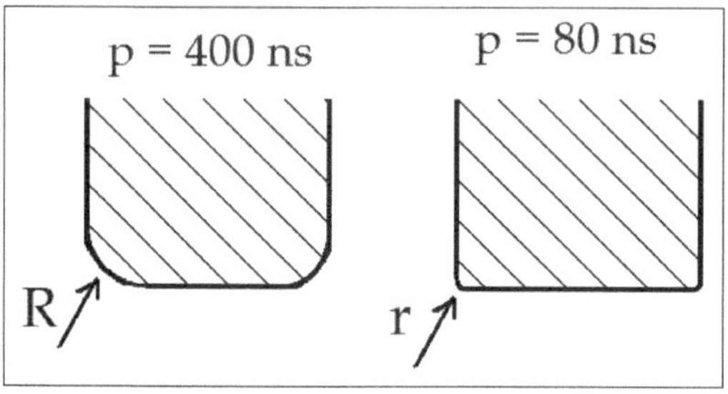

Fig. 11. Influence of the pulse width on the radius on the top of the tool

3.3 Manufacturing of nickel

Nickel is a hard (Mohs hardness: 3,8) and ductile metal with a silvery-white and slightly golden shine. Nickel is apart from chrome and molybdenum an important element for the refinement of steel. The ferromagnetic metal is corrosion-resistant. Nickels protective oxide surface resists most acids and alkalis. The corrosion-resistance is one of the most important characteristics of parts in laboratory environments or health care, therefore nickel is the common material in those branches. For the electrochemical manufacturing of nickel the electrolyte hydrochloride acid (HCl) is used. HCl deactivates the passive surface of nickel and renders the material processable. The following experiments were done to find the optimal processing parameters for the manufacturing of products and special surfaces made of nickel. To evaluate the outcome of the experiments, the produced structures were measured with a high-end optical measuring device. Also optical considerations through a light microscope helped to evaluate the following characteristics of the produced surfaces:

- shape / geometry
- topology (smoothness of the bottom surface)
- shine of the surface
- edge rounding

3.3.1 Pulse width (p) and amplitude (A)

In the first experiment the pulse width and the amplitude of the pulse were varied in order to see which effects the adjustment of these parameters cause. The experimental setup is a block with five parallel grooves. Every groove is made with different pulse widths from 400 ns to 80 ns. A sketch of the groove geometry is illustrated in figure 12. Overall four of these blocks with different amplitudes were manufactured. The range of the amplitudes was from

3000 mV to 2100 mV in 300 mV steps. After measuring the width of every groove, the working gap can be calculated via formula (3).

$$D + 2a = B \tag{3}$$

D tool diameter in µm
a working gap in µm
B measured width of the groove in µm

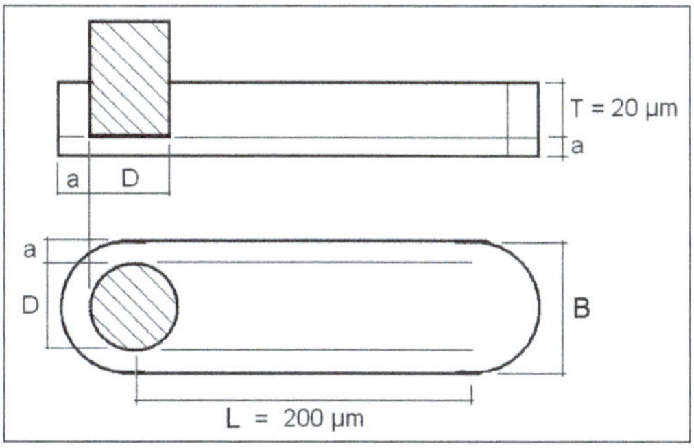

Fig. 12. Sketch of the produced groove

The diagram in figure 13 shows that a smaller pulse width reduces the working gap. The optical estimation shows that grooves made with lower pulse widths have much better optical qualities (figure 13, left). This outcome can be explained by the localization of the manufacturing reactions. Smaller voltage pulses lead to a spatial confinement of the electrochemical reactions so that the working gap shrinks and the geometry gets more precise which is confirmed in figure 13, right. As a consequence, the pulse width is the most important parameter for the machining precision. Dependent on the machine, the minimal pulse width of p = 80 ns is further used in the experiments to produce grooves in high quality. The adjusted electrochemical parameters for this experiment are indicated in table 3.

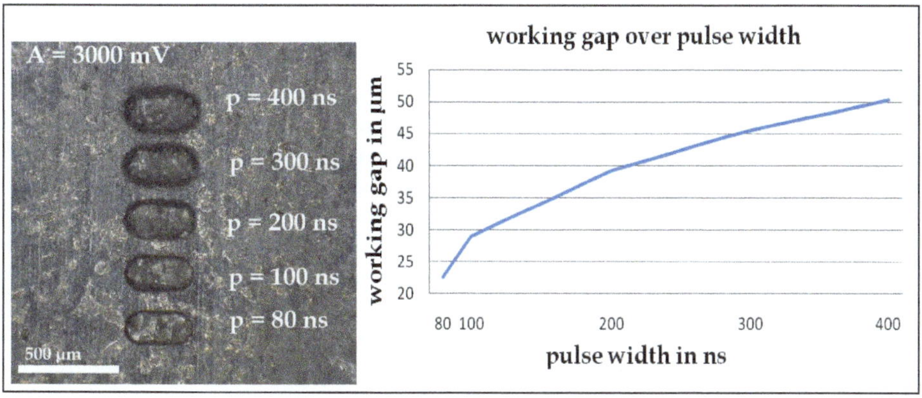

Fig. 13. Illustration of grooves (left) - from top downwards different pulse widths were used. Diagram of the appurtenant working gaps over pulse widths (right).

A = 3000 mV	p = varied	T = 200 mV	E = 1M HCl
I = 1000 μA	ppr = 1/8	D = 150 μm	

Table 3. Adjustments for the experiment of figure 13

Figure 14 shows that similar to the pulse width the reduction of the amplitude causes a reduction of the working gap. At a pulse width of p = 80 ns an amplitude of less than 3000 mV does not lead to a removal of material, due to the fact that the double layers cannot be sufficiently charged with the provided energy. Equally the provided energy of 2400 mV amplitude and 100 ns pulse width is not sufficiently for production. The overview of the production parameters for these experiments is mentioned in table 4.

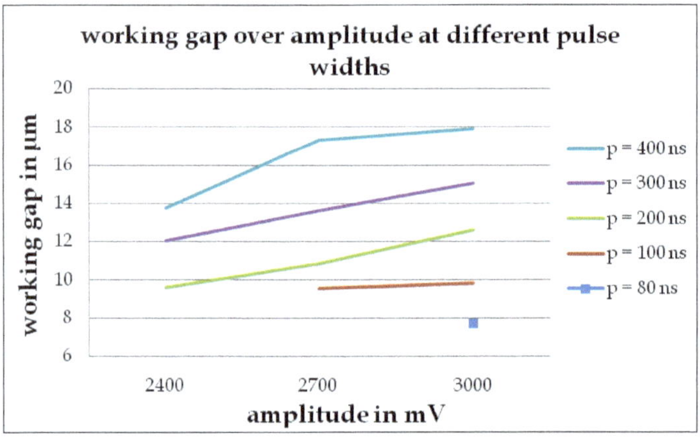

Fig. 14.Working gaps over amplitude at different pulse widths.

A = varied	p = varied	T = 200 mV	E = 0,2M HCl
I = 1000 μA	ppr = 1/8	D = 75 μm	

Table 4. Adjustments for the experiment of figure 14

3.3.2 Electrolyte-concentration

The concentration of the electrolyte is a very important parameter for the electrochemical processing. In the equivalent circuit diagram of the electrochemical cell, the electrolyte is equal to an ohmic resistor. For this experiment hydrochloric acid (HCl) in three different concentrations was used to explore the correlation between the electrolyte-concentration and the working gap. The diagram in figure 15 shows that the reduction of the electrolyte concentration leads to smaller working gaps. This outcome can be explained by the reduced conductivity of the electrolyte and the following localization of the reactions.

A reduction of the concentration increases the resistance because of the lack of ions in the aqueous solution. In such solutions ions are the charge carriers and therefore responsible for the electric conductivity. The illustration in figure 15 shows the optical differences of changed electrolyte concentrations. The processing parameters for this experiment are indicated in table 5.

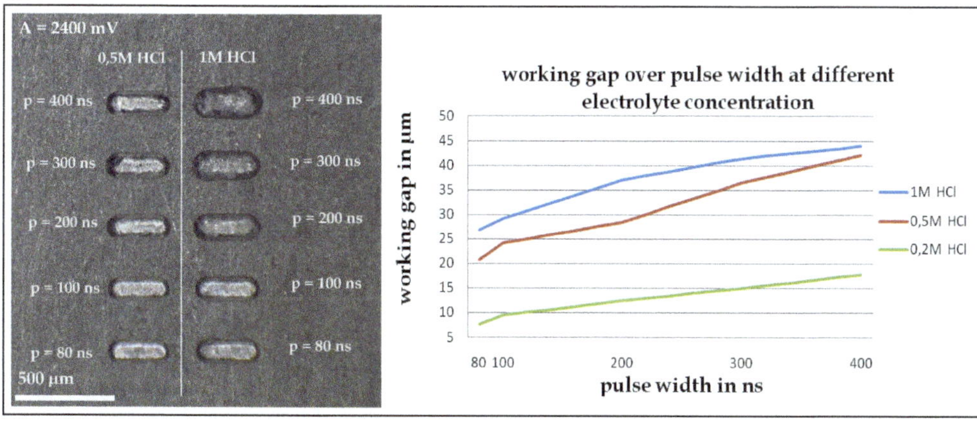

Fig. 15. Image of grooves made with 0,5M HCl and 1M HCl for A = 2400 mV (left). Working gap over pulse width at different electrolyte concentrations for A = 3000 mV (right)

A = varied	p = varied	T = 200 mV	E = varied
I = 1000 µA	ppr = 1/8	D = 75 µm	

Table 5. Adjustments for the experiment of figure 15

3.3.3 Current through the backing electrode (I)

To investigate the influence of the current through the backing electrode, the current was varied between 500 µA and 4000 µA. The results in figure 16 (left) show an increased processing time at higher currents. The minimal working gaps are in the range of 2000 to 3000 µA, as illustrated in figure 16 (right). Because of the optical criteria and the working gap a current of I = 2000 µA was used for further experiments. The illustration in figure 17 shows the difference between a high-quality and a low-quality groove. The electrochemical parameters for this experiment are shown in table 6.

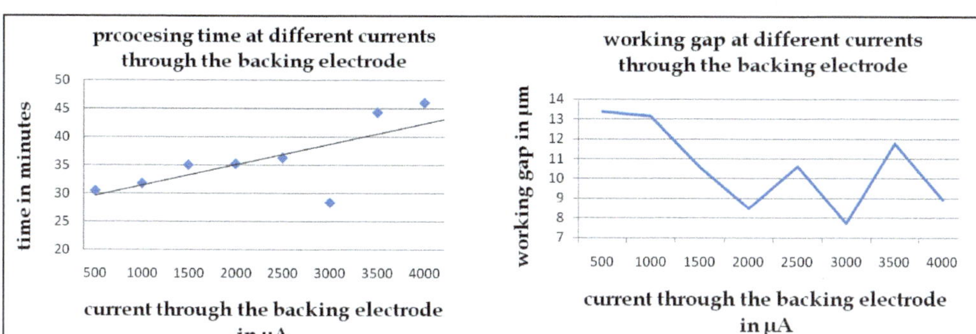

Fig. 16. Processing time at different currents (left) and working gap at different currents (right)

Fig. 17. Image of grooves with I = 2000 µA (above) and I = 500 µA (below)

A = 3000 mV	p = 80 ns	T = 200 mV	E = 0,5M HCl
I = varied	ppr = 1/8	D = 75 µm	

Table 6. Adjustments for the experiment of figure 17.

3.3.4 Tool voltage (T)

For successful application of ultra short voltage pulses for electrochemical machining, the electrochemical conditions, e.g. the average electric potentials of the tool (T) and the work piece have to be precisely controlled. These potentials are independently adjusted by a low-frequency bipotentiostat and a platinium backing electrode. [Kock M.]

To investigate the influence of T, seven grooves with different tool voltages were produced. The production parameters for this manufacturing are indicated in table 7. After the measurement and evaluation of the working gap via formula (3), the results show that between -100 mV and + 100 mV the working gap reaches a minimum (figure 18, left). The optical appearance of these grooves has also the highest quality (figure 18, right). Another advantage is that the processing time decreases with lower tool voltages. For the further experimental work a tool voltage of +100 mV was used.

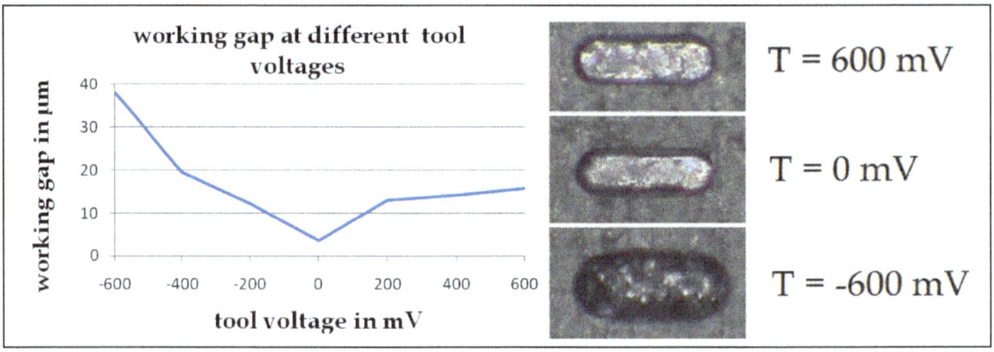

Fig. 18. Working gap at different tool voltages (left), image of grooves with T = 600 mV, 0 mV and -600 mV

A = 3000 mV	p = 80 ns	T = varied	E = 0,5M HCl
I = 2000 µA	ppr = 1/8	D = 75 µm	

Table 7. Adjustments for the experiment of figure 18

3.3.5 Pulse-pause ratio

The pulse-pause ratio is an important parameter that influences the electrochemical reactions. To ensure a precise and fast dissolution of the material, the ratio of pulse time to pause time should be correctly chosen. Every single pulse that charges the electrochemical double layer dissolves a monolayer of atoms from the material into the electrolyte solution. Due to the fact, that one monolayer of atoms is a very small amount of material the pulses must be applied with very high frequency to solvate the material in a reasonable rate. If the ratio is too high, the process time is unnecessarily lengthened as these rates obey an exponential law (Butler-Volmer equation). To find an appropriate pulse-pause ratio, five grooves with a different ppr-parameter were produced. Figure 19 shows a decreased removal rate at higher pulse-pause ratios for the drilling and milling processes. All of these grooves have the same working gap with negligible deviations in the range of maximal 5 µm. There is great potential to speed up the process by reducing the pulse-pause ratio without losing much precision. The used parameters for the experiment are specified in table 8. Considering the optical estimations, a pulse-pause ratio between 1/6 and 1/8 is recommended.

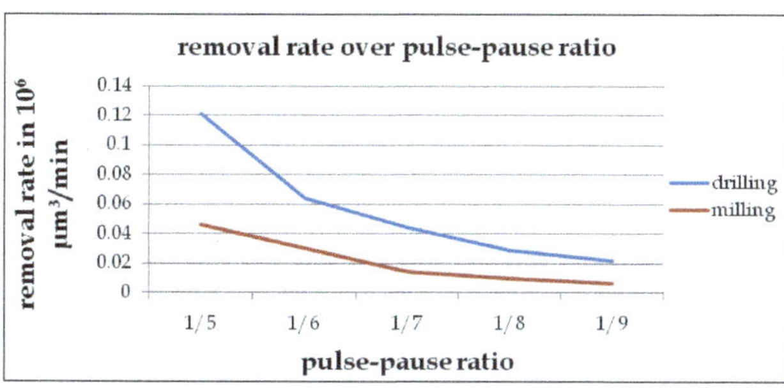

Fig. 19. Removal rate over pulse-pause ratio

A = 3000 mV	p = 80 ns	T = 100 mV	E = 0,5M HCl
I = 2000 µA	ppr = 1/8	D = 75 µm	

Table 8. Adjustments for the experiment of figure 19

3.3.6 Drilling with µPECM

In this experiment the maximum possible drilling depth should be found. The drilling process works without any problems to a depth of 140 µm. All over the removal speed slows down slightly. At a depth of 140 µm the drilling speed slows down rapidly and the experiment has to be stopped. An explanation is that in this depth the exchange of electrolyte is not sufficient, so the dissolved metal ions saturate the electrolyte in the drilled hole and prevent any further metal dissolution. This can be disabled by an alternately up and down movement of the tool to realize a kind of flushing (pulsed mechanical movement). In figure 20 the removal speed over drilling depth is shown. Table 9 indicates the drilling parameters for the process.

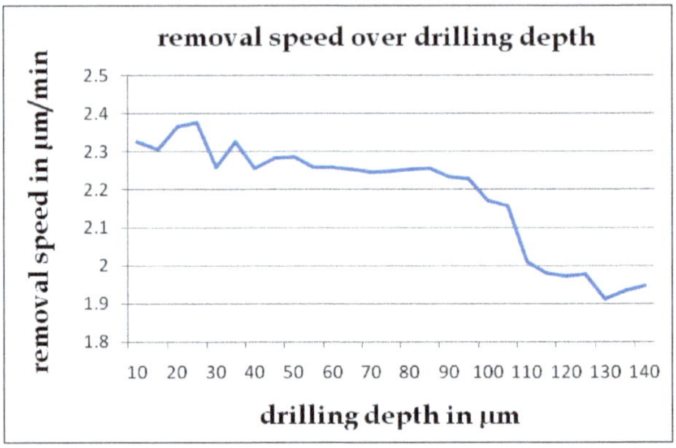

Fig. 20. Removal speed over drilling depth

A = 3000 mV	p = 80 ns	T = 100 mV	E = 0,5M HCl
I = 2000 µA	ppr = 1/8	D = 75 µm	

Table 9. Adjustments for the experiment of figure 20

3.3.7 Dwelling time

For this experiment the tool was positioned 4 µm above the nickel surface and remained at this position for different time periods. At the first position the dwelling time was 0 seconds. On each position the dwelling time was doubled to finally 640 seconds. The longer the pulses are applied, the more material is removed (figure 21). At 0 seconds only a scratch was produced. At higher dwelling times the holes are deeper. Finally, the removal rate decreases and a maximum gap will be developed. The electrical resistance between tool and work piece grows with the distance of them, until finally no more reaction/dissolution is possible. A referential groove was produced for the measurement. It is very important to adjust an optimized machine feed rate, because longer dwelling times lead to enlarged working gaps. The Adjustments for this experiment are illustrated in the table 10.

A = 3000 mV	p = 80 ns	T = 100 mV	E = 0,5M HCl
I = 2000 µA	ppr = 1/8	D = 75 µm	

Table 10. Adjustments for the experiment of figure 21

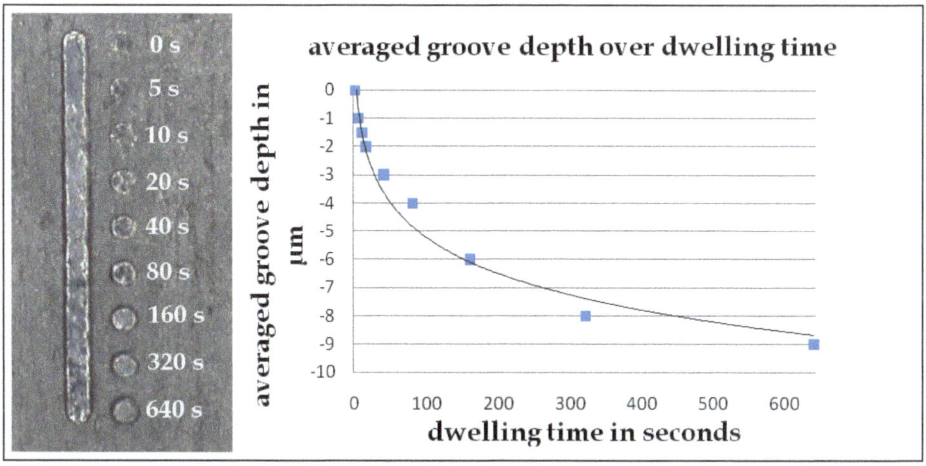

Fig. 21. Averaged groove depth over dwelling time

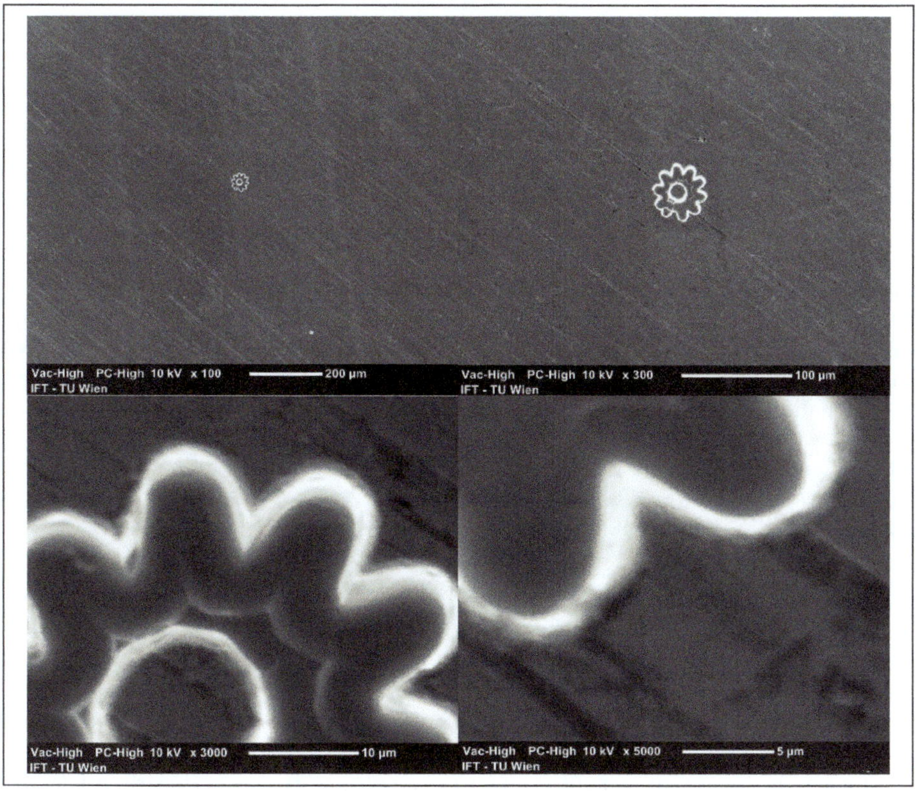

Fig. 22. Images of the microstructure, photographed with a scanning electron microscope (SEM) at different resolutions

3.3.8 Part production (micro injection mould)

The manufactured microstructure in figure 22 has an overall diameter of less than 50 μm, is 15 μm deep, and approximately shaped like a gearwheel. This microstructure was manufactured in 4 hours, with an electrolyte concentration of 0,2M HCl. The tool for this experiment (figure 23) was made out of a tungsten wire with diameter D = 150 μm by successively reducing the diameter in the tooling basin to < 5 μm. The magnification of 45 in a light microscope was not sufficient to examine the structure; therefore, a scanning electron microscope has to be used. The experiment shows that the production of a micro injection mould in a range < 100 μm is possible with the IFT´s machine.

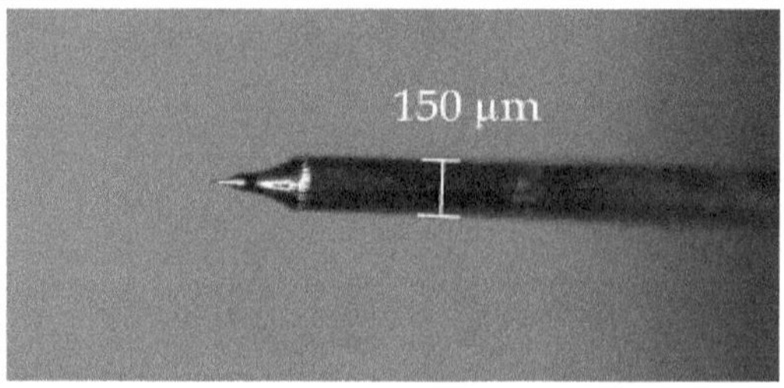

Fig. 23. Image of the tool to produce the micro injection mould with a top of D < 5μm.

A = 3000 mV	p = 80 ns	T = 100 mV	E = 0,2M HCl
I = 2000 μA	ppr = 1/8	D < 5 μm	

Table 11. Adjustments for the experiment to produce a micro injection mould

3.4 Manufacturing of steel (1.4301)

1.4301 steel is the most widely used non corroding steel and it has a very broad scope of application. The need of micro-structuring of such a standard material is continually growing. A solution of hydrofluoric acid and hydrochloric acid was used as electrolyte. The exact designation of this electrolyte solution is 3% HF/3M HCl. As previously mentioned, four criteria were used for the optical consideration of the grooves. These are:

- shape/ geometry
- topology/ smoothness of the bottom surface
- shine of the surface
- edge rounding

The experiments on 1.4301 were the same as on nickel with the difference that the electrolyte was not changed.

3.4.1 Pulse width (p) and amplitude (A)

Grooves with a length of 200 μm and a depth of 20 μm were manufactured. Thereon the amplitudes and the pulse widths were varied and the optical consideration of the grooves was performed to classify the results. The spatial resolution is almost linearly related to the pulse width. [Kock M.]. Figure 24 confirms this as the working gap shrinks with the reduction of the pulse width. The combination with the highest manufacturing precision was A = 2800 mV and p = 100 ns. The production with shorter pulse widths with the tool diameter of 150 µm was not possible. The energy applied by shorter pulse widths or lower amplitudes was not sufficient to recharge the double layer in order to realize material removal. By increasing the amplitude it was possible to finish grooves made with a pulse width of 80 ns, but the overall result was not favorable. The overview of the used parameters for the experiment shown in figure 24 is illustrated in table 12.

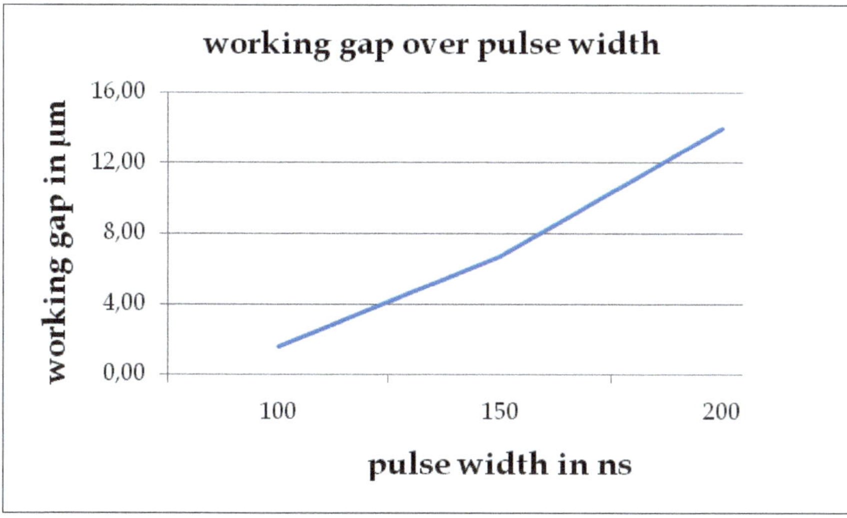

Fig. 24. Working gap over pulse width for A = 2800 mV

A = varied	p = varied	T = 100 mV	E = 3% HF/3M HCl
I = 1500 µA	ppr = 1/8	D = 150 µm	

Table 12. Adjustments for the experiment of figure 24

3.4.2 Current through the backing electrode (I)

This experiment was performed to show the influence of the cathodic protection-current on the process. The applied current protects the work piece in the electrolyte from corrosion or any other reactions. Eight grooves with the same dimensions as in the experiment before were made with I from 4000 to 500 µA. An obvious trend of how the cathodic protection-current influences the process could not be observed from the series of grooves. The results show that I from 3000 to 4000 µA achieves the smallest working gap and the best surface condition. Figure 25 shows two grooves with an obvious optical difference. Topology of the ground, sharpness of the edges, and form of the groove is much better with I = 3000 µA. Therefore, I has to be fixed at 3000 µA for the next attempts. All other electrochemical parameters for this experiment are indicated in table 13. During this phase of the experiments, the choice of which of the parameters to fix was dedicated by the optical assessment and the working gap measurement and not yet by the removal rate.

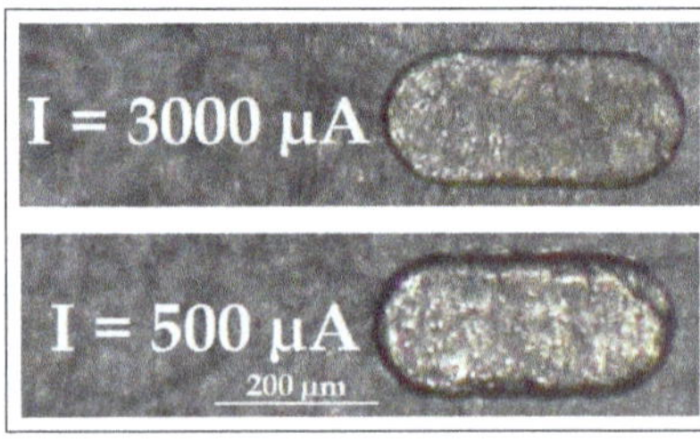

Fig. 25. Image of grooves made with I = 3000 µA above, respectively I = 500 µA below

A = 2800 mV	p = 100 ns	T = 100 mV	E = 3% HF/3M HCl
I = varied	ppr = 1/8	D = 150 µm	

Table 13. Adjustments for the experiment of figure 25

3.4.3 Pulse-pause ratio

The idea of this experiment was a variation of the pulse-pause ratio from 1/5 to 1/11. Figure 26 shows the manufactured grooves of the ppr experiment. The manufacturing parameters of this process are illustrated in table 14. For this experiment the voltage at the tool was zero. An experiment with the potential at the tool has shown that a very low voltage leads to the best results in case of the optical considerations. But these low tool voltage could bring up some problems.
When the drilling depth is higher, it can happen that the positive ions from the work piece treatment deposit at the tool. This deposition starts with a slight change of the tool geometry and can lead to a kind of ion based short circuit bridge between tool and work piece. Such a short circuit disrupt the manufacturing process. For the further experimental work the tool voltage was set at 100 mV to avoid any unwanted occurances.

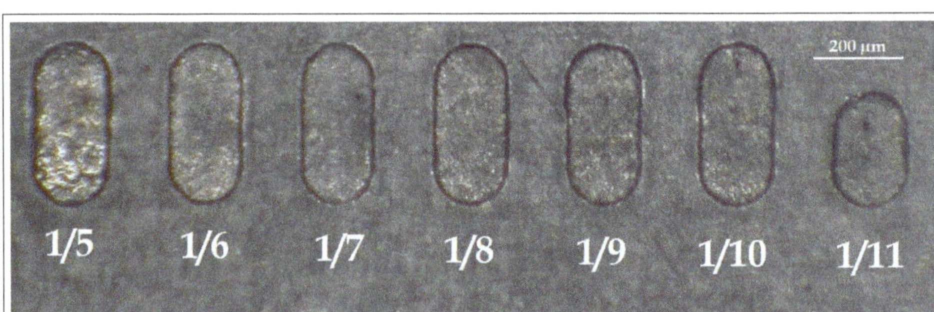

Fig. 26. Grooves produced for the pulse–pause ratio experiment

Figure 27 shows that the higher the pulse-pause ratio, the lower the removal rate. If within a period of time fewer pulses are applied, the charging and discharging of the electrochemical double layer also occurs less frequently. This is the obvious explanation for the low manufacturing speed of the groove made with a ppr of 1/11. For this ratio the manufacturing process was stopped because economic material removal could not be realized.

The best combination of the optical quality of the surface and the removal rate was detected from a pulse-pause ratio of 1/7. The consequence was to fix this parameter for the next experiments. Based on the optical result, the pulse-pause ratio of 1/5 was not viewed in the evaluation.

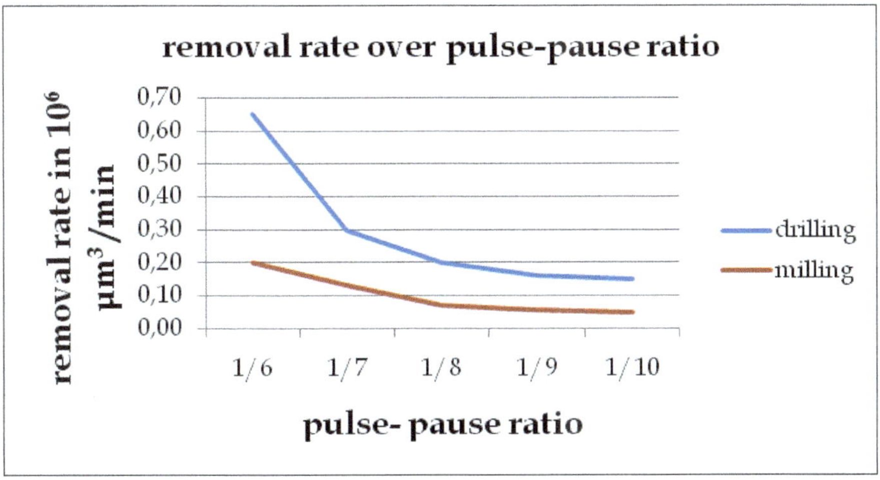

Fig. 27. Removal rate over pulse-pause ratio

A = 2800 mV	p = 100 ns	T = 0 mV	E = 3% HF/3M HCl
I = 3000 µA	ppr = varied	D = 150 µm	

Table 14. Adjustments for the experiment of figure 26 and 27

3.4.4 Drilling with µPECM

To this point in the series of experiments all grooves were manufactured with an adjusted depth of 20 µm. This experiment was done to show how the manufacturing depth influences the process. Figure 28 shows that at a depth between 125 - 175 µm the speed of removal rapidly reduces from above 35 to less than 10 µm per minute. A possible explanation is that the electrolyte is not sufficiently available in the drilled hole. The electrolyte is sated in such depth, so the transport of new solved ions out of the bore slows down and the removal speed reduces. After the depth of around 425 µm was reached, the process was stopped, because it was no longer possible to manufacture the work piece. To prepare sufficient electrolyte solution in such depth and thus realize better transport of the solved ions out of the bore, the mechanical movement of the tool inside the drilled hole could be pulsed to get a kind of flushing and reach higher depths. The manufacturing parameters of this experiment are illustrated in table 15.

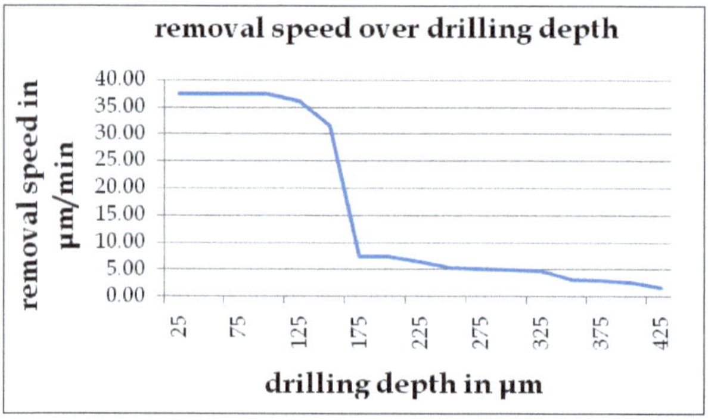

Fig. 28. Removal speed over drilling depth

A = 2800 mV	p = 100 ns	T = 100 mV	E = 3% HF/3M HCl
I = 3000 µA	ppr = 1/7	D = 75 µm	

Table 15. Adjustments for the experiment of figure 28

3.4.5 Dwelling time

Figure 29 shows the effect of the dwelling time during the process. In this experiment a tool with a diameter of 150 µm was positioned 4 µm above the work piece´s surface. The parameters of the experiment are shown in table 16. The tool was stopped at eight different positions. On the first position the dwelling time was about 0 s, and afterwards it was doubled on each position from 5 s to 640 s. With the maximum depth of around -10 µm at the longest dwelling time this experiment confirmed the relevance of the dwelling time for the manufactured geometry. If the manufacturing feed rate is chosen too low, the precision of the manufactured geometry shrinks - caused by the time-dependent development of the working gap. This is one of the effects, which has to be controlled in industrial usage of the µPECM technology. Table 16 gives a overview of the process parameters for the dwelling time experiment.

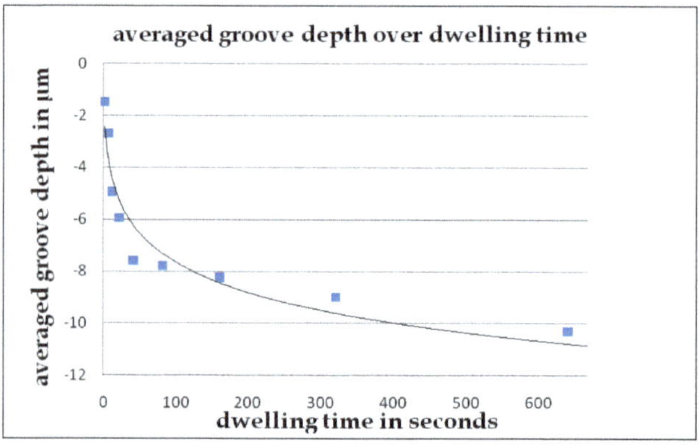

Fig. 29. Averaged groove depth over dwelling time

A = 2800 mV	p = 100 ns	T = 100 mV	E = 3% HF/3M HCl
I = 3000 μA	ppr = 1/7	D = 150 μm	

Table 16. Adjustments for the experiment of figure 29

3.4.6 Manufacturing of the Institute´s logo with μPECM

The goal of the last experiment was to produce a micro structure with the knowledge of the described experimental work. So, the emblem of the Institute for Production Engineering and Laser Technology was chosen to be machined in a small steel plate. The first step, as in all other experiments, was to provide an appropriate tool to produce a high quality result. To manufacture grooves with a maximum width of 30 μm a tool diameter of about 20 μm is necessary. In a special tooling basin the diameter reduction from 150 μm to 20 μm was realised. Figure 30 shows the result of the tooling process.

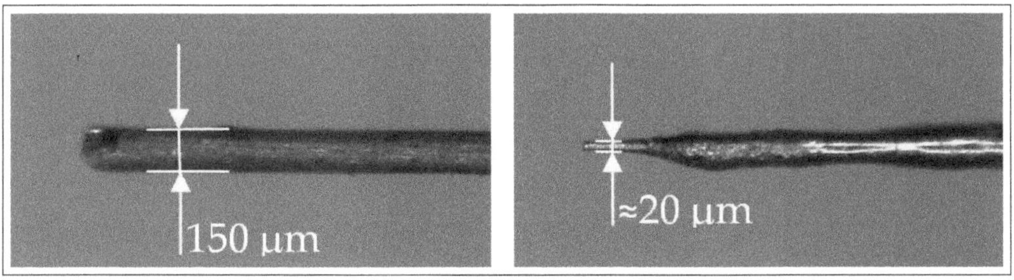

Fig. 30. Tool before (diameter 150 μm - left) and after the tooling process (diameter ≈20 μm - right)

Figure 31 shows the result seen through a light microscope with forty-five-fold magnification and table 17 illustrates the used processing parameters. To get an idea of the dimensions of the emblem, a human hair was attached for comparison. The total removal time to produce this logo was 03:04:44 (hh:mm:ss). The groove 0-1 has an adjusted length of 322,5 μm and an adjusted depth of 30 μm. The manufacturing time was 11,02 minutes and the width is 26,3 μm. This leads to a removal rate of 0,027 10^6 μm^3/min.

A = 2300 mV	p = 80 ns	T = 100 mV	E = 3% HF/3M HCl
I = 3000 μA	ppr = 1/7	D ≈ 20 μm	

Table 17. Adjustments for the manufacturing of the Institute's logo

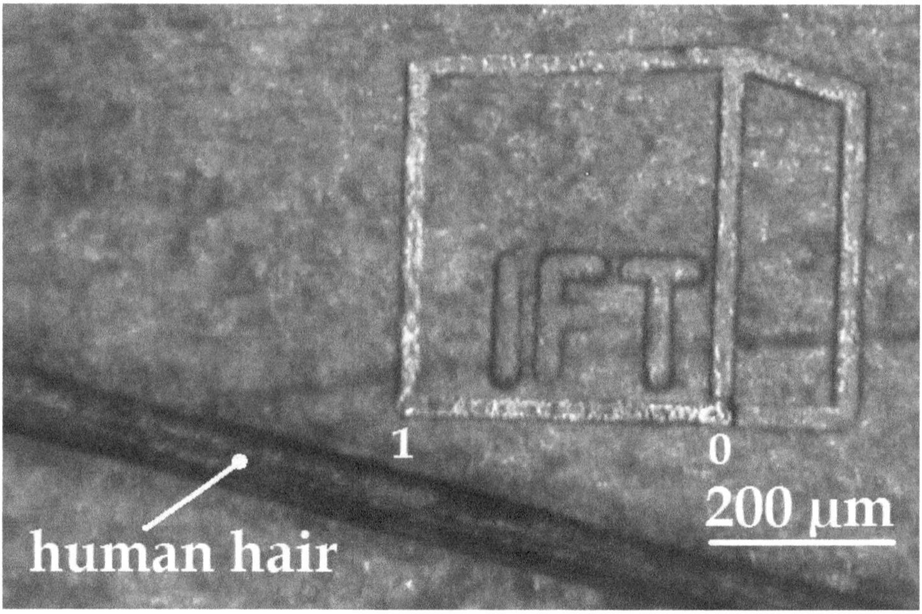

Fig. 31. Logo of the Institute in comparison to a human hair (diameter ≈ 50 μm)

4. Conclusion

The technology of electrochemical micromachining with ultra short voltage pulses has successfully displayed the many applications especially for prototype building or for the manufacturing of special products where there is no other technology which can combine a very high manufacturing precision for special materials without any mechanical forces or thermal influences. [Zemann R.] In principal, it can be applied to all electrochemically active materials, including semiconductors. [Schuster R.] Also, the use of applicable effects on process accuracy and material removal rate of difficult to machine materials offers a wide range of possible applications for μPECM technologies in the future. The occurring electrochemical problems are tradable and topics at the IFT, as well as the micromachining of many different materials like nickel, tungsten, titanium, non-corroding steels, or hard metals. As already mentioned, the machine at the IFT is simple constructed and very easy to maintain, so it is adequate for industrial use. However, a more complex machine structure would enable to reach highest precision requirements, but needs more maintenance and a higher financial investment. The experiments on the IFT´s machine proved that electrochemical micromachining is achievable for SME's. With the parameter sets in table 18 and 19 appropriate results were manufactured. Appropriate results means, that with these parameters, the grooves deliver adequate working gaps and optical results - geometry, topology, sharpness of the edges, and shine of the ground. Other parameters would perhaps reach higher removal rates, but on the other side lose quality with regard to precision.

A = 3000 mV	p = 80 ns	T = 100 mV	E = 0,2M HCl
I = 2000 μA	ppr = 1/8	D = 75 μm	

Table 18. Adjustments to achieve appropriate results working on nickel

Caused by the complexity of this technology, the variation of one of the adjustable parameters could significantly affect the result. Therefore at this point of research it is not definitely possible to give tangible instructions on how to reach requested results. It is very much experience necessary to interpret the proceedings at the machine correctly and to enhance the manufacturing process. Due to the multidisciplinary nature of this technology, intensified cooperation with other experts and an extensive research study has to be done; before a reasonable forecast for the processing parameters of a specific manufacturing process can be done.

A = 2800 mV	p = 100 ns	T = 100 mV	E = 3% HF/3M HCl
I = 3000 μA	ppr = 1/7	D = 150 μm	

Table 19. Adjustments to achieve appropriate results working on steel (1.4301)

5. Prospects

In the course of the experiments, it was also tried to treat carbide metal by electrochemical micromachining with ultra short pulses. The work piece used for experimental work was a K40FF. This carbide metal consists of a 12% cobalt matrix with 88% tungsten-carbide as stengthener. The electrolytes used were 3% HF/3M HCl and 2M NaOH. Both electrolytes were found to be unsuitable in combination with this carbide metal. A major challenge is to find new material-electrolyte combinations to apply electrochemical micromachining with ultra short pulses. The IFT has some tangible visions to realize treatment of carbide metal. A prospectively area for application of this technology could be protection of plagiarism. Technical devices and parts could be branded with the μPECM technology so, that only the producer can find the printed serial number, due to the small size of it.

6. References

Buhlert, M. (2009). *Elektropolieren*, Eugen G. Leuze Verlag, ISBN 978-3-87480-249-9, Saulgau, Germany

Hamann, C.H. & Vielstich, W. (2005). *Elektrochemie*, 4. vollständig überarbeitete und aktualisierte Auflage, WILEY-VCH Verlag GmbH & Co. KGaA, ISBN 3-527-31068-1, Weinheim, Germany

Kirchner, V. (2001). *Elektrochemische Mikrostrukturierung mit ultrakurzen Spannungsimpulsen*, Dissertation - Freie Universität Berlin, Berlin, Germany

Kock, M. (2004). *Grenzen der Möglichkeiten der elektrochemischen Mikrostrukturierung mit ultrakurzen Spannungspulsen*, Dissertation - Freie Universität Berlin, Berlin, Germany

Schuster, R., Kirchner V., Allongue, P. (2000). *Electrochemical Micromachining*, SCIENCE Vol 289, sciencemag, 7. July 2000, p. 98-101

Zemann, R. (2010). Electrochemical Milling, *Annals of DAAAM for 2010 & Proceedings of the 21st International DAAAM Symposium "Intelligent Manufacturing & Automation: Focus on Interdisciplinary Solutions"*, 20-23rd October 2010, Zadar, Croatia, B. Katalinic, ISSN 1726-9679, ISBN 978-3-901509-73-5, S. 843 - 844, DAAAM International, Vienna, Austria

3

A New Pre-Wet Sizing Process – Yes or No?

Ivana Gudlin Schwarz and Stana Kovačević
University of Zagreb,
Faculty of Textile Technology,
Department of Textile Design and Management,
Croatia

1. Introduction

As one of the most complex steps in fabric production, sizing plays a very important role in the weaving process. The primary purpose of the sizing process is to obtain the warp threads that can successfully be woven without major damages which occur during the yarn passage through sliding metal parts of the weaving machine (Lord, 2003). It applies to the improvement of physical and mechanical parameters of warp threads, primarily to increase strength and abrasion resistance and thus to reduce the number of warp breaks to a minimum in order to achieve the maximum efficiency of weaving machines and energy savings. Also, the goal of sizing is to keep the fibers in the yarn in a position where they were before sizing, with minimal yarn deformations during weaving. The success of the weaving process depends on the complexity of several factors including the characteristics of the desired material, the sizing process, the sizing ingredients and yarn properties, but also the extensive knowledge of a textile technologist (chemistry, rheology, electronics, mechanical engineering, physics, mechanics, mathematics, etc...), which makes this process more difficult and more important for the overall process of making woven fabric (Adanur, 2001). Today's achievements in all engineering branches enable an exceptional progress of the sizing processes to achieve a very high quality of sizing that meets the needs of today's modern weaving. However, the sizing costs, despite the complete automation of the regulation and control of the most important sizing process parameters, are still very high. Their reduction is possible by reducing the consumption of sizing agents and energy, as well as by modernization and development of machinery and technology, and all without consequence on the quality of the sized yarn. The choice of sizing agents plays the most important role in meeting all the requirements placed on the sizing process and sized yarn as well as optimizing and keeping the sizing process conditions and the size pick-up constant (Kovačević et al., 2006). Even today the optimization of the size pick-up applied to the yarn presents a major problem in the sizing process, despite the high degree of automation and high quality sizing agents. Influential parameters in the optimization of size pick-up are defined with the substance balance that enters and exits the size box (Equation 1). The requirement of keeping the size pick-up optimized and constant can be achieved by continuous measuring and keeping temperature and yarn moisture, size concentration in the size box constant, as well as automatic regulation of squeezing force and sizing speed (Pleva & Rieger, 1992; Soliman, 1995).

$$Sp = \frac{W_{Sp} - W_H}{\frac{100}{C} - 1 - \frac{W_{Sp}}{100}} \quad (\%) \tag{1}$$

Where: S_p - size pick-up, W_H - warp moisture at the box entry (%),W_{Sp} - warp moisture at the box exit (%), C - size concentration in the box (%)

Both, sizing conditions and yarn parameters affect yarn size pick-up. If the yarn is in some sections "more closed" with fewer interspaces among fibers or with more twists, the absorption of size in this section will be lower, resulting in a lower size pick-up on the yarn, despite constant sizing conditions. During the sizing process two types of forces are needed to be overcome: the forces of surface tension (wetting) and the forces of diffusion (Goswami, 2004). Penetration of the liquid (wetting) occurs in two phases:

1. penetration of the liquid into the capillary spaces between the fibers in the yarn (during which two forces have to be overcome - the difference between the pressure of enclosed air and surrounding liquid, and tension forces of the interface between fiber and water)
2. penetration of liquids into the fiber, i.e. extrusion of air bubbles and filling those spaces with a liquid.

Also a big unknown in size pick-up optimization is the sizing of wet warp, as well as the entire pre-wet sizing process. All previous knowledge of pre-wet sizing points to the obtainment of outstanding results, relevant physical-mechanical properties, reduction of consumption in sizing agents and energy, and an increase in weaving productivity (Hyrenbach, 2002; Rozelle, 1999, 2001; Sherrer, 2000). Therefore, this chapter aims to bring knowledge of the pre-wet sizing process by its analysis and by making a comparative analysis of the standard sizing process. The goal is to prove that there are a number of justified reasons for a new technological process, and to highlight the advantages and also disadvantages and possible improvements of better physical-mechanical properties of the yarn, reduction of size and energy consumption and increase in weaving productivity (Gudlin Schwarz et al. 2010, 2011).

2. Sizing machine

The pre-wet sizing process is still a rather unexplored area and not confirmed by scientific research. The most important reason why this research area has remained unexplored and with such a poor representation of this topic in scientific work is aggravated laboratory samples processing. Thanks to the laboratory sizing machine (Fig. 1) designed and constructed at the Faculty of Textile Technology, University of Zagreb, Croatia (whose segments are sizing box and dryer - which represents two consensual patents No.: PK20070247 and PK20070248, registered at the Croatian State Intellectual Property Office) both sizing processes - standard sizing process and pre-wet sizing process were able to be carried out. It consists of a creel for cross wound bobbins, with the possibility of tension regulation, two boxes - a box for pre-wetting with hot water and a size box, and a dryer. The pre-wetting box consists of a pair of immersion rollers and a pair of rollers for squeezing out excess water. The size box consists of a working box with two pairs of immersion rollers and two pairs of rollers for size squeezing, as well as a pre-box that allows to keep size levels in the working box constant, namely continuous size circulation from the working box to the pre-box with natural flow, and from the pre-box to the working box using a pump. During the sizing process it is possible to keep water temperature constant in the pre-wetting box and size temperature in the size box with integrated heaters and thermostats, which indirectly warm up the water and size through the walls of the boxes. Thread tension was measured during the sizing process at the box entry, while warp moisture was measured at all important places: at the box entry, between two boxes - the pre-wetting box and size box, at the size box exit and after the dryer. Drying the sized yarn is preformed by contact, moving it across the two heated cylinders of the contact dryer. It is also possible to regulate and keep the sizing speed constant using the winder of the sized and dried yarn, as well as the speed regulator (Gudlin Schwarz et al. 2010, 2011).

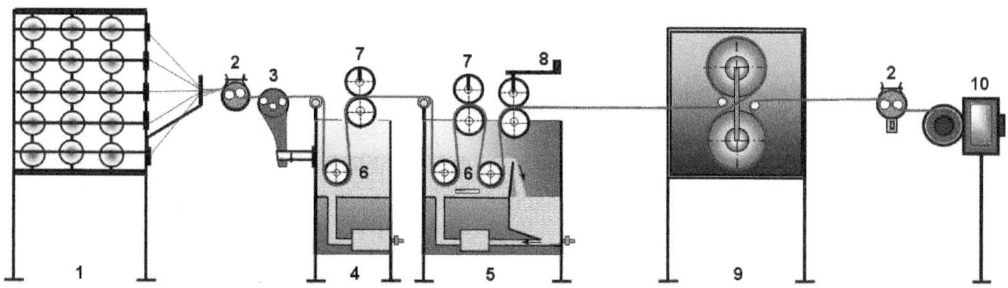

Fig. 1. Laboratory sizing machine (constructed on Faculty of Textile Technology, University of Zagreb, Croatia): 1 - creel for cross wound bobbins, 2 – moisture contact measuring device, 3 - thread tension measuring device, 4 - pre-wetting box, 5 - size box, 6 - rollers for immersing yarn into water and size, 7 - rollers for water and size squeezing, 8 - regulation of the pressure of the last squeezing roller, 9 - contact dryer, 10 - winder of the sized yarn

3. Testing methods and materials (yarn and size)

Sizing is a process which can be carried out on yarns of different raw material composition, and in this case it was 100% cotton carded ring spun yarn with a nominal count of 20 tex and with certain properties shown in Table 1.

Parameters			Yarn
Tt_n (tex)			20.00
Tt_r (tex)			18.55
Parameters of unevenness (400 m/min)	CV (%)		16.01
	Thin places		11.50
	Thick places		95.50
	Neps		208.00
T (twists/m)		$\bar{x}$	913.04
		CV	5.20
H (number of protruding fibers)		$\bar{x}$	20428.00
		CV	2.30
A (number of cycle)		$\bar{x}$	80.76
		CV	9.39
F (cN)		$\bar{x}$	281.69
		CV	6.78

Parameters		Yarn
E (%)	$\overline{x}$	3.83
	CV	7.26
W (cNx cm)	$\overline{x}$	292.75
	CV	13.08
σ (cN/tex)	$\overline{x}$	15.19
	CV	6.78

Table 1. Properties of unsized yarn; where : Tt_n – nominal count (tex), Tt_r – real count (tex), T – twist (twist/m), H – hairiness (No. of protruding fibers longer of 1mm from the yarn surface), A – abrasion resistance (No. of cycles), F – breaking force (cN), ε – elongation at break (%), W – work to rapture (cN×tex), σ – tenacity (cN/tex), CV – coefficient of variation, $\overline{x}$ - mean value

As it was mentioned above, for a successful sizing process right choice of sizing agents and size preparation are of great importance, depending on yarn (fiber) type, origin of the sizing agent, different sizing auxiliaries, and the requirements of the sizing process itself. (Vassallo, 2005; Zhu, 2003). In the presented example and based on these needs, two different recipes with different concentrations were used in both sizing processes, as shown in Table 2.

Compounds	Recipe	Concentration
1. Water 2. Sizing agent based on polyvinilalcochol (PVA) 3. Sizing auxiliaries composed of natural fats and waxes with a specific emulsifier system	Recipe 1 - R 1	7.5%
	Recipe 2 - R 2	5.0%

Table 2. Characteristics of the sizing agents and auxiliaries, and size recipes

Sizing conditions are exactly defined and held constant during both sizing processes, and are shown in Table 3.

Sizing condition	
Thread tension between creel for cross wound bobbins and pre-wetting box	40 cN
Temperature of water in the pre-wetting box	65°C
Size temperature in the size box	75°C
Sizing speed	3 m/min
Pressure on the last pair of the rollers for squeezing excess size	19.1 N/cm^2
Temperature on the cylinders of the contact dryer	140°C
Output moisture	6 %

Table 3. Sizing conditions

To test the samples before and after sizing, standardized methods were used. Thus, the real count of yarn was tested according to HRN ISO 2060:2003, while yarn unevenness was tested on an Unevenness tester 80, type B, Keisokki Company. The breaking properties (breaking force, elongation at break, work to rapture and tenacity of yarns) were tested on a Statimat M made by Textechno according to ISO 2062. Yarn hairiness was tested before and after sizing by recording the fibers protruding from the yarn structure using a Zweigle G 565 hairiness meter according to ASTM D 5674-01, while the twists were tested by means of a MesdanLab Twist tester according to ISO 17202. Abrasion resistance tests were performed on a Zweigle G 551 abrasion tester before and after sizing, where each of 20 types of thread loaded with a weight of 20g was subjected to the abrasion process until thread breakage. The movement of the cylinder coated with emery paper (fineness 600): left - right and its rotation around its axis achieves certain abrasion intensity in the yarn and emery paper. During the process the yarn weakens, and at the moment when the mass of the weights hung on the yarn overcomes the yarn strength, a break occurs, and the number of roller movements until breaking the yarn is recorded.

The determination of the size pick-up can be performed in several ways, but for the purposes of this study the gravimetric method was used. The implementation process of this method is as follows: before sizing the samples were dried to absolutely dry, after which they were weighed, and then returned to climatic conditions and sized; after sizing the samples were again dried to absolutely dry and weighed (Kovačević et al., 2002). The amount of size pick-up is calculated using Equation 2:

$$Sp = \frac{G_S(g) - G_U(g)}{G_U(g)} \cdot 100 \ (\%) \tag{2}$$

Where: Sp (%) - amount of size pick-up
G_S (g) - mass of absolutely dry sized yarn
G_U (g) - mass of absolutely dry unsized yarn

4. Testing methods and materials (yarn and size)

4.1 Yarn breaking properties

The most prominent mechanical properties of yarn are primarily breaking properties, which include: breaking force, elongation at break, work to rapture and tenacity. These parameters show us some of the most important yarn characteristics for the weaving process, and the positive impact of sizing to those properties is also one of the most relevant role of the sizing process. Sizing pursued to a greater increase in breaking force and at the same time in a less decrease in elongation at break, which in turn depends on the size pick-up and size distribution in the yarn (Gudlin Schwarz et al. 2011; Kovačević & Penava, 2004).

Figure 2 shows an F-E diagram of unsized yarn and yarns sized with two different size recipes R1 and R2, where the differences in force (F) and elongation (ε) between the yarns subjected to different processing methods can be clearly seen.

The values of breaking force of the tested samples are shown in Figure 3a, where a very small difference between the sized yarns occurs, with an average increase of almost 40%

compared to the unsized yarn. The only yarn that shows a deviation from the others in the form of a small increase in breaking force (of only 32%) is the yarn sized with R2 and the pre-wet sizing process.

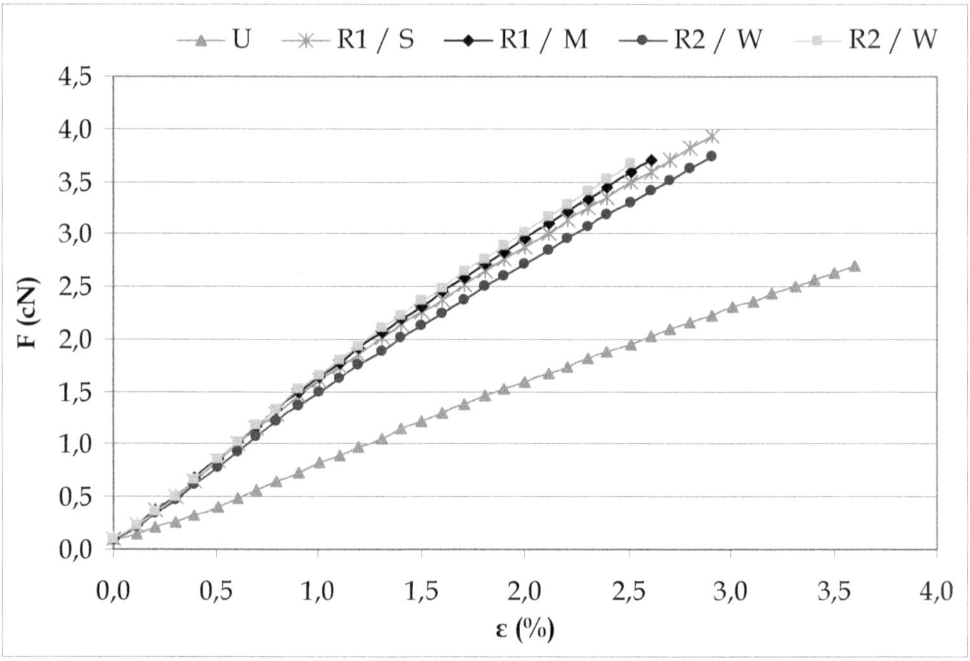

Fig. 2. F-E diagram of unsized yarn and yarns sized with recipes 1 and 2, by standard sizing process and pre-wetting sizing process; where: U - unsized yarn, R1 - yarn sized with recipe 1, R2 - yarn sized with recipe 2; S – standard sizing process, W – pre-wetting sizing process

Generally, by using the sizing process the elongation at break reduces and therefore represents the disadvantage of this process. The values of the elongation at break of the tested yarns are shown in Figure 3b, where the results are divided into two groups with almost identical values: the yarns sized with the standard process, which shows a decrease of almost 20%, and those sized with the pre-wet sizing process, with a decrease of almost 25%, compared to the unsized yarn.

In Figure 3c, which shows the results of work to rapture, are presented the uniform values within one sizing process, where a larger increase by 12% is recorded in the yarns sized with the standard process than in the yarns sized with the pre-wet sizing process, where the values increase by 4% for the yarn sized with R1, while the yarn sized with R2 records even a slight drop by only 3% compared to the unsized yarn.

Tenacity is a parameter that brings into relation yarn finesses and force, and the values obtained are shown in Figure 3d. The values of the sized yarns are quite consistent with an increase by nearly 33% for the yarn sized with the standard process, and by 31% for the yarn sized with the pre-wet sizing process.

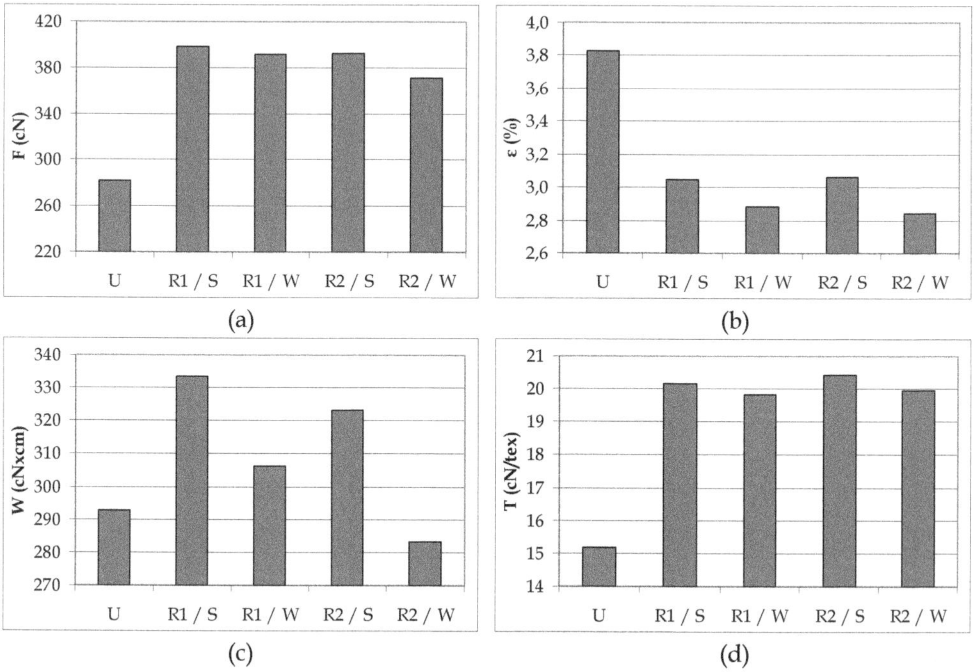

Fig. 3. Diagrams of breaking properties of the unsized yarn and the yarns sized with recipe 1 and 2 by the standard sizing process and the pre-wetting sizing process

4.2 Yarn hairiness and abrasion resistance

Yarn hairiness and abrasion resistance are parameters which are extremely important for the weaving process, which are greatly improved by a successful sizing process. Hairiness, i.e. the number of protruding fibers, is reduced by sizing, and the abrasion resistance is increased, which affects the reduction in friction resulting from the thread passing through the metal elements of the weaving machine and, therefore, the number of thread breaks in the weaving process (Gudlin Schwarz, 2011).

Figure 4a shows the values of the tested hairiness for the unsized yarn and the yarns sized with both processes and with both recipes. Yarn hairiness reduction sized with both processes is very similar, and amounts to 78% for the yarns sized with the standard process, and 81% for the yarns sized with the pre-wet sizing process compared to the unsized yarn.

The value diagram of the abrasion resistance of the unsized yarn and the yarn sized with both procedures and with both recipes is shown in Figure 4b. It is interesting that the yarns sized with both processes but with a higher concentration of size (R1) show good results in terms of increasing the abrasion resistance compared to the unsized yarn. The yarns sized with the standard process recorded an increase by even 68%, while the yarns sized with the pre-wet process showed almost half an increase by only 36%. Regarding the samples sized with a smaller size concentration (R2) with both processes, a notable decrease in abrasion resistance compared to the unsized yarn is recorded, namely by 4% for the yarns sized with the standard process, and by 14% for the yarn sized with pre-wet process. This phenomenon, in spite of the size pick-up which strengthens the yarn, is attributed to the yarn extension that occurs during sizing, and it is greater in sizing with a lower size concentration (R2).

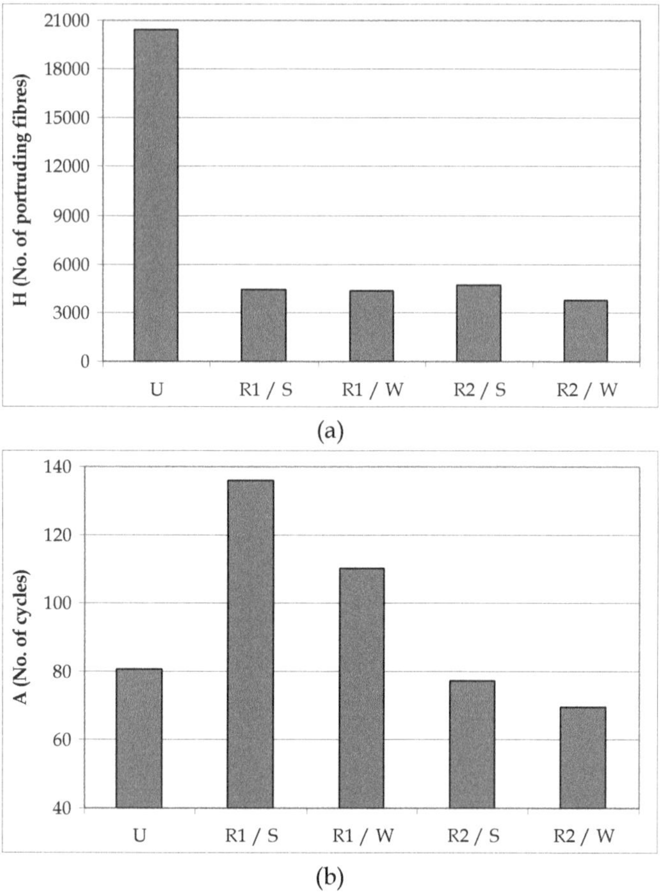

Fig. 4. Diagram of hairiness and abrasion resistance of the unsized yarn and the yarns sized with recipe 1 and 2 by the standard sizing process and pre-wet sizing process

4.3 Yarn extension

In the sizing process warp tension is a very important and unavoidable parameter, which in turn causes extension (visible even at a minimum tension) and thus the deformation shown by changes of mechanical properties. The appearance of yarn extension during the sizing process is unfortunately a reality that can not be avoided despite the minimum warp tension in segments when the warp is the most sensitive and that is in the wet state. The sensitivity of wet yarn begins with the entry into the size box and lasts until the exit from the dryer. The greater the yarn tension and its length in those segments, the higher is the yarn extension. During the pre-wet sizing process the yarn length in the wet state additionally increases between the pre-wetting box and the size box, which further increases tension sensitivity, and thus susceptibility to extension and deformation. Yarn unevenness also affects extension properties to a great extent, where thin and thick places represent weaker yarn parts which are more sensitive to tension, especially in the wet state (Gudlin Schwarz, 2010). Similarly, the values shown in Figure 5 indicate that the elongation is higher during sizing with a lower size concentration (R2) in both sizing processes.

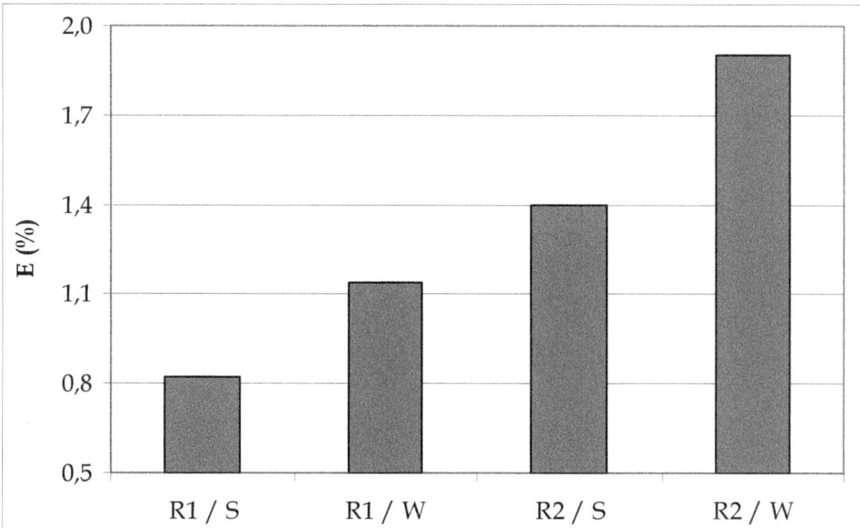

Fig. 5. Diagram of extension of yarns sized with recipe 1 and 2 by standard sizing process and pre-wet sizing process

4.4 Size pick-up of yarn

As stated in the introduction, optimizing the size pick-up to achieve the maximum utilization of the sizing process represent the biggest challenge in the whole process (Kovačević et al., 2002). The obtained results indicate that the amount of size pick-up and its distribution determine many features of sized yarn properties.

Figure 6 shows the value of the amount of size pick-up on the yarn. A small difference between the yarns sized with R1 by both sizing processes is easily observable. Significant differences are evident in the yarn sized with R2 in both processes, where an almost equal difference in reduced size pick-up (an average of 50%) between the standard sizing process (R2/S) and the pre-wet sizing process (R2/W) is maintained.

The yarn sized with lower size concentrations showed very good results in all important properties in terms of no significant deviations (in spite of a lower amount of size pick-up on the yarn) in relation to the yarn sized with a higher size concentration. This phenomenon is particularly interesting for the yarn sized with the pre-wet sizing process, where the amount of size pick-up on the yarn is considerably lower (due to water filling the interior of the yarn and the different distribution of the size pick-up on the yarn), than on the yarn sized with the standard process, indicating a reduced consumption of sizing agents and resulting in great savings (Sejri et al., 2008, 2011).

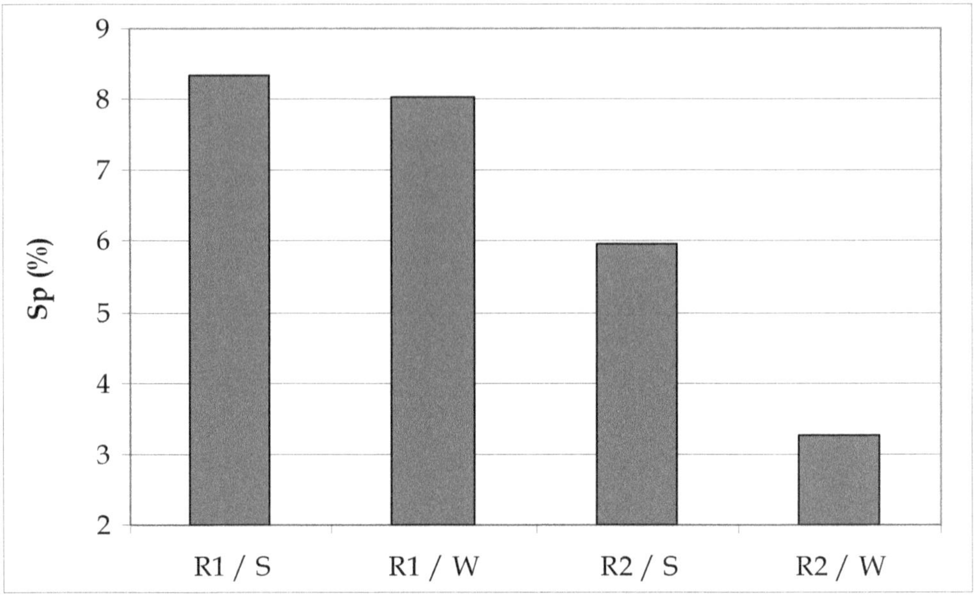

Fig. 6. Diagram of size pick-up (Sp) on the yarns sized with recipe 1 and 2 by the standard sizing process and the pre-wet sizing process

4.4.1 Distribution of size pick-up on yarn

As already mentioned, the pre-wetting sizing process differs from the standard sizing process in the construction of the sizing machine, where another pre-wet box with hot water is installed in the front of size box. The importance of the pre-wetting box is in soaking the yarn in hot water (60-70°C) before entering the size box, which enables the dissolving and removal of grease and other impurities and additives present in the raw yarn. Furthermore, in the phase of pre-wetting, it comes to wetting the yarn in water, i.e. to fill interstitial spaces in the interior of the yarn with water, and after squeezing the excess water the yarn remains wet and partially filled with water. As such, it enters the size box with much higher humidity than it is the case with the yarn in the standard sizing process where it enters the size box dry. Therefore, the contact of the retained water in the yarn with size leads to very rapid mutual bonding, allowing faster and easier penetration and diffusion of the size into the yarn. However, the size concentration in the interior of the yarn is lower than the size concentration in the size box, because the water remained in the interior of the yarn after wetting diluted it. Therefore, the greater part of the size remains on the surface of the yarn. When sizing the dry yarn, penetration of size into the interstitial yarn spaces is not as rapid as in the case of wet yarn, and thus the inner part remains almost unfilled with size, while around or on its periphery a solid size coat is formed, unlike the yarns sized with the pre-wetting sizing process, where the size pick-up on the yarn periphery does not form such an intensive solid size coat (Fig. 7-9) (Gudlin Schwarz, 2011; Johnen, 2005).

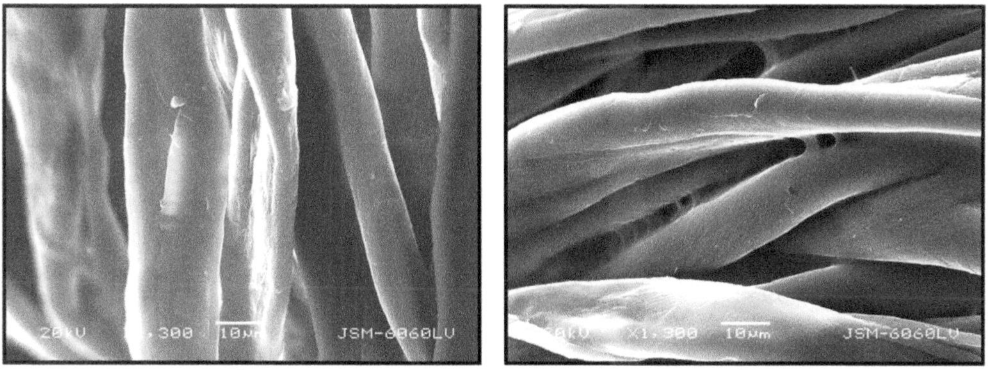

A B

Fig. 7. Microscopic longitudinal-section image of the yarn sized with the standard sizing process (A) and with the pre-wetting sizing process (B); SEM microscopy JSM - 6060LV

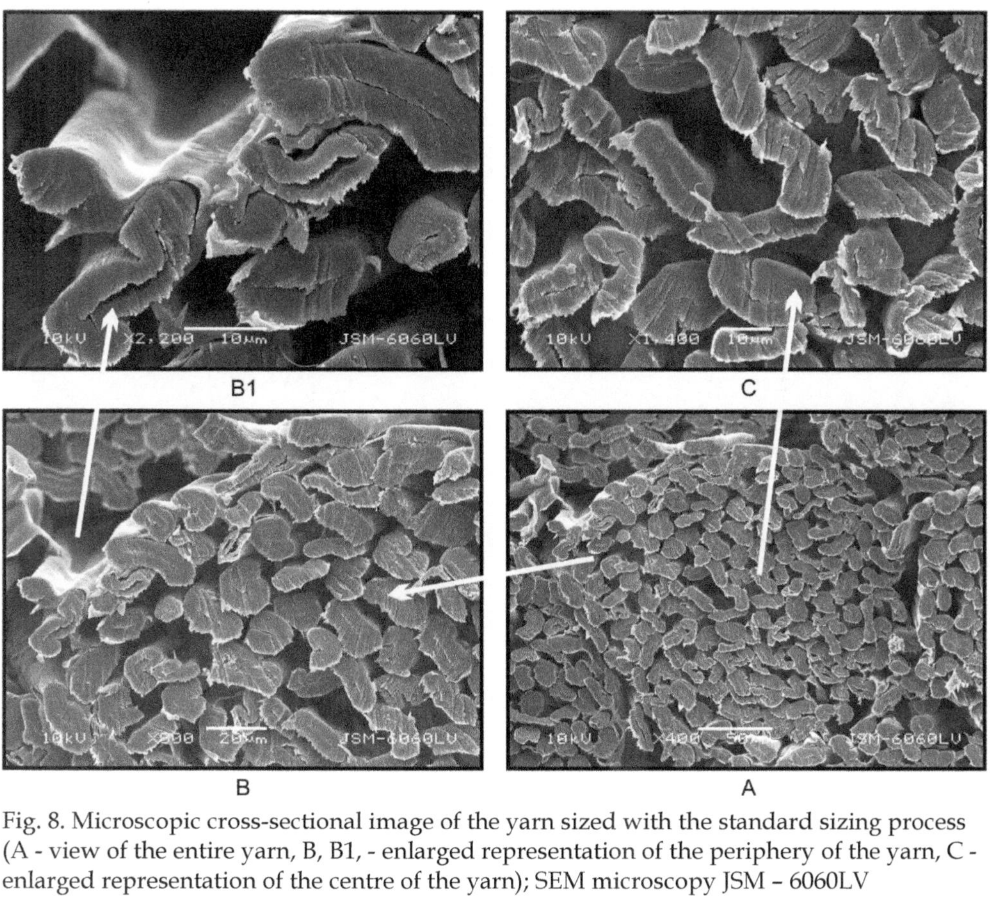

B1 C

B A

Fig. 8. Microscopic cross-sectional image of the yarn sized with the standard sizing process (A - view of the entire yarn, B, B1, - enlarged representation of the periphery of the yarn, C - enlarged representation of the centre of the yarn); SEM microscopy JSM - 6060LV

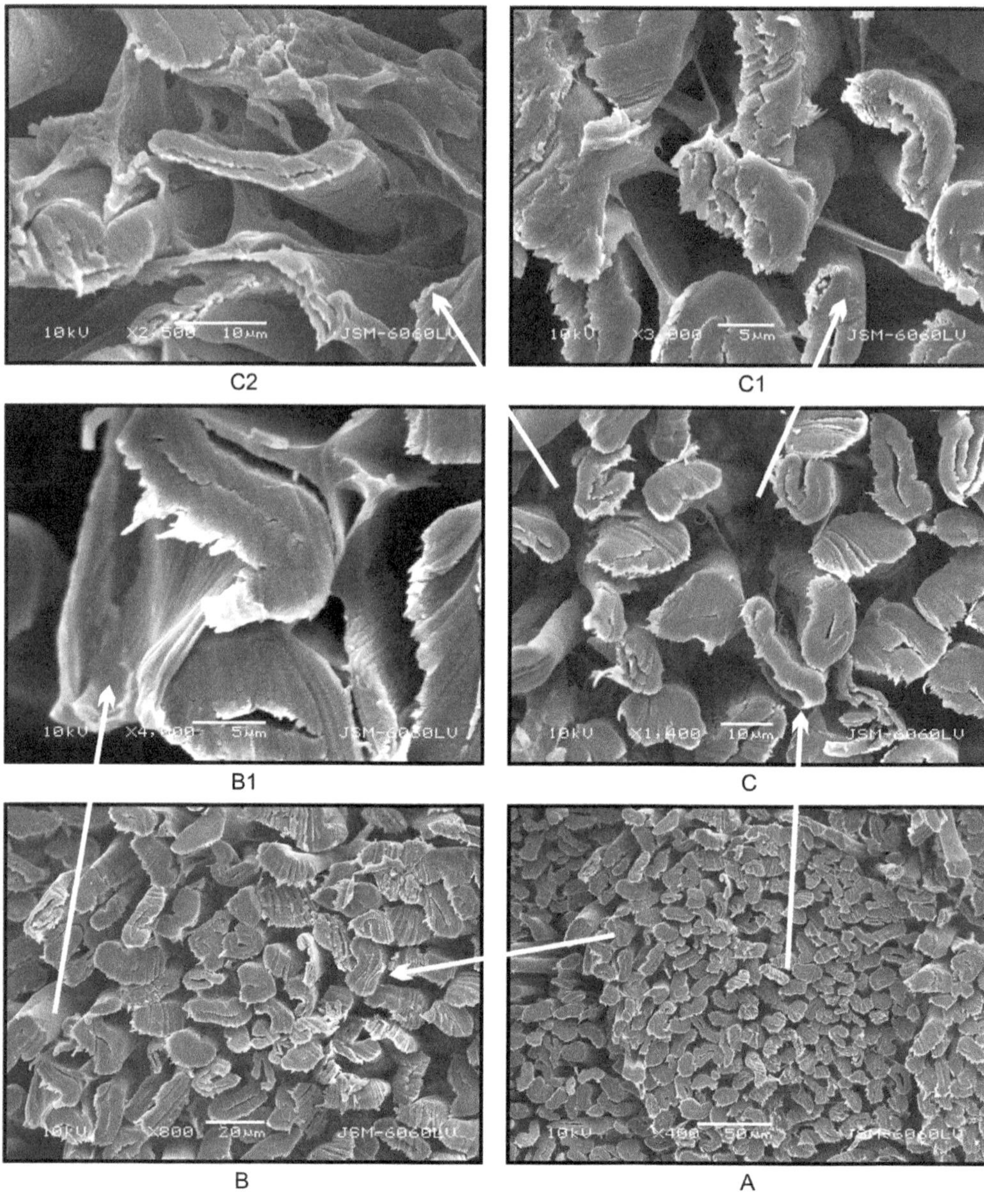

Fig. 9. Microscopic cross-sectional image of the yarn sized with the pre-wetting sizing process (A - view of the entire yarn, B, B1, - enlarged representation of the yarn periphery, C, C1, C2 - enlarged representation of the yarn centre); SEM microscopy JSM – 6060LV

5. Conclusion

Existing knowledge as well as the results of the conducted research shows that there is no big difference in the sized yarn properties obtained with the two processes, relevant for further process of making woven fabrics, while some properties of the yarn sized with the pre-wet sizing process are even noticeably better. Each of these processes brings certain advantages and disadvantages. The standard sizing process is a generally well- known, accepted and ubiquitous process in the textile industry. But of great significance is the fact that the replacement of the standard process in terms of upgrading and installing a part of the sizing range necessary for the implementation of the pre-wet sizing process does not require complex procedures or large financial expenditure. All these indicators are of great importance for the new sizing process, which give this process a priority over the standard sizing process in view of the possibility to reduce costs (size, water and energy costs - both for the sizing process and desizing process), with no negative impact on the properties of the sizing process nor on the quality of the sized yarn, including exceptional significance for the environmental aspect of the overall process.

6. Acknowledgment

The knowledge and results shown in this chapter came from the extensive research carried out within the project "Advanced Technical Textiles and Processes", code: 117-0000000-1376, Faculty of Textile Technology, University of Zagreb, Croatia, conducted with the support of the Ministry of Science, Education and Sports of the Republic of Croatia.

7. References

Adanur S. (2001). *Handbook of weaving*, Technomic Publishing Company, ISBN: 1- 58716-013-7, Basel, Switzerland.

Goswami, B.C.& Anandjiwala R.D., Hall D.M. (2004). *Textile Sizing*, Marcel Dekker Inc., Basel, ISBN: 0-8247-5053-5, New York

Gudlin Schwarz I., Kovacevic S. & Dimitrovski K. (2011). Comparative Analysis of the Standard Sizing Process and the Pre-wet Sizing Process, *Fibres & Textiles in Eastern Europe*, 19, 4 (87), 131-137

Gudlin Schwarz I., Kovačević S. & Dimitrovski K. (2011). Analysis Of Changes In Mechanical And Deformation Yarn Properties by Sizing, *Textile research journal*, 81, 5, 545-555

Gudlin Schwarz I., Kovačević S., Dimitrovski K. & Katović D. (2010). Istraživanje parametara pređa škrobljenih postupkom s prednamakanjem, *Tekstil*, 59, 279-286

Hyrenbach H. (2002). Partical experience with the prewetting proces in sizing, *Melliand International*, 8, December, 251-252

Johnen A. (2005). Experiences in wet-in-wet sizing, *Melliand International*, 11, March, 34-36

Kovačević S. & Penava Ž. (2004). Impact of Sizing on Physico-mechanical Properties of Yarn, *Fibres & Textiles in Eastern Europe*, 48, 4, 32-36

Kovačević S. Grancarić A.M. & Stipančić M. (2002). Determination of the Size Coat; *Fibres & textiles in Eastern Europe*, 10, 3, 63-67

Kovačević S., Penava Ž. & Oljača M. (2006). Optimisation of Production Costs and Fabric Quality, *Fibres & Textiles in Eastern Europe*, 14, 2, 40-48

Lord P.R. (2003). *Handbook of yarn production: technology, science and economics*, The Textile Institute, Woodhead Publishing, ISBN: 1855736969, Manchester, UK

Maletschek F. (2002). New developments in weaving preparatory processes, *Melliand International*, May, 8, 108-110

Pleva R. & Rieger W. (1992). Measurement and Optimization of Size Pick-up, *Textile Praxis International*, 47, 3, 230-232

Rozelle W.N. (1999). Pre-wet: New Money Maker in Warp Sizing Operations, *Textile World*, 149, 5, May, 73-79

Rozelle W.N. (2001). Pre-wet Sizing System Bases on Water Atomization, *Textile World*, 151, 3, March, 28-30

Sejri N., Harzallah O., Viallier P., Amar S.B. & Nasrallah S.B. (2011). Influence of wetting phenomenon on the characteristics of a sized yarn, *Textile Research Journal*, 81, 3, 280–289

Sejri N., Harzallah O., Viallier P., Amar S.B. & Nasrallah S.B. (2008). Influence of Pre-wetting on the Characteristics of a Sized Yarn, *Textile Research Journal*, 78, 326-335

Sherrer A. (2000). Benninger: SaveSize Pre-Wet Warp Sizing, *Textile World* 4, April, 42-43

Soliman H.A. (1995). Evaluation of Sizing as Conntrolling Parameter in the Tendency to Yarn Entangling, *ITB Garn-und Flachenherstellung*, 41, 2, 42-44

Vassallo J.C. (2005). Spun Sizing Chemicals and Their Selection. *AATCC Review*, December 2005, 25-28

Zhu Z. (2003). Starch mono-phosphorylation for enhancing the stability of starch/PVA blend pastes for warp sizing, *Carbohydrate Polymers*, 54, 115-118

Part 2

Control Systems, Automation

4

Applying the Technology of Wireless Sensor Network in Environment Monitoring

Constantin Volosencu
"Politehnica" University of Timisoara
Romania

1. Introduction

This chapter presents some considerations related to the applications in environment monitoring of some concepts as: estimation, fault detection and diagnosis, theory of distributed parameter systems and artificial intelligence based on the modern technology of wireless sensor networks. All these concepts allow treatment of large, complex, non-linear and multivariable system of the environment by learning and extrapolation. The environment may be seen as a complex ensemble of different distributed parameter systems, described with partial differential equations.

Sensor networks (Akyildiz & all, 2002) have large and successful applications in monitoring the environment, they been capable to measure, as a distributed sensor, the physical variables, on a large area, which are characterizing the environment, and also to communicate at long distance the measured values, from the distributed parameter environmental processes. A lot of papers and books have been published in the fields of using sensor networks in environment monitoring in the last years. Some related work is surveyed as follows. The paper (Cuiyun & all., 2006) presents some research consideration related the changes of urban spatial thermal environment, for sustainable urban development, to improve the quality of human habitation environment. The urban thermal phenomenon is revealed using thermal remote sensing imagery, based on the instantaneous radiant temperature of the land surfaces. An architecture of sensor network for environment is presented in (Lan & all, 2008). Environmental pollution and meteorological processes may be studied using various kinds of environmental sensor networks. The modern intelligent sensor networks comprise automatic sensor nodes and communication systems which communicate their data to a sensor network server, where these data are integrated with other environmental information. The paper (Giannopoulos & all, 2009) presents the design and implementaion of a wireless sensor network for monitoring environmental variables and evaluates its effectiveness. It has application in environment variable monitoring such as: temperature, humidity, barometric pressure, soil moisture and ambient light, for research in agriculture, habitat monitoring, weather monitoring and so on. In order to improve the capacity of the environmental sensor networks different techniques may be used. The paper (Talukder & all, 2008) is using a model predictive control for optimal resource management in environment sensor networks, for with application at spatio-temporal events of a coastal monitoring and forecast system. The paper (Dardari & all, 2007) presents and application at the estimation of atmospheric pressure using a wireless sensor network, which is randomly distributes. The estimation error is dicussed and a design criterion is proposed. The author has contribution in the field of monitoring distributed parameter systems based on sensor networks and estimation using adaptive-network-based fuzzy inference (Volosencu, 2010), (Volosencu & Curiac, 2010).

Using the modern intelligent wireless sensor networks multivariable estimation techniques may be applied in environment monitoring, seen as distributed parameter systems. Based on these concepts, environment monitoring becomes more easily and more performing (Fig. 1).

The chapter presents a methodology of how to use the above mentioned topics in the problem of the environmental monitoring, as follows: - principles and technical data of modern sensor networks, - examples of distributed parameter systems, with their mathematical models, useful in environment description, - examples of modeling and simulation of environmental temperature variation, - technical data of the sensor network used in practical experiments, - a case study of environmental temperature estimation base on auto-regression and neuro-fuzzy inference engine.

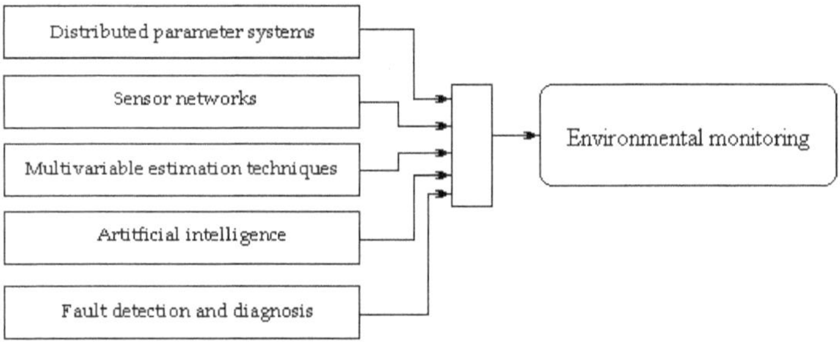

Fig. 1. Scientific domains for environmental monitoring

The most important domains of applications are: the processes of heat conduction, with propagation of heat in anisotropy medium: propagation of heat in a porous medium, processes of transference of heat between a solid wall and a flow of hot gas; applications related to electricity domain as electrostatic charges in atmosphere; the motion of fluid, the processes of cooling and drying, phenomenon of diffusion. Other applications are: the growing of the gas particles in a fluid, the temperature modification in the air mass.

The chapter presents a short survey of the main characteristics of the above topics involved in the problem of the environmental monitoring, some principles and technical data of modern sensor networks, some examples of distributed parameter systems, with their mathematical models, useful in environment description. The second paragraph presents some equation useful in modelling environmental processes. The third paragraph presents some estimation algorithms useful in environment monitoring, for future estimation of changing in physical variables of the medium. The fourth paragraph presents some examples of modelling and simulation of environmental temperature variation. The fifth paragraph presents some technical data of the sensor network used in practical experiments. The sixth paragraph presents the monitoring structure, the monitoring method and the estimation mechanism. The seventh paragraph presents an example of expert system useful in environment monitoring, based on environment knowledge. The eighth paragraph presents a case study. The ninth paragraph presents a technical solution of implementation of the monitoring system based on virtual instrumentation. The main results and future perspectives are presented in conclusion.

2. Equations for environmental systems

2.1 Primary physical and mathematical models

The environment systems, which are complex heterogeneous systems of distributed parameter systems, may be described using partial differential equations. These equations are used to formulate problems involving functions of several variables, such as the propagation of sound or heat, electrostatics, electrodynamics, fluid flow. Some examples of distributed parameter systems are presented as follow (Rosculet & Craiu, 1979). Diverse categories of systems have specific characteristics that are important in their investigation, simulation, prediction, monitoring and diagnosis. One of the most important domains of applications is represented by the process of heat conduction, with propagation of heat in anisotropy medium. In the field of motion of fluid there are: plane motion of viscous fluids, running of viscous fluids in medium as a tube or running of gases. The processes of cooling and drying are also met in environment systems. Phenomenons of diffusion could be: diffusion flow for chemical reactions, the flames diffusion, the density repartition of particles loading by the meteorites. Other applications in the environment could be: estimation of the ice height covering the snow the arctic seas, motion of underground waters, the growing of the gas particles in a fluid, the temperature modification in the air mass. For some of the above processes some equations are given as follows.
The function of the object's temperature is $\theta(P, t)$, at the time moment t, where P is a point in the space. If different points of object have different temperatures, $\theta(P, t) \neq ct.$, then a heat transfer will take place, from the warmer parts to the less warm parts. The vector *grad* θ has its direction along the normal at the level surface for $\theta = ct.$, in the sense of θ rising. The law of heat propagation through an object in which there are no heat sources:

$$\gamma\rho\frac{\partial\theta}{\partial t} = \frac{\partial}{\partial x}\left(k\frac{\partial\theta}{\partial x}\right) + \frac{\partial}{\partial y}\left(k\frac{\partial\theta}{\partial y}\right) + \frac{\partial}{\partial z}\left(k\frac{\partial\theta}{\partial z}\right) \quad (1)$$

The heat sources in the object have a distribution given by the function:

$$F(t,P) = F(t,x,y,z) \quad (2)$$

If the object is homogenous $a = \sqrt{k/\gamma/\rho} = ct.$ and the equation (2) is written:

$$\frac{1}{a^2}\frac{\partial\theta}{\partial t} = \frac{\partial}{\partial x}\left(\frac{\partial\theta}{\partial x}\right) + \frac{\partial}{\partial y}\left(\frac{\partial\theta}{\partial y}\right) + \frac{\partial}{\partial z}\left(\frac{\partial\theta}{\partial z}\right) \quad (3)$$

The initial conditions or of the limit conditions have physical significance. They are given by the equation:

$$\theta(x,y,z,t)\big|_{t=0} = f(x,y,z) \quad (4)$$

Running of viscous fluids in rectilinear medium may be analyzed with he following equations. Let it be a rectilinear medium, which is leading a viscous liquid. The ax of medium, seen as a tube is Oz. Let us consider the movement of a part of the liquid between two transversal sections z_1 and z_1+h. If A is the transversal section area supposed to be constant and ρ is the fluid density, the movement equation is

$$\rho A h \frac{\partial v}{\partial t} = A(p_1 - p_2) - R \tag{5}$$

where p_1 and p_2 are the pressures in the two sections and R is the force on the tube wall. If v is the fluid speed in the direction of Oz axis, v is independent of z if the liquid is incompressible

$$v = v(x, y, t) \tag{6}$$

The partial derivative equation is

$$\rho \frac{\partial v}{\partial t} = -\frac{\partial p}{\partial z} + \mu \left(\frac{\partial^2 v}{\partial x^2} + \frac{\partial^2 v}{\partial y^2} \right) \tag{7}$$

if the pressure p is constant

$$\frac{1}{a} \frac{\partial v}{\partial t} = \left(\frac{\partial^2 v}{\partial x^2} + \frac{\partial^2 v}{\partial y^2} \right) \tag{8}$$

which it is the equation of heat propagation in plane, where $a=\mu/\rho$.
For the plane motion of viscous fluids let's consider an incompressible, viscous fluid of constant density ρ, in a plane movement. If (v_x, v_y) are the speed components in the point P(x, y) of the plane at the time moment t , the movement equations are

$$\begin{aligned} \frac{\partial v_x}{\partial t} + v_x \frac{\partial v_x}{\partial x} + v_y \frac{\partial v_x}{\partial y} &= -\frac{1}{\rho} \frac{\partial p}{\partial x} + \nu \Delta v_x \\ \frac{\partial v_y}{\partial t} + v_x \frac{\partial v_y}{\partial x} + v_y \frac{\partial v_y}{\partial y} &= -\frac{1}{\rho} \frac{\partial p}{\partial y} + \nu \Delta v_y \end{aligned} \tag{9}$$

where p is the pressure in this point, $\nu = \frac{\mu}{\rho}$, μ is the viscosity coefficient. At the equations (5) the equations of continuity are added

$$\frac{\partial v_x}{\partial x} + \frac{\partial v_y}{\partial y} = 0 \tag{10}$$

The current function is introduced

$$v_x = -\frac{\partial \varphi}{\partial y}, \quad v_y = -\frac{\partial \varphi}{\partial x} \tag{11}$$

Analyze of the no stationary heat in subterranean could be done when a series of problems arise at the calculation of heat losses in conditions of a heat change no stationary. For determining the no stationary heat losses in the subterranean, the next equation is used

$$\frac{\partial \theta}{\partial t} = a\left[\frac{\partial^2 \theta_s}{\partial x^2} + \frac{\partial^2 \theta_s}{\partial y^2}\right] \tag{12}$$

where θ is the temperature of the material, θ_s is the soil temperature, t is the time, x, y are the Cartesians coordinates and a is a coefficient what is characterizing soil thermal diffusion.
Suddenly in practice a great importance is to analyze the running of gases, to establish the pressure in a certainly point of a medium. The no stationary running of a gas is defined by the system

$$\begin{aligned} \frac{\partial p}{\partial x} &= \rho\left(\frac{\partial v}{\partial t} + \frac{\lambda v^2}{2d}\right) \\ \frac{\partial \rho}{\partial t} &= \frac{\partial(\rho v)}{\partial x} \end{aligned} \tag{13}$$

where p is the pressure, v is the speed related at a section, d is the medium diameter and λ is the friction coefficient.
A method used to determine the ice height of the arctic seas is the radiometry. Radiometry is based on registration of the heat radiation of which intensity varies with temperature and the radiation coefficient of the objects. The value of the radiations will characterize the relation between ice heights in their different stages. The temperature of the ice surface is determined from the heat equation, which describes the heat repartition in snow and ice

$$c_j \rho_j \frac{\partial \theta_j}{\partial t} = \lambda_j \frac{\partial^2 \theta}{\partial z^2}, \; j = 1,2,3 \tag{14}$$

where c_j is the specific heat, ρ_j is the density, λ_j is the thermal conductivity coefficient, θ_j is the temperature, t is the time, z is the height coordinate. The indices j=1,2,3 correspond to the there medium: air, snow and ice. At the frontiers there are the conditions of equilibrium.

2.2 General equations for modeling environment system seen as distributed parameter systems

The distributed parameter systems have general mathematical models in continuous time and space as partial differential equation, of parabolic or hyperbolic form, as:

$$\frac{\partial \theta}{\partial t} = c_1 \nabla(c_2 \nabla \theta) + c_3 \theta + Q \tag{15}$$

$$\frac{\partial^2 \theta}{\partial t^2} = c_1 \nabla(c_2 \nabla \theta) + c_3 \theta + Q \tag{16}$$

where the variables $\theta(\zeta, t)$ are depending on time $t \geq 0$ and on space $\zeta \in V$, where ζ is x for one axis, (x, y) for two axis or (x, y, z) for three axis, c_1, c_2 and c_3 are coefficients, which could be also time variant and $Q(\zeta, t)$ is an exterior excitation, variable on time and space.
So, in the general case, an implicit equation may be written:

$$f\left(\frac{\partial \theta}{\partial t}, \frac{\partial^2 \theta}{\partial t^2} \cdot \frac{\partial \theta}{\partial \zeta}, \frac{\partial^2 \theta}{\partial \zeta^2}, \ldots\right) = 0 \tag{17}$$

For the partial differential equations (1, 2) some boundary conditions may be imposed to establish a solution. So, when the variable value of the boundary is specified, there are Dirichlet conditions:

$$c_4\theta = q \tag{18}$$

And, when the variable flux and transfer coefficient are specified, there are Neumann conditions:

$$c_5\nabla\theta + c_6\theta = 0 \tag{19}$$

In the practical application case studies limits and initial conditions of the equation (1) are imposed:

$$\begin{aligned}&\theta(0,t) = \theta_{\zeta 0},\ t \in [0,T], \theta(\zeta,0) = 0,\ \zeta \in [0,l],\\ &\theta(l,t) = \theta_{\zeta l},\ t \in [0,T]\end{aligned} \tag{20}$$

A system with finite differences may be associated to the equations (1) and (2). For this purpose the space S is divided into small dimension pieces l_p:

$$l_p = l / n \tag{21}$$

In each small piece S_{pi}, $i=1,\dots,n$ of the space S the variable θ could be measured at each moment t_k, using a sensor from the sensor network, in a characteristic point $P_i(\zeta_i)$, of coordinate ζ_i. Let it be θ_i^k the variable value in the point $P_i(\zeta_i)$ at the moment t_k.
The points from the space in which the phenomenon is happening are denoted P_i, with the coordinate z_i. For a bi-dimensional space in a system coordinate xOy $z_i=(x_i,y_i)$. The phenomenon as distributed system is monitored with a sensor network with n sensors S_i, $i=1,\dots,n$, placed in n points P_i from the space, like in Fig. 2.

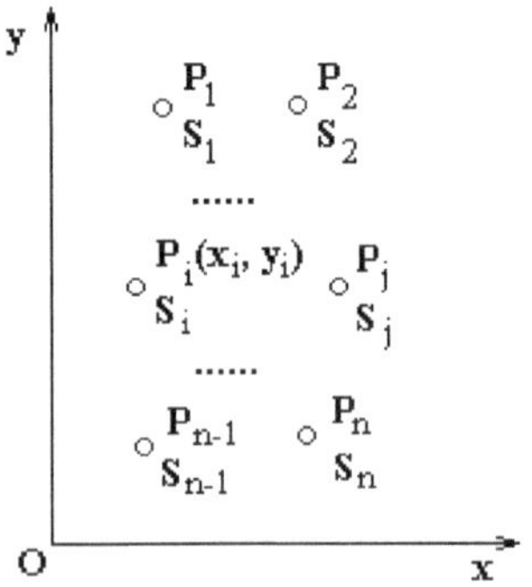

Fig. 2. Space monitoring scheme

It is a general known method to approximate the derivatives of a variable with small variations. In the equation with partial derivatives there are derivatives of first order, in time, and derivatives of first and second order in space. So, theoretically, we may approximate the variable derivatives in time with small variations in time, with the following relations:

$$\frac{\partial \theta}{\partial t} = \frac{\theta_i^{k+1} - \theta_i^k}{t_{k+1} - t_k} \tag{22}$$

$$\frac{\partial^2 \theta}{\partial t^2} = \frac{\theta_i^{k+1} - 2\theta_i^k + \theta_i^{k-1}}{(t_{k+1} - t_k)^2} \tag{23}$$

The first and the second derivatives in space may be approximated with small variations in space to obtain the following relations. For the *x*-axis we may write the following equations:

$$\frac{\partial \theta}{\partial x} = \frac{\theta_i^k - \theta_{i-1}^k}{l_p} \tag{24}$$

$$\frac{\partial^2 \theta}{\partial x^2} = \frac{\theta_{i+1}^k - 2\theta_i^k + \theta_{i-1}^k}{l_p^2} \tag{25}$$

The same equations may be written also for the *y* and also *z*-axis. Of course, an equation with variables written in vectors could be written.

We may consider the variable is measured as the sample $\theta_i^k = \theta(\zeta_i, t_k)$, $\zeta_i \in V$, at equal time intervals with the value:

$$h = t_{k+1} - t_k \tag{26}$$

called sample period, in a sampling procedure, with a digital equipment, at the sample time moments $t_k = k.h$.

For the above equation, a linear approximate system of derivative equations of first degree may be used:

$$\frac{d\Psi}{dt} = A\Psi + BQ \tag{27}$$

where, this time, ψ is a vector containing the values of the variable $\theta(\zeta, t)$ in different points of the space and at different time moments.

Combining the equations (17, 22, 24) in equation (15), a system of equations with differences results for the parabolic equation:

$$f_p(\theta_i^k, \theta_{i-1}^k, \theta_i^{k+1}, \theta_{i+1}^k) = 0 \tag{28}$$

and, combining the equations (17, 23, 25) in equation (16), an equivalent system with differences results as a model for the hyperbolic equation:

$$f_h(\theta_i^k, \theta_{i-1}^k, \theta_{i+1}^k, \theta_{i-1}^{k+1}, \theta_i^{k-1}, \theta_{i-1}^{k-1}) = 0 \tag{29}$$

Taking account of equations (28, 29), it is obvious that several estimation algorithms may be developed as follows, based on the discrete models of the partial derivative equations. These algorithms of estimation are presented as it follows.

3. Algorithms of estimation

3.1 Parabolic systems

Estimation algorithm 1. It estimates the value of the variable θ_i^{k+1} at the moment t_{k+1}, measuring the values of the variables $\theta_{i-1}^k, \theta_{i+1}^k, \theta_i^k$ at the anterior moment t_k:

$$\theta_i^{k+1} = f_1\left(\theta_{i-1}^k, \theta_{i+1}^k, \theta_i^k\right) \tag{30}$$

This is a multivariable estimation algorithm, based on the adjacent nodes.

Estimation algorithm 2. It estimates the value of the variable θ_i^{k+1} at the moment t_{k+1}, measuring the values of the same variable $\theta_i^k, \theta_i^{k-1}, \theta_i^{k-2}, \theta_i^{k-3}$, but at four anterior moments t_k, t_{k-1}, t_{k-2} and t_{k-3}.

$$\theta_i^{k+1} = f_2\left(\theta_i^k, \theta_i^{k-1}, \theta_i^{k-2}, \theta_i^{k-3}\right) \tag{31}$$

This is an autoregressive algorithm, using the values from the same node.

3.2 Hyperbolic systems

Estimation algorithm 3. It estimates the value of the variable θ_i^{k+1} at the moment t_{k+1}, measuring the values of the variables $\theta_{i-1}^k, \theta_{i+1}^k, \theta_i^k$ at the anterior time moment t_k and t_{k-1}:

$$\theta_i^{k+1} = f_1\left(\theta_{i-1}^k, \theta_{i+1}^k, \theta_i^k, \theta_{i-1}^{k-1}, \theta_{i+1}^{k-1}, \theta_i^{k-1}\right) \tag{32}$$

This is a multivariable estimation algorithm, based on the adjacent nodes and 2 time anterior moments.

Estimation algorithm 4. It estimates the value of the variable θ_i^{k+1} at the moment t_{k+1}, measuring the values of the same variable (the same node) $\theta_i^k, \theta_i^{k-1}, \theta_i^{k-2}, \theta_i^{k-3}, \theta_i^{k-4}, \theta_i^{k-5}$, but at six anterior moments t_k, t_{k-1}, t_{k-2}, t_{k-3}, $t_{k-4}$and t_{k-3}:

$$\theta_i^{k+1} = f_2\left(\theta_i^k, \theta_i^{k-1}, \theta_i^{k-2}, \theta_i^{k-3}, \theta_i^{k-4}, \theta_i^{k-5}\right) \tag{33}$$

4. Modeling and simulation

Environment behavior may be modeled with the equation from the above paragraph. Using these models, some analysis in time and space domains may be accomplished. Some transient characteristics of the temperature are there presented for 101 samples. The nodes and meshes structure for a sensor network with reduced number of sensor, in this case 13, is presented in Fig. 3.

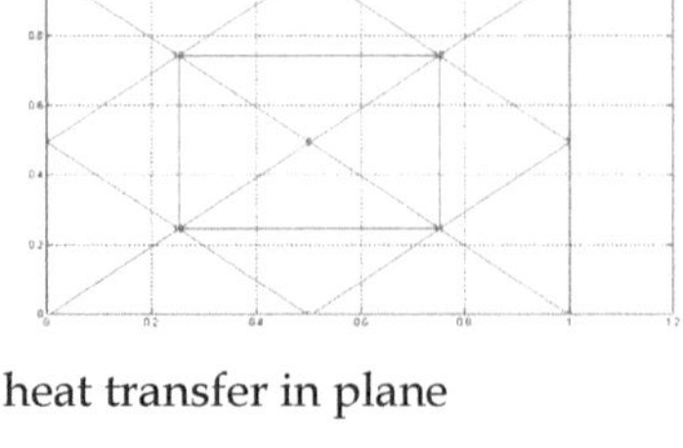

Fig. 3. Nodes and meshes for heat transfer in plane

The temperature variation in 3D is presented in Fig. 4, at a certain time moment.

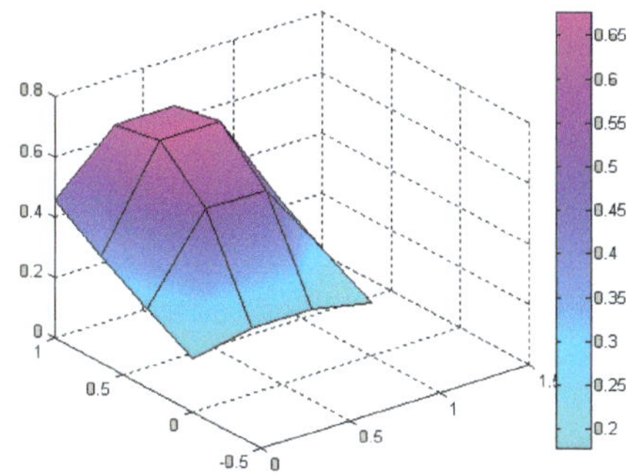

Fig. 4. Temperature variation in space

Temperature isotherms in plane are presented in Fig. 5.
Identical characteristics may be obtained for other distributed parameter systems involved in environmental modeling.

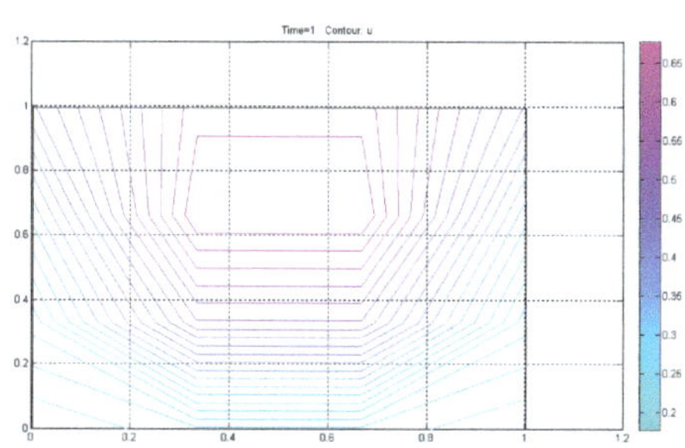

Fig. 5. Temperature isotherms

5. Sensor network

The modern sensors are smart, small, lightweight and portable devices, with a communication infrastructure intended to monitor and record specific parameters like temperature, humidity, pressure, wind direction and speed, illumination intensity, vibration intensity, sound intensity, power-line voltage, chemical concentrations and pollutant levels at diverse locations. The sensor number in a network is over hundreds or thousands of ad hoc tiny sensor nodes spread across different area. Thus, the network actively participates in creating a smart environment. With them we may developed low cost wireless platforms, including integrated radio and microprocessors. The sensors are adequate for autonomous operation in highly dynamic environments as distributed parameter systems. We may add sensors when they fail. They require distributed computation and communication protocols. They insure scalability, where the quality can be traded for system lifetime. They insure Internet connections via satellite.
The structure of a modern sensor is presented in Fig. 6.

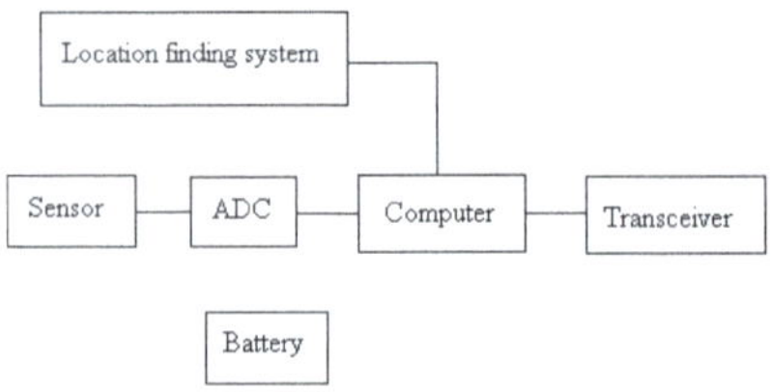

Fig. 6. The structure of a modern sensor

The constructive and functional representation of a sensor network is presented in Fig. 7.

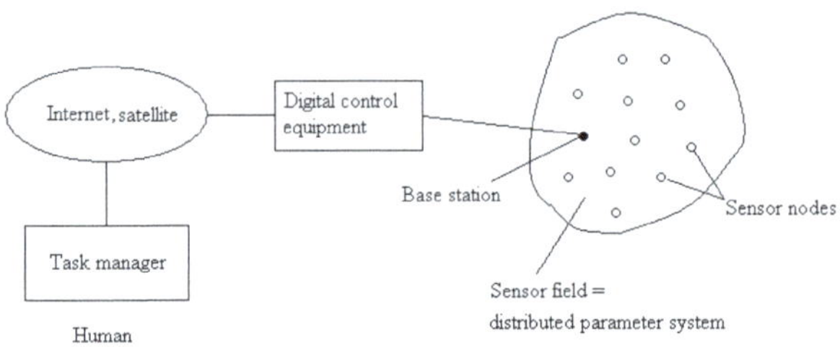

Fig. 7. Sensor network

The sensor S_A measures the temperature θ_A in a point in this space. The sensor S_A measures the temperature θ_A in a point in this space.
There have been used in practice: a Memsic eKo Outdoor Wireless Monitoring System with 4 eKo sensor nodes EN2100, an eKo base radio EB2110, an eKo gateway w/ built-in eKoView web application. The eKo Wireless Sensor Nodes form wireless mesh network with communication range from several hundred meters, accepting up to four sensor inputs. Solar cell or rechargeable batteries powered them. The eKo base radio provides connection between eKo sensor nodes and eKo gateway via USB interface for data transfer. pThere had been used an eKo weather station sensor suite with wind speed, wind direction, rain gauge, ambient temp/humidity, barometric pressure and solar radiation. Each node has a temperature and humidity sensor to measure the ambient relative humidity and air temperature and to calculate the dew point. The base station is wireless, with computing energy and communication resources, which is acting like an access gate between the sensor nodes and the end user. The senor nodes have two components. The processor/radio modules are activating the measuring system of small power.

Fig. 8. Components of the sensor network used in practice

They are working at the frequency of 2.4 GHz. The sensor network is also provided with a software for data acquisition, which is reading data from a data base. The sensor network is working in real time with a driver which insures data acquisition from the base station.

6. Monitoring application

6.1 Monitoring structure

The estimation model describes the evolution of a variable measured over the same sample period as a non-linear function of past evolutions. This kind of systems evolves due to its "non-linear memory", generating internal dynamics. The estimation model definition is:

$$y(t) = f(u_1(t),...,u_n(t)) \quad (34)$$

where u(t) is a vector of the series under investigation (in our case is the series of values measured by the sensors from the network):

$$u = \begin{bmatrix} u_1 & u_2 & \dots & u_n \end{bmatrix}^T \quad (35)$$

and f is the non-linear estimation function of non-linear regression, n is the order of the regression. By convention all the components u_1(t),...,u_n(t) of the multivariable time series u(t) are assumed to be zero mean. The function f may be estimated in case that the time series u(t), u(t-1),..., u(t-n) is known (recursive parameter estimation), either predict future value in case that the function f and past values u(t-1),..., u(t-n) are known (AR prediction). The method uses the time series of measured data provided by each sensor and relies on an (auto)-regressive multivariable predictor placed in base stations as it is presented in Fig. 9.

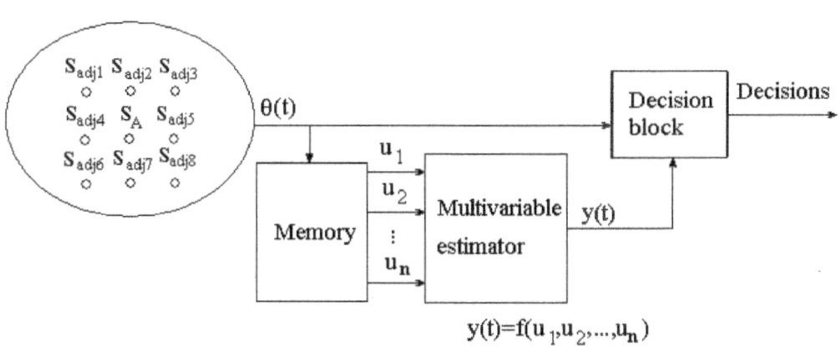

Fig. 9. Estimation and detection structure

The principle of the estimation is: the sensor nodes will be identified by comparing their output values θ(t) with the values $y(t)$ predicted using past/present values provided by the same sensors or adjacent sensors (adj). After this initialization, at every instant time t the estimated values are computed relying only on past values θ_A(t-1), ..., θ_A(0) and both parameter estimation and prediction are used. First, the parameters of the function f are estimated using training from measured values with a training algorithm as back-propagation, for example. After that, the present values $\theta_A(t)$ measured by the sensor nodes may be compared with their estimated values $y(t)$ by computing the errors:

$$e_A(t) = |\theta_A(t) - y(t)| \quad (36)$$

If these errors are higher than the thresholds ε_A at the sensor measuring point, a fault occurs. Here, based on a database containing the known models, on a knowledge-based system, we may see the case as a multi-agent system, which can do critics, learning and changes, taking decision based on node analysis from network topology. Two parameters can influence the decision: the type of the distributed parameter system, which is offering the data measured by sensors and the computing limitations. Because both of them are a priori known, an off-line methodology is proposed. Realistic values are situated between 3 and 6.

6.2 Estimator mechanism

The estimator is a non-linear one, described by the function $y=f(u_1, u_2, ..., u_n)$, using the adaptive-network-based fuzzy inference. Its general structure is presented in Fig. 10.

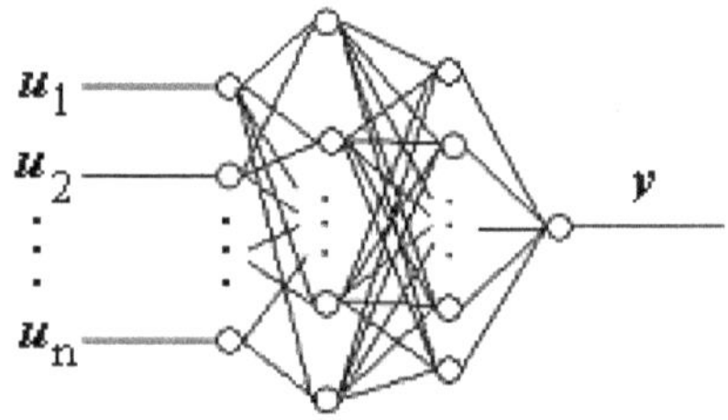

Fig. 10. The estimator input-output general structure

The number of inputs depends on the estimation algorithm, on the specific position in space of the measuring points, on the conditions of determination. The ANFIS procedure is well known and it may use a hybrid learning algorithm to identify the membership function parameters of the adaptive system. A combination of least-squares and back-propagation gradient descent methods may be used for training membership function parameters, modeling a given set of input/output data.

6.3 Monitoring method

The following method is according to the objectives of monitoring of defined distributed parameter system from the practical application in the real world, as heat distribution, wave propagation. These systems have known mathematical model as a partial differential equation as a primary model from physics, with well-defined boundary and initial conditions for the system in practice. These represent the basic knowledge for a reference model from real data observation. The primary physical model must be meshed, in order to obtain a mathematical model as a multi input - multi output state space model. The unstructured meshes may be generated. The sensors must be placed in the field, according to the meshes structured under the form of nodes and triangles. A scenario for practical applications could be chosen and simulated. The simulation and the practical measurements are producing transient regime characteristics. Those transient characteristics are due to the system dynamics in a training process. In steady state we cannot train the neural model. On these transient characteristics, seen as times series, the estimation algorithms may be applied. ANFIS is used to implement the non-linear estimation algorithms. With these algorithms, future states of the process may be estimated. Possible fault in the system are chosen and strategies for detection may be developed, to identify and to diagnose them, based on the state estimation. In practice, applying the method presumes the following steps: -placing a sensor network in the field of the distributed parameter system; -acquiring data, in time, from the sensor nodes, for the system variables; -using measured data to determine an estimation model based on ANFIS; -using measured data to estimate the future values of the system variables; -imposing an error threshold for the system variables; -comparing the measured data with the estimated values; -if the determined error is greater then the threshold, a default occurs; -diagnosing the default, based on estimated data, determining its place in the sensor network and in the distribute parameter system field.

7. Expert system

7.1 Process knowledge

Knowledge that may be determinate from measurements upon the process variables made using sensor networks is as it follows:

- the value v_i of the phenomenon at a time moment, in a point of the space P_i, which is the value provided by the sensor Si, place in the point P_i, at the time moment t: $\theta_i(t)$, temperature in this case;
- the speed of the phenomenon s_i, which is the derivative in time of the variables measured by sensor S_i, in the point P_i, at two consecutive time moments t and t-h: $\frac{d\theta_i(t)}{dt} \approx \frac{\theta_i(t) - \theta_i(t-h)}{h}$, where the discrete time approximation is used, for a constant sample period h;
- the value of the difference in space d_{ij}, from two adjacent sensor variables: $\Delta\theta_{ij}(t) = \theta_i(t)-\theta_j(t)$, given by the sensors S_i and S_j, place din the points P_i and P_j; -the difference in space is given the sense in which the phenomenon is happening. The positive sense is considering from S_i to S_j. This difference is proportional to the space between two sensors S_i to S_j, or points P_i and P_j $l_{ij}=|z_i-z_j|$, where z_i and z_j are the space coordinates of the two points. For a bi-dimensional space, the coordinates are for $P_i(x_i,y_i)$ and $P_j(x_i, x_j)$.
- the speed sd_{ij} of difference variation between two adjacent sensors S_i and S_j, place din the points P_i and P_j, as time derivative of space difference $\frac{d\Delta\theta_{ij}(t)}{dt} = \frac{\Delta\theta_{ij}(t) - \Delta\theta_{ij}(t-h)}{h}$, at two onsecutive time moments t and t-h. The speed of the difference in space is given the speed of the space displacement in a sense in which the phenomenon is happening.

We may use also the variables obtained as estimation, as it follows.

- the estimated value $\hat{v}_i$ of the phenomenon at a time moment, in a point of the space P_i, which is the value provided by the estimator E_i for the point P_i, at the time moment t: $\hat{\theta}_i(t)$;
- the speed of the estimated phenomenon $\hat{s}_i$, which is the derivative in time of the estimated variables provided by the estimator E_i for the point P_i, at two consecutive

time moments t and t-h: $\frac{d\hat{\theta}_i(t)}{dt} = \frac{\hat{\theta}_i(t) - \hat{\theta}_i(t-h)}{h}$, where the discrete time approximation is used, for a constant sample period h;

the estimated difference in space $\hat{d}_{ij}$, from two values of two adjacent sensor variables: $\Delta\hat{\theta}_{ij}(t) = \hat{\theta}_i(t) - \hat{\theta}_j(t)$, given by the estimators E_i and E_j, for the points P_i and P_j; The estimated difference in space is given the estimated sense in which the phenomenon is estimated to take place.

- the estimated speed $\hat{sd}_{ij}$ of the estimated difference variation between two estimators E_i and E_j, for two adjacent points P_i and P_j, as time derivative of estimated space difference $\frac{d\Delta\hat{\theta}_{ij}(t)}{dt} = \frac{\Delta\hat{\theta}_{ij}(t) - \Delta\hat{\theta}_{ij}(t-h)}{h}$, at two consecutive time moments t and t-h. The speed of the difference of estimates in space is given the speed of the estimate of the space displacement in a sense in which the phenomenon is estimated to happen.

Some errors between the estimates and the actual variables may be introduced: $e_v = v - \hat{v}$ - the error at the process value; $e_s = s - \hat{s}$ - the error in speed of phenomenon happening in some field point; $e_d = d - \hat{d}$ -the error in space difference of two adjacent points and $e_{sd} = sd - \hat{sd}$ - the error of speed of phenomenon propagation in space.
In order to make estimations, we may use the values provided by the sensors.

7.2 Expert system structure

For these process variables v, s, d, sd and for the estimated variables $\hat{v}, \hat{s}, \hat{d}, \hat{sd}$ some values may be defined as negative N and positive P or around zero Z, with some degrees: small S, medium M or big B. So, we may have the following combinations put on an axis: NB, NM, NS, Z, PS, PM, PB. To emphasize a non-linear character of the process, the usage of only three fuzzy values is recommended.
The reasoning is as it follows: -If the derivatives are negative, we may say the phenomenon is decreasing. -If the derivative are positive, the phenomenon is increasing; -If the differences are negative, the phenomenon sense is opposite from the two sensors and measuring points. - If the speed of the difference is positive, the space becomes to be not homogenous, something is happening in the space between the two sensors.
The expert system is developed using a backward chaining. Some rules from the rule base for this expert system are: (1) IF v is Z THEN the process is supressed (c_f = 10 %); (2) IF v is NOT Z THEN the process is NOT supressed (c_f = 90 %); (3) IF s is Z THEN the process is NOT in course (c_f = 10 %); (4) IF s is NOT Z THEN the process is in course (c_f = 90 %), and so on. Many other rules may be developed according to the above considerations.
The application may be framed in so called "goal driven methods". In the real distributed parameter systems there are phenomena with small certainty and their opposite seems to be

true. When an exert system is developed for monitoring distributed parameter systems, it is necessary to test both, to see what it is happening in the field.

8. Case study

There is presented a basic case study consisting in a heat distribution flux through a plane square surface of dimensions l=1, with Dirichlet boundary conditions as constant temperature on three margins:

$$h_\theta \theta = r \tag{37}$$

with r=0, and a Neumann boundary condition as a flux temperature from a source

$$nk\nabla\theta + q\theta = g \tag{38}$$

where q is the heat transfer coefficient q=0, g=0, h_θ=1.
The heat equation, of a parabolic type, is:

$$\rho C\frac{\partial\theta}{\partial t} = \nabla(k\nabla\theta) + Q + h_\theta(\theta_{ext} - \theta) \tag{39}$$

where ρ is the density of the medium, C is the thermal (heat) capacity, k is the thermal conductivity, coefficient of heat conduction, Q is the heat source, h_θ is the convective heat transfer coefficient, θ_{ext} is the external temperature. Relative values are chosen for the equation parameters: ρC=1, Q=10, k=1.
In the case of study, a small sensor network with only 13 nodes had been used in laboratory tests. The number of sensor is equivalent to a reduced number of nodes and meshes, as it is in the position scheme from Fig. 11.
In the case study, we are choosing the nodes 8, 13, 12 5 and 11 in order to apply the estimation method. These nodes are marked with bold characters on figure.
The transient characteristics of the temperature (in relative values) are presented in Fig. 12, for 101 samples.
The transient characteristics of the 12th and 13th nods are the same, so they are plotted one over the other, and in the Fig. 12 there are only four characteristics instead of five.

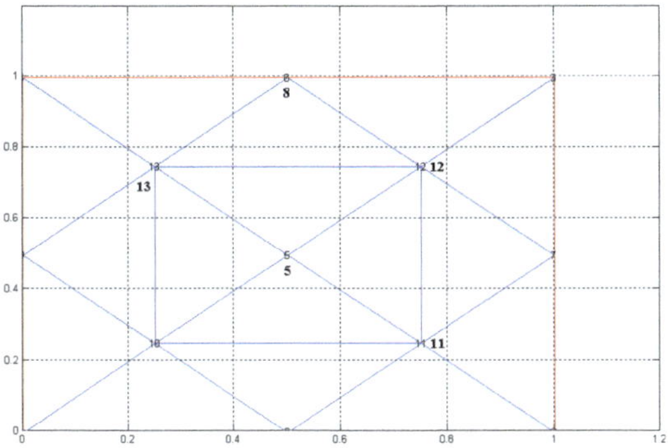

Fig. 11. Sensor network position in the field

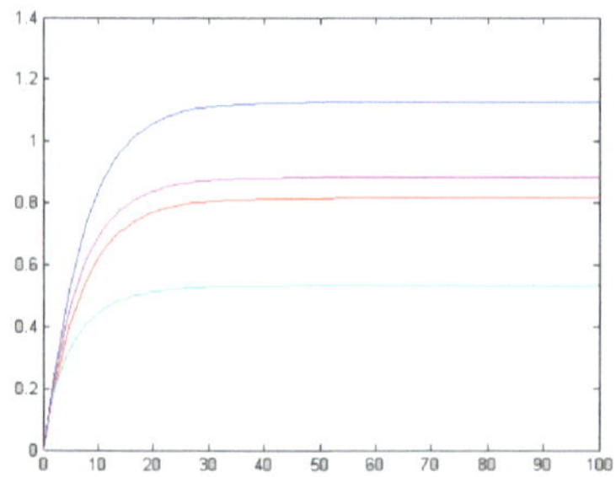

Fig. 12. Transient characteristics

We are presenting as an example the estimation for the 5th node. It is the node of the estimated variable, based on the first recursive algorithm:

$$\theta_5^{k+1} = f(\theta_8^k, \theta_{13}^k, \theta_{12}^k, \theta_{11}^k) \quad (40)$$

The fuzzy inference system structure is presented in Fig. 13.

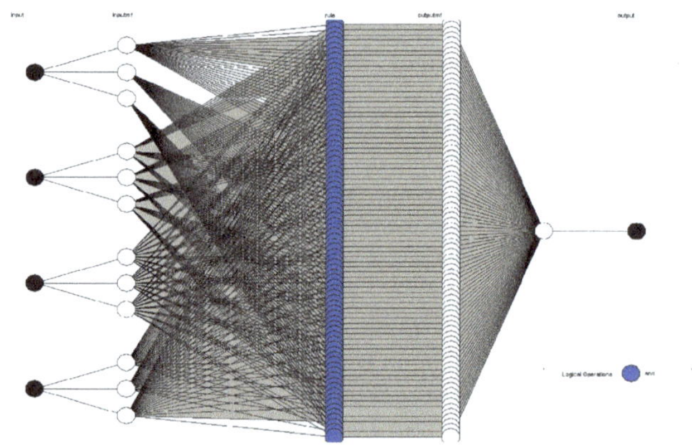

Fig. 13. FIS structure

A short description about the ANFIS and its function approximating property is provided as it follows. The number of inputs depends on the algorithm type. For the 1st and 2nd algorithms there are 4 inputs, because of the first order derivation in time of the parabolic model. For the 3rd and the 4th algorithms there are 6 inputs, because of the second order derivation in time of the hyperbolic model. The ANFIS procedure may use a hybrid learning algorithm to identify the membership function parameters of single-output, Sugeno type fuzzy inference system. A combination of least-squares and back-propagation gradient descent methods may be used for training membership function parameters, modeling a given set of input/output data.

In the inference method *and*, there may be implemented with product or minimum, *or* with maximum or summation, implication with product or minimum and aggregation with maximum or arithmetic media. The first layer is the input layer. The second layer represents the input membership or fuzzification layer. The neurons represent fuzzy sets used in the

antecedents of fuzzy rules determine the membership degree of the input. The activation function represents the membership functions. The 3rd layer represents the fuzzy rule base layer. Each neuron corresponds to a single fuzzy rule from the rule base. The inference is in this case the sum-prod inference method, the conjunction of the rule antecedents being made with product. The weights of the 3rd and 4th layers are the normalized degree of confidence of the corresponding fuzzy rules. These weights are obtained by training in the learning process. The 4th layer represents the output membership function. The activation function is the output membership function. The 5th layer represents the defuzzification layer, with single output, and the defuzzification method is the centre of gravity.
The comparison transient characteristics for training and testing output data are presented in Fig. 14.

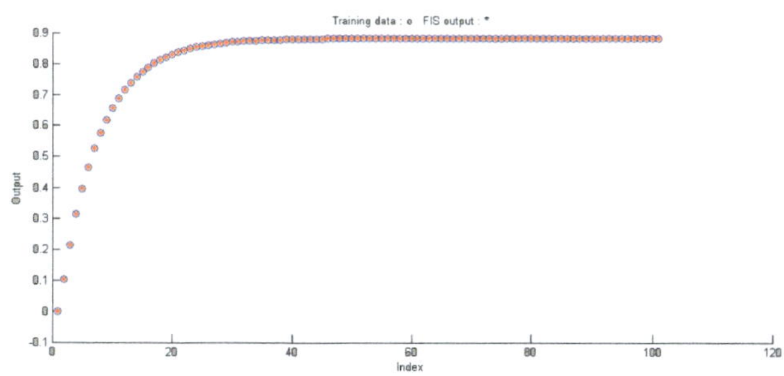

Fig. 14. Comparison between training and testing output

The characteristics are plotted two on the same graph, to show that there is no significant difference. The characteristic for the training data is plotted with °. The characteristic for the FIS output is there plotted with *. The difference between the training case and the testing case is very small. The plotting signs ° and * are on the same points for the both characteristics. The average testing error is $2{,}017.10^{-5}$. The number of training epochs was 3.
If a fault appears at a sensor, for example at the time moment of the 50th sample, an error occurs in estimation, as it is in Fig. 15.

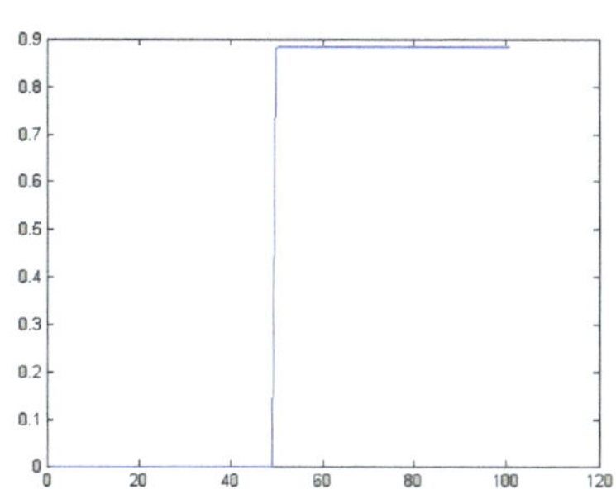

Fig. 15. Error at the fifth node for a fault in the network

Detection of this error is equivalent to a default at this sensor, from other point of view in the place of the monitored sensor in the space of the distributed parameter systems and in the heat flow around the sensor.

9. Implementation using virtual instrumentation

Virtual instrumentation, based on National Instruments technology, had been used for sensor network monitoring. A virtual instrument for sensor network monitoring was built on a personal computer [11]. It includes: data acquisition and processing, estimator, data base, results table and an Excel data base. The control panel is presented in Fig. 16.

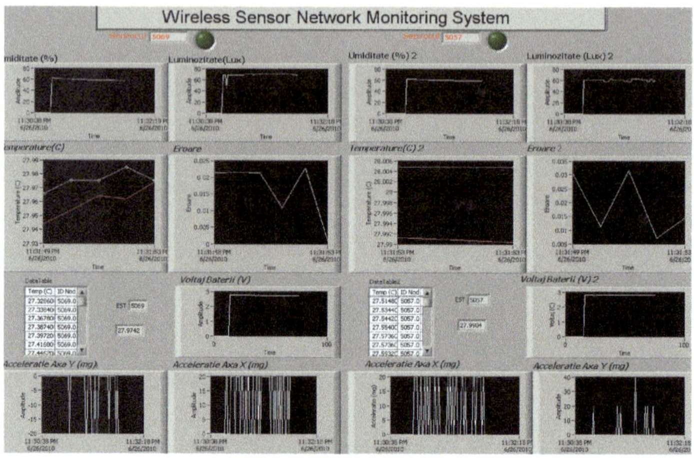

Fig. 16. The control panel of the virtual instrument

The block diagram of the virtual instruments for sensor network monitoring is presented in Fig. 17.

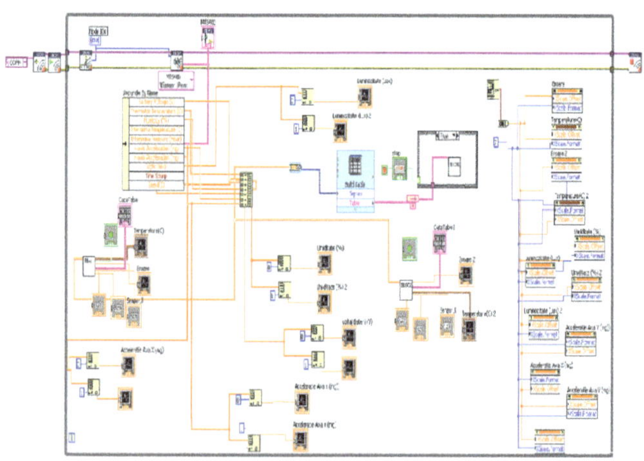

Fig. 17. The block diagram of the virtual instrument

The block diagram is built using sub-VIs, input-output virtual instruments and estimation sub-VIs. In this block diagram, the rules may be introduced and computed using inference and confidence factors. The driver assures data manipulation with a very small delay.
A long distance monitoring is allowed, using a web page, presented in Fig. 18.

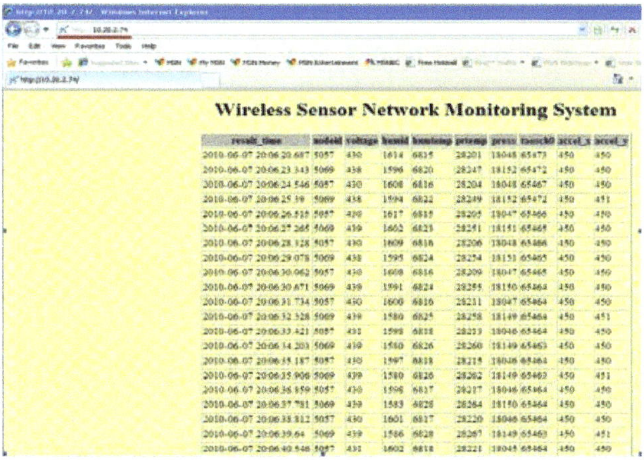

Fig. 18. Web page for monitoring

10. Conclusion

This chapter presents some considerations on environmental monitoring using sensor networks and estimation techniques based on ANFIS, one of the main tools of artificial intelligence.

There are presented four algorithms for estimation and one method for fault detection and diagnosis of distributed parameter systems. The algorithms are based on non-linear exogenous models with regression and auto-regression. The firsts are using the values provided by the adjacent nodes of the sensor network. The seconds are using the values from anterior time moments of the same node. The non-linear adaptive network based fuzzy inference scheme (ANFIS) is used for system identification based on time series data acquired from an autonomous wireless intelligent sensor network. There is presented an application expert systems for environment monitoring, based on distributed parameter system theory, with exemplification at the process of heat transfer. There are used: the knowledge on distributed parameter system, the measured variables acquired from the system using a sensor network and some estimates obtained with estimation techniques. The sensor network is seen as a distributed sensor, placed in the measuring field of the distributed parameter system. The positioning of sensors in the field may be done according to optimal nodes and triangular meshes of a modelling and simulation of the environmental process based on distributed parameter system theory. There is presented an example of generated meshes and estimated temperature. The method offers the way of how to use all these concepts for fault detection and diagnosis in environment systems, based on the measured values provided by the sensors and the estimated values computed by the ANFIS estimator, calculating an error and detecting the fault based on a decision taken after a threshold comparison. The usage of virtual instrumentation on personal computers offers a good user interface. This methodology can be efficiently implemented on sensor network base stations, so there is no need for other hardware resources. The research results are presented in the frame of a practical case study, with tests, which are validating the theory. The key point of the chapter is the development of a methodology for environment monitoring, based on some summated concepts: estimation techniques, the theory of distributed parameter systems, expert systems and wireless sensor networks. A negative aspect is the lack of information related to the error of measuring data, for different environment applications in practice. In future, some researches may be done in order to respond to this question related to the accuracy of measurements for different practical cases. Future applications could be done in computing interpolative values in inaccessible places from the sensor area, in the control of distributed parameter systems, and other.

11. Acknowledgement

This work was made in the frame of the CNCSIS - UEFISCSU, PNII -IDEI_PCE_ID923 grant.

12. References

Akyildiz, F.; Su, W.; Sankarasubramaniam, Y. & Cayirci, E., Wireless Sensor Networks: A Survey. *Computer Networks*, 38(4), March, 2002.

Cuiyun, W.; Naiang, W.; Xibao, X.; Feng, Z. & Yinzhou, H. Study on Spatial Thermal Environment in Lanzhou City Based on Remote Sensing and GIS, *IEEE Int. Conf. On Geoscience and remote Sensing Symposium*, July, 31, 2006, Denver, pp. 2466-1468.

Dardari, D.; Conti, A.; Buratti, C. & Verdone, R. Mathematical Evaluation of Environmental Monitoring Estimation Error through Energy-Efficient Wireless Sensor Networks, IEEE Trans. on Mobile Computing, July, 2007, Vol. 6, Issue 7, p. 790-802.

Giannopoulos, N; Goumopoulos, & Kameas, A. Design Guidelines for Building a Wireless Sensor Network for Environmental Monitoring, *PCI'09 13th Panhellenic Conf. on Informatics*, 10-12 Sept. 2009, Corfu, p. 148-152.

Lan, S.; Qilong, M. & Du, J. Architecture of Wireless Sensor Networks for Environmental Monitoring, *GRS 2008 Int. Workshop on Geoscience and Remote Sensing*, 21-22 Dec. 2008, Shanghai, Vol. 1, p. 579-582.

Rosculet, M.N. & M. Craiu, M. *Differential applicative equations*, RSR Academy Publishing, Bucharest,, 1979.

Talukder, A.; Panangadan, A. & Herrington, A.T. Autonomous Adaptive Resource Management in Sensor Network Systems for Environmental Monitoring, *2008 IEEE Aerospace Conf.*, 1-8 March 2008, Big Sky, MT, pp. 1-9.

Volosencu, C., Environmental Monitoring Based on Sensor Networks and Artificial Intelligence, *Development, Energy, Environment, Economics (DEEE '10)*, Puerto De La Cruz, Tenerife, Nov. 30- Dec. 2, 2010, p. 79 – 83.

Volosencu, C., Algorithms for estimation in distributed parameter systems based on sensor networks and ANFIS, *WSEAS Transactions on Systems*, Volume 9, Issue 3, March 2010, pag. 283- 294.

Tracking Players in Indoor Sports Using a Vision System Inspired in Fuzzy and Parallel Processing

Catarina B. Santiago[1], Lobinho Gomes[1], Armando Sousa[1],
Luis Paulo Reis[2] and Maria Luisa Estriga[3]
[1]*Faculty of Engineering, University of Porto and INESC-Porto, Porto*
[2]*School of Engineering, University of Minho and LIACC, University of Porto, Porto*
[3]*Faculty of Sports, University of Porto and CIFI2D, Porto*
Portugal

1. Introduction

Sports are an important part of nowadays society and there is an increasing interest by the sports' community on having mechanisms that allow them to better understand the dynamics of teams (their own and their opponents). This information is frequently extracted manually by operators that, after the game, visualize game recordings (frequently TV footages) and perform hand annotation, which is a time consuming and error prone task. There is a clear necessity for developing automatic mechanisms and methodologies which allow performing these tasks much faster and systematically. The importance of such systems was first highlighted in the late 80's by Franks et al. ((Franks & Nagelkerke, 1988; Franks et al., 1987)).
In this chapter, we present an automatic and intelligent visual system for detecting and tracking handball players based on two cameras that cover the entire playing area. The followed methodology includes the identification of foreground pixels using dynamic background subtraction, the definition of colour subspaces for each team using a Fuzzy inspired model that allows detecting the players based on the colour properties of their clothes. Player tracking is further improved by using one Kalman Filter per player (object to track). The resulting information is aggregated in an undistorted image view of the entire field that is very interesting and meaningful to the target end-user. The generation of the video is a demanding computational task that takes advantage of using parallel computing. The resulting videos are interesting and include several important informations to the human end-user. Tests were conducted on videos collected during the Handball Portuguese SuperCup competition, where the best six Portuguese teams competed for the trophy of the year of 2010/11.
The chapter structure is as follow: the next section presents some relevant information on image segmentation methodologies, parallel processing and implementations of automatic visual systems for detecting and tracking players and describes the main methodologies used. Section 3 discusses the proposed architecture principles, providing an overview of the methodology used. Section 4 presents the results achieved and finally section 5 concludes this chapter with the main conclusions.

2. Related research

This section focus on the three main areas involved on the implementation of the system, and therefore presents some of the most common used methodologies for video/image segmentation and for parallel processing as well as an overview of systems that are able to detect and track players in indoor team sports.

2.1 Video/image segmentation

Video and inherently image segmentation is the first step and probably the most critical step in any vision system. Video segmentation can be subdivided into temporal segmentation and spatial segmentation.

Temporal segmentation corresponds to segmenting the video into meaningful temporal sequences. This kind of segmentation is usually used as the first step of video annotation and tries to segment the video taking into account similarities/dissimilarities between successive frames (Koprinska & Carrato, 2001). On the other hand, spatial segmentation analyses the content of each frame, and divides it into homogeneous regions that correspond to independent objects. The focus of this work is more on spatial segmentation, therefore for a detailed survey on temporal video segmentation please refer to (Koprinska & Carrato, 2001).

Spatial video segmentation may be performed using the methodologies that are used for image segmentation and further enhanced using the temporal characteristics of video. In addition, when performing colour analysis there is also the need to choose a colour space.

Regarding colour image segmentation, a detailed survey is provided by (Cheng et al., 2001). As they state, most of the existing colour image segmentation methodologies have their origins on grey scale image segmentation with the addition of a proper colour space choice. The main categories of image segmentation methodologies (Cheng et al., 2001) are summarized on Table 1.

Nowadays there is the tendency to apply techniques from different categories in order to achieve better results. A good example of this tendency is the JSEG algorithm (Deng & Manjunath, 2001) which initially clusters colours into several representative classes, afterwards replaces each pixel by their corresponding colour class label and at the end employs a region growing process directly to the class map in order to identify homogeneous regions.

On videos, contrary to static images, besides the physical (x and y coordinates) and colour information there is also the time component. Using this property it is possible to segment images based on motion along time. There are two main approaches to perform this task: background subtraction and optical flow.

Background subtraction is usually used in situations where a more or less fixed background exists and subdivides the image into foreground and background regions. Several background subtraction techniques have been proposed in literature. The main issue on these methods is to obtain a good estimate of the background.

The simplest method to model the background is to use a single static image without objects, however its efficiency is low, because it does not take into account changes such as light effects or shadows that may occur in the background. More robust methods include estimating the background model using a moving average (Heikkila & Silvén, 1999), median or even a mixture of Gaussians (Grimson et al., 1998).

Category	Description
Histogram Thresholding	Determine the peaks or modes of the multi-dimensional histogram of a colour image.
Feature Space Clustering	Groups the image feature space into a set of meaningful groups or classes based on intensity, colour or texture characteristics of pixels and not on the spatial relation among them.
Region based	Includes region growing, Watershed transform and region split and merge. These methods try to divide the image domain based on the fact that adjacent pixels in a same region have similar visual features (colour, intensity, texture or motion).
Edge Detection	Segment the image by finding the edges of each region using one of the well known edge detectors (Canny (Canny, 1986), Sobel (Sobel, 1970), Roberts (Roberts, 1963)).
Fuzzy	These methods allow classes and regions to have a certain uncertainty and ambiguity which is generally the case in image processing.
Neural Networks	Allow parallel processing and the incorporation of non-linearities. They can be used either to pattern recognition, classification or clustering.

Table 1. Overview of image segmentation categories

Optical flow (Barron & Thacker, 2005) is based on the fact that when an object moves in front of a camera, there is a corresponding change on the image, however it assumes small displacements during time.

Following the tendency, we propose, for the video segmentation step, a methodology that combines background subtraction for detecting foreground regions, region growing and a Fuzzy inspired categorization for colour calibration.

2.2 Parallel processing

Initially, computing power was pure sequential processing, that is, sequential operations over time in a single dedicated processor. The hunger for usefulness and processing power lead to advanced solutions such as concurrency and distributed computing. Given large "workloads", concurrency is interleaving processing over time for the multiple "workloads" (even in a single processor). Distributed computing is sharing "workloads" (or parts) over different processors linked over a network. Recent advances in computer architecture introduced multi core architectures that enable true parallel processing of several "workloads" in parallel in the same computer. The challenge remains unaltered: to maximize the usefulness of a computer systems and harness as much computing power as possible, possibly circumventing technical limitations of an architecture, whilst maximizing performance but still keeping cost interesting.

When considering a single complex (large) computational job, parallel processing starts by:

- (i) dividing work into several pieces (this can be done by the programmer, at run time or a mixture of both situations) then

- (ii) "transfers" work parts to collaborators (several processors inside the same computer or not) and later
- (iii) completes the job by assembling all intermediate results into a meaningful final solution.

There are a number of complexities with parallel computing:

- Management of parallelization - the additional computational cost (overhead) to start and maintain parallel execution (sharing program, data, managing requests, extra memory required, etc);
- Communication overhead - additional computational cost to communicate to several other processors (examples: transmission of initial relevant data or fundamental intermediate calculations);
- Synchronization problems - the explicit need for sequential operations caused by interdependence among several work parts or using shared resources;
- Load Imbalancing - the difficulty in sharing workloads as even amounts of work with the likely problem that, if one of the intermediate calculations takes longer to process than the others, optimization is not as perfect as one could wish for.

In parallel computing, the Scalability curve is defined as the speed-up in processing (performance gain), over the number of available processors. Generically, it is not likely at all that a job done in a single processor in a time t_1 could be done in N processors in t_1/N of that time, that is, time to solve a problem in N processors is very frequently $t_N > t_1/N$. By adding too many processors (large N), the scalability curve will drop heavily when too many work parts are generated and computational overhead for parallel processing outweighs the benefits of having many "workers" (processors). Additionally, adding processors will likely have severe financial impact on the final overall cost of the computational solution.

OpenMP ((OpenMP, 2011)) is a technological framework for cross platform, shared-memory application programming interface (API) for taking advantage of run time distribution of work parts into several processors ((Chapman et al., 2007)). OpenMP (OMP) was introduced formally in 1997 and is maintained by the Architecture Review Board (ARB), a consortium of industry and academia and its licence is essentially free. Advertised benefits include, among others, good scalability, implicit communication and programming at (somewhat) high level. The technological limitations for performance include transferring code and data among several types of memory and assigning work parts (threads) to available resources (processors).

CUDA ((NVIDIA, 2011)) stands for Compute Unified Device Architecture and is a software platform for massively parallel high-performance computing using NVidia's video graphics hardware. The CUDA software toolkit is proprietary from NVidia, it essentially allows a free licence of use and was formally introduced in 2006. A large portion of the (recent, high-end) hardware from that manufacturer is usable with CUDA. The intent is to turn the resources of the video card into a number of generic processors and memory ((Halfhill, 2008)). The actual number of CPUs and memory available for generic use in the video card is hardware dependent and a highly volatile issue over time. Sometimes 16 or more independent processors are available for custom parallel programming (working frequencies frequently below 1 GHz) and about 512 MB of memory is also frequently usable. These resources are usable if the video card is used for showing 2D images. Communication with the video card has inherent technological limitations as code and data must be transferred into the video card, an "external" element when compared to CPUs and Memory Systems. While using the video card(s) as generic processors presents limitations to hardware changes it is most interesting for application sharing lots of data, with the added benefit of freeing up the main processor(s) for other tasks. High-end video cards have risen in complexity, performance and cost and may currently be more expensive than the "main" general purpose CPU of the computer.

The reader should be aware that at, the time of writing this article, parallel programming is still at its infancy. A promising framework exists, however: OpenCL (Open Computing Language) is open, royalty-free, standard for general-purpose parallel programming of heterogeneous systems across different hardware (Group, 2011b). OpenCL provides a uniform programming environment for software developers to write efficient, portable code for high-performance computing using a diverse mix of multi-core CPUs, GPUs, etc. OpenCL is recent and its first introduction is in late 2009 by the Kronos Group (Group, 2011a), a consortium of companies Intel, AMD, ATI, ARM, NVidia, among others. The performance will likely be inferior to pure CUDA as OpenCL calls CUDA when possible. These topics are currently issues of large interest in the Scientific and Technological communities because they promise great performance benefits. The most important topics and dominant toolkits were, however, briefly addressed in this section.

Prospective users of parallel computing should take into consideration that there is a non-negligible effort in learning the concepts at stake and even more in porting large conventional algorithms to take advantage of parallel processing. While OpenMP is not far to common programming techniques, CUDA approach of offering massive parallelism with common data requires different approaches and most likely the new algorithm will demand very large portions of code to be written almost from start. As in other optimization techniques, profiling the application is of great interest to find which part of the code takes more time to execute and what part(s) of the code will benefit from parallel computing.

2.3 Player detection and tracking using vision systems

Although there are devices to detect and track players using other methodologies besides vision (Radio Frequency Identifiers, Local Position Measurement (Stelzer & Fischer, 2004) or Global Positions Systems) these other methodologies are beyond the scope of this paper and will not be addressed.

The sport that has deserved the highest attention by the scientific community has been soccer, nevertheless the focus of this section will be on the work developed in indoor sports since our object of study is handball and also because the challenges posed by indoor/outdoor sports are quite different. Outdoor sports usually receive more attention by the media and therefore it is possible to use images provided by TV broadcast cameras, however the light conditions are much worse and the environment is less controllable. On the other hand, indoor sports usually need a dedicated camera system and despite being a more controlled environment, players tend to be closer to each other (the playing area is smaller) which brings added difficulties since merging and occlusion situations occur more often.

On indoor team sports it is possible to find works on basketball, handball and indoor soccer, using either broadcast video footages or dedicated cameras placed at strategic places and several methodologies for player detection and tracking.

Concerning basketball games we can find (Hu et al., 2011) using broadcast videos. Players are detect by extracting the field (through dominant colour detection) and generating a player mask. Afterwards players are tracked in image coordinates using a CamShift-based tracking algorithm and their positions are converted into real world coordinates using an automatic calibration methodology. Results are presented for 48 consecutive frames and show high precision and recall percentages, 91.38% and 91.34%, respectively. However occlusion and merging between players of the same team affect the detection and are not taken into consideration.

A very promising project is the Autonomous Production of Images based on Distributed and Intelligent Sensing (APIDIS), (APIDIS, 2008), that equipped a basketball court with an acquisition network composed of microphones, conventional and (arrays of) omnidirectional cameras in order to provide a basketball dataset. This dataset was used by (Alahi et al., 2009; Delannay et al., 2009).

Taking advantage of the setup flexibility and amount of cameras (Alahi et al., 2009) were able to minimize occlusion and merging problems and determine the 3D positions of the players. The player detection is performed via a sparsity constrained binary occupancy map based on severely degraded silhouettes. These silhouettes are obtained through a basic background subtraction method. Results show that the usage of more than one camera can tremendously increase both the precision and recall rates for the player detection, from 57% and 62% in a single camera system to 76% and 72% if an omnidirectional camera is added.

(Delannay et al., 2009) detect the players from the foreground activity masks determined on multiple views. They take into consideration players occupy a 3D space and therefore sum the cumulative projections of the multiple views' foreground activity masks on a set of planes that are parallel to the ground plane. Afterwards, regions with larger projection values are considered to be a player and scanned for digits in order to detect the player's number. The tracking propagation is performed frame by frame and is based on the Munkres general assignment algorithm (Munkres, 1957). They compare their methodology with methods that project the activity masks into a single plane and conclude that projecting into multiple planes increases the detection rate and minimizes the shadows effects.

We could find two works (Kristan et al., 2009; Monier et al., 2009) that explore both basketball and handball, use closed world assumptions and two dedicated cameras placed at the ceiling of the sports hall. Placing the cameras at the ceiling has the advantage of minimizing occlusion problems.

On the first work, players are detected by applying a background mask image obtained from thresholding the differences between the background (background is obtained by a simple method, for example a median filter) and the current frame with a dynamic threshold specific for each player. The tracking algorithm is formulated as a closed world problem based on a Boot-Strap Particle Filter and each tracker is initialized manually. The multiple players' tracking is achieved by partitioning the world into several Voronoi cells, one for each player, that are updated at each time step. Results indicate low failure rates per player per minute, less than 1.1 in the worst test case.

The second work follows some of the assumptions of the one from (Kristan et al., 2009), however the foreground pixels are detected by creating a dynamic background model rather than acting on the threshold level or generating a background mask and afterwards apply template matching to track the players. The templates for each player are manually initialized by the user and the tracking is performed by searching in a region of interest determined by image resolution and players' speed restrictions. The multiple tracking is also achieved by partitioning the world into Voronoi regions where each tracker acts. The fusion between the two images is performed in real world coordinates. Results indicate average correction rates between 0.0019 and 0.00677 correction/frame/player.

Regarding indoor soccer, there are the works of (Needham & Boyle, 2001) and (Kasiri-Bidhendi & Safabakhsh, 2009). (Needham & Boyle, 2001) use a single stationary camera system and apply a multiple object condensation algorithm. Initially a bounding box of each player is detected, then, through a propagation algorithm, the fitness of each bounding box is evaluated and adjusted. The prediction stage of the condensation algorithm takes into consideration position estimates from a Kalman filter. They report on trajectory based results compared with hand annotated values and indicate distance errors less than 1 meter, which is a quite high value given that an indoor soccer minimum field measures 15x25 meters.

On the other hand, (Kasiri-Bidhendi & Safabakhsh, 2009) use TV broadcast images. Initially, the background colour is determined through clustering, afterwards the entire field region is extracted using the Mahalanobis distance between pixels on the frame and on the background colour distribution. Once the image is background free the lines of the field are extracted using a Canny edge operator. The remainder of the information stored in the image corresponds to the players and the ball that are further refined using physical constraints of area and roundness. The player and ball tracking is based on level set contours. By using this tracking technique occluded players are tracked by a single contour instead of two during the occlusion period and are again correctly detected after splitting.

This work follows (Santiago et al., 2011), that present a methodology based on colour in order to detect handball players. In addition, we have included a dynamic background model in order to take into consideration light changes and a Fuzzy inspired methodology to calibrate the colour teams. Moreover, a vector of Kalman filters is used for the tracking process and an HyperVideo with the resulting information is generated.

3. Architecture

3.1 Engineering solution

The challenges of defining a vision system able to identify and track the players in a team game are quite huge due to the dynamic and spatial characteristics of the game itself. Usually indoor team games are quite dynamic with high physical contact among players and rapid movements, for example a player can achieve velocities of more than 5 m/s. These characteristics impose a careful choice on the system's architecture which includes choosing not only the cameras and their disposition but also defining the software design.

In order to minimize crowd interference, merging and occlusion situations and have a good view of the entire field we propose using two cameras placed at the ceiling of the sport's hall to cover each half of the field (this solution was also adopted by (Kristan et al., 2009; Monier et al., 2009)).

The cameras chosen are GigEthernet (DFK 31BG03.H model from Imaging Source) which allows placing the recording unit far away from the cameras and maintaining the signal integrity, have a resolution of 1024x768 pixels and can supply images at 30 frames per second. The images collected from the cameras are shown on Fig. 1 .

We propose a two module software system, one responsible for acquiring the images from the two cameras (Acquisition System) and another responsible for the offline processing of

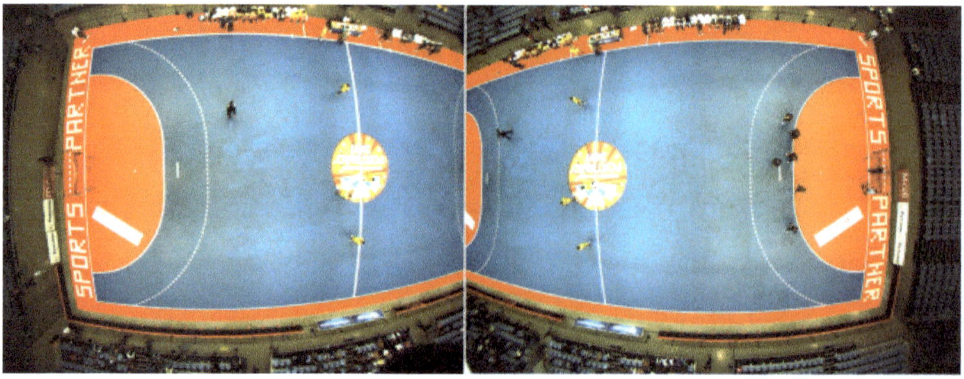

Fig. 1. A single frame from the two video streams.

the two video streams (Processing System). This last module is responsible for detecting and tracking the players as well as for generating an HyperVideo that consists on an unified image of the two streams with the positions of the players. Additionally it generates log files with the players' positions so that they can be used by the sports community to perform game analysis and infer game statistics. Figure 2 presents a scheme of the proposed architecture.

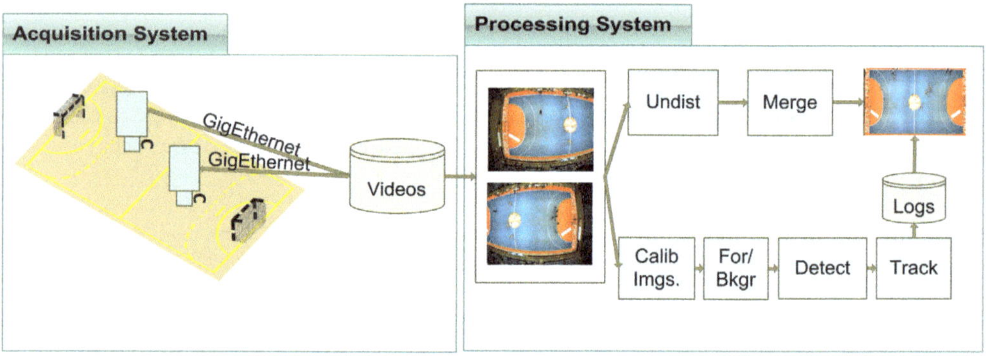

Fig. 2. System's architecture.

3.2 Player detection

Following previous work ((Santiago et al., 2011)) the player detection is achieved through colour identification and is composed of three steps. The first step consists on the colour calibration of each team using a region growing method allied with a Fuzzy inspired categorization methodology. This calibration is responsible for subdividing the colour space into subspaces, which are not necessary disjoint since there may be colours that are common to both teams (for example there are many teams that have uniforms with white stripes).
Afterwards, the user manually indicates the location of each player on the field by clicking on the images. Despite some works make this initialization automatically we choose this approach because both handball and basketball allow unlimited players' substitutions, and this manual initialization enables the possibility to always give the same number to the player and also discard the player when he/she leaves the field.

The second step consists on detecting foreground regions through a dynamic background subtraction method, which uses an empty image of the field and a dynamic threshold that is continuously and locally updated at each new frame (as will be described later on this chapter).
After the foreground pixels are identified, their colour is compared against the colour subspaces and classified into one of the teams. In case there is a belonging tie between teams, the adjacent pixels are searched in order to break the tie. Additionally the teams' colour subspaces are updated with new information.
Finally, pixels are aggregated to form blobs and categorized into player or no player according to size and density restrictions. The centre of mass of the blob is considered the player's position that is afterwards transformed into real world coordinates (court coordinates) using the cameras' homographies.

3.2.1 Colour calibration

The colour calibration is performed under the supervision of the user and is achieved using a region growing method (Santiago et al., 2011).
Let us define colour subspace S_c as the set of RGB colour triplets that are tagged as having the colours of the vests of the team c. The initial colour seeds $C(x_s, y_s)$ for each colour subspace S_c are set manually using the mouse to click on the objects that will be segmented, afterwards the surrounded pixels' colours $C(x_a, y_a)$ are agglomerate around these seeds using colour distance criteria. The colour expansion is performed on the HSL (Hue, Saturation and Luminance) colour space in order to minimize the effects of shadows and light variations.
The regions growth is performed in all directions (using a 8 neighbour mask n_8) in a recursive way until reaching a pixel that in terms of colour is away from the seed more than a global threshold (C_{ThresG}) or from its previous neighbour $C(x_p, y_p)$ more than a local threshold (C_{ThresL}), according to the following definition (both thresholds are user definable):

$$C(x_a, y_a) \in S_c \Leftrightarrow \forall (x_a, y_a) \in n_8(x_p, y_p):$$
$$\Delta(C(x_a, y_a), C(x_p, y_p)) < C_{ThresL} \wedge \Delta(C(x_a, y_a), C(x_s, y_s)) < C_{ThresG}$$

,where

- $C(x, y)$ is the HSL colour of pixel at location (x,y),
- $n_8(x, y)$ are the eight neighbours of pixel at location (x,y),
- $\Delta(C_1, C_2)$ is a configurable weighed distance function, involving HSL components of colours C_1 and C_2.

During the colour expansion each colour value is attributed a given belonging degree to the subspace being calibrated. This value is stored in a lookup table that contains for each colour triplet the belonging degree to each subspace. Despite the expansion is performed on the HSL colour space, the colour lookup table is built on the RGB (red, green, blue) colour space as well as the remainder image processing operations.
The fuzzy belonging $\mu()$ of the colour $C()$ of a pixel P of coordinates (x_P, y_P) to a given colour subspace Sc is $\mu_{Sc}(C(x_P, y_P))$ and can assume four levels: when not belonging to the colour subspace $\mu() = 0$ and $B() = C_0$, by default and before the calibration takes place, all the colours are categorized with no belong degree to every subspace; for low belong degree to the colour subspace $\mu() = 0.5$ and $B() = C_L$; for full belong degree to the colour subspace $\mu() = 1$

and $B() = C_F$ and additionally a full belong degree with the characteristic of also being a colour seed $\mu() = 1$ and $B() = C_S$. The B_{Sc} function maps the four colour categories (C_0, C_L, C_F and C_S) into the fuzzy belonging according to Table 2.

Colour	B_{Sc}	μ_{Sc}
Not the colour	C_0	0
Resembles the colour	C_L	0.5
Is the colour	C_F	1
Is a seed colour	C_S	1

Table 2. Mapping of B_{Sc} to fuzzy belonging $\mu()$.

In order to determine the belonging degree of the colour triplet the following rules are applied sequentially during the region growing process:

- **Rule1** - if the pixel was assigned to the subspace, is physically quite close to the initial seed pixel and the colour distance to the initial seed pixel is less than one fifth the maximum allowed distance for the growing process (less than $\frac{1}{5}C_{ThresG}$) then it is also assumed to be a seed pixel with a full belonging degree.

$$B_{S_c}(C(x_a, y_a)) = C_S \Leftarrow C(x_a, y_a) \in S_c \wedge \Delta((x_a, y_a), (x_s, y_s)) < \frac{1}{5}C_{ThresG}$$

- **Rule2** - if the colour distance to the initial seed pixel is less than two fifths the maximum allowed distance for the growing process than the pixel is categorized with a full belonging degree but without being a seed.

$$B_{S_c}(C(x_a, y_a)) = C_F \Leftarrow C(x_a, y_a) \in S_c \wedge \Delta(C(x_a, y_a), C(x_s, y_s)) < \frac{2}{5}C_{ThresG}$$

- **Rule3** otherwise and in case the pixel obeys to the region growing conditions it is categorized with a low belong degree (C_L).

By the end of the calibration process the colour space is subdivided into subspaces, which are not necessary disjoint since the same colour can belong to different subspaces. The motivation for allowing non-disjoint subspaces is that teams frequently share colours, for example uniforms with white stripes are common and thus the exact same well known colour belongs to the two opposing teams. Hence the Fuzzy Logic inspiration methodology, however implementation issues make it not interesting to allow for continuous degrees of belonging. The belonging degrees attributed to each colour triplet, as will be seen later, will allow to break ties but also to generate dynamic subspaces that can adapt (either grow or shrink) during the game. Subspaces do not have nor ever create any predefined specific shape as they are created from user-selected seeds on the image, that have been grown in the user selected video frames and in colour space.

3.2.2 Background subtraction

Since the background is more or less static, due to the semi-controlled environment of an indoor game, the subtraction is performed using an empty image of the viewed scene and only the threshold used to distinguish between foreground and background pixels is dynamic and specific for each pixel.

The background subtraction is performed on the RGB colour space, because tests showed that for some pixels a small difference between the RGB colour components of the background and the processed images corresponded to a large difference on the Hue component (HSL colour space). In fact, non-linear colour spaces suffer from the non removable singularity problem as stated by (Cheng et al., 2001).
Also in order to make the processing time shorter, the subtraction is executed locally and not to the entire image. In other words, only predefined regions, which are defined by the Kalman Filter predictive stage, suffer this process.
The threshold applied to each pixel is only updated if the pixel is classified as background, otherwise its value remains unchanged. The update obeys to equation 1 and the value is never allowed to go below 4% or above 23.5% of the entire colour range (0-255) for each colour component. These values were obtained experimentally.

$$\sigma_{t+1}^{c}(x,y) = \begin{cases} \alpha(I_t^c(x,y) - B^c(x,y)) + (1-\alpha)\,\sigma_t^c(x,y)\,, \text{ if } I_t(x,y) \in B(x,y) \\ \\ \sigma_t^c(x,y) \quad ,otherwise \end{cases} \tag{1}$$

,where

σ is the threshold of the pixel at position *(x,y)*, time *t+1* and colour component *c*,

I is the colour intensity of the pixel at position *(x,y)*, time *t* and colour component *c*,

B is the background colour intensity of the pixel at position *(x,y)* and colour component *c*,

α is a learning constant, that for our specific case was set to 0.02.

Pixels whose colour difference to the background image is less than the respective threshold are labelled as background, the others are labelled as foreground.

3.2.3 Team identification

After the foreground pixels are identified, their colour is compared against the colour lookup table that resulted from the calibration process (3.2.1) and classified into one of the teams.
Since the same colour can belong to different teams (subspaces) it may occur that a pixel is classified into more than one team. To break this tie, information not only from the belonging degree itself but also from adjacent pixels that have already been classified is used.
Spatial information is used by counting the number of adjacent pixels that belong to each team, afterwards if the sum of the team with the highest value is 1.5 times the value of the lowest team then that pixel is assigned a weight of 2 to that team otherwise it is assigned a value of 1 to both teams. Colour calibration information is used by adding to the previous weight the corresponding fuzzy belong values (as shown in Table 2).
The team with the highest final weight is the one assigned to the pixel. This way, it is possible that, although the belonging degree of a pixel to a team based on the colour calibration information is higher than the belonging degree to the other team it may be the winner due to the neighbourhood characteristics.
Additionally, if the winning team has a full belong to that colour triplet and corresponds to a seed colour ($B() = C_S$) then a region growing process is triggered and the colour lookup table that contains the information concerning the colour subspaces is updated. This auto expansion is more restrictive than the one performed during the manual initialization and is performed at time intervals (t_{expans}), an adjustable setting. This setting may change to take

into consideration the speed at which light changes in the pavilion and yields better overall performance to the application.

In order for this update to add not only colour triples to the subspaces but also to remove them (otherwise subspaces would grow too much), each colour triplet has associated a persistence ($p_{S_c}(R,G,B)$) to that subspace. Colours with lower belonging have lower persistence and colours with higher belonging have higher persistence. The initial persistence given to the colour is proportional to the time between auto expansions according to Eq.2.

$$\begin{cases} p_{S_c}(R,G,B) & = \frac{1}{8} t_{expans} & \text{, if } B_{S_c}(R,G,B) = C_L \\ p_{S_c}(R,G,B) & = \frac{1}{4} t_{expans} & \text{, if } B_{S_c}(R,G,B) = C_F \\ p_{S_c}(R,G,B) & = \infty & \text{, if } B_{S_c}(R,G,B) = C_S \end{cases} \tag{2}$$

The persistence is maximum (with the values defined in Eq. 2) when the colour is added to the subspace and diminishes whenever it is not detected in a frame, however seed colours have infinite persistence and will therefore always remain in the subspace. Whenever the persistence value reaches zero the colour triplet is removed from the subspace.

With the introduction of this dynamic it is possible to have mutable subspaces that adapt to light changes either occurring at different regions of the same frame or between frames.

At the same time the foreground pixels are classified, they are also aggregated horizontally to form run length encoding (RLE) structures characterized by the y, x_{min} and x_{max} positions of the RLE. Finally the RLEs are merged vertically to form blobs. A full description of this pixel aggregation can be found in (Santiago et al., 2011) with the particularity that small horizontal RLEs are ignored and do not pass to the vertical merging process in order to minimize noise.

The blobs resulting from this pixel aggregation are further refined according to size and colour density constraints. Therefore, blobs that are too small or too large or blobs that have low colour density are discarded as being players. The colour density is measured as the percentage of pixels inside the bounding box of the blob that belong to the team divided by the total number of pixels of the bounding box. The remainder blobs are considered players that belong to a given team (S_c) and have an (x,y) position on image and world coordinates.

The position on image coordinates is calculated as being the center of mass of the blob according to Eq.3 .

$$(x_{cm_{blob}}, y_{cm_{blob}}) = \left(\frac{\sum_x \sum_y \mu_{Sc}(C(x,y))\, x}{\sum_x \sum_y \mu_{Sc}(C(x,y))}, \frac{\sum_x \sum_y \mu_{Sc}(C(x,y))\, y}{\sum_x \sum_y \mu_{Sc}(C(x,y))}\right) \tag{3}$$

, where c is the team the blob belongs to.

The world coordinates[1] are obtained by first removing the barrel effect produced by the lens (only radial effect was considered, since the tangential component can, in most cases, be neglected) using Eq. 4 . The unknowns in these equations (k_1 , k_2 , k_3 , x_c and y_c) are determined using the information extracted from the field lines.

$$\begin{cases} x_u = x_d + (x_d - x_c)(k_1 r^2 + k_2 r^4 + k_3 r^6) \\ y_u = y_d + (y_d - y_c)(k_1 r^2 + k_2 r^4 + k_3 r^6) \end{cases} \tag{4}$$

[1] We would like to thank professor Paulo Costa and Paulo Malheiros from the University of Porto, Faculty of Engineering for the specific camera calibration software.

, where

- $r^2 = (x_d - x_c)^2 + (y_d - y_c)^2$,
- (x_u, y_u) are the undistorted coordinates,
- (x_d, y_d) are the distorted coordinates,
- (x_c, y_c) are the coordinates of the center of distortion of the lens,
- k_1, k_2 and k_3 are the radial coefficients for barrel distortion.

Fig. 3 illustrates the images before and after removing the barrel effect for the two cameras. Once the barrel effect is removed from the players' positions, it is possible to apply the pinhole camera model in order to obtain the world coordinates of the players. This model uses intrinsic parameters (K) and extrinsic parameters (R and T) to map image coordinates (X) into world coordinates (x) according to Eq.5.

$$x = K[R|T]X \Leftrightarrow x = H_w X \tag{5}$$

The H matrix is defined according to Eq.6.

$$H_w = \begin{bmatrix} f & 0 & c_x \\ 0 & f & c_y \\ 0 & 0 & 1 \end{bmatrix} \begin{bmatrix} \cos\phi\cos\alpha & \sin\omega\sin\phi\cos\alpha - \cos\omega\sin\alpha & \cos\omega\sin\phi\cos\alpha + \sin\omega\sin\omega\sin\alpha & T_x \\ \cos\phi\sin\alpha & \sin\omega\sin\phi\sin\alpha + \cos\omega\cos\alpha & \cos\omega\sin\phi\sin\alpha - \sin\omega\cos\alpha & T_y \\ -\sin\phi & \sin\omega\cos\phi & \cos\omega\cos\phi & T_z \end{bmatrix} \tag{6}$$

, where

- f is the focal length,
- c_x and c_y are the coordinates of the optical center,
- ϕ, ω and α are the rotations around the x, y and x axis, respective,
- T_x, T_y, T_z are the translations on x, y and z directions.

The coordinates are projected at 1.2m from the ground which corresponds to a best effort of the height of the center of mass of an average person, when seen in most frequent positions of the field. This projection allows to have a more correct measure of the players' positions and enables the information fusion between cameras.

3.3 Player tracking

The player tracking is based on a vector of Kalman filters (Kalman & Others, 1960; Welch & Bishop, 2002) (one per player) with state x_k (Eq. 7) and measure z_k (Eq. 8) at instant time k.

$$x_k = [\, x \; y \; v_x \; v_y \,]^T \tag{7}$$

$$z_k = [\, x \; y \,]^T \tag{8}$$

, where

- x and y are the player center of mass position in real world coordinates,
- v_x and v_y are the player velocity in real world coordinates.

And modelled according to the following linear stochastic difference equations (Eq. 9, 10).

$$x_k = Ax_{k-1} + w_{k-1} \tag{9}$$

$$z_k = Hx_k + v_k \tag{10}$$

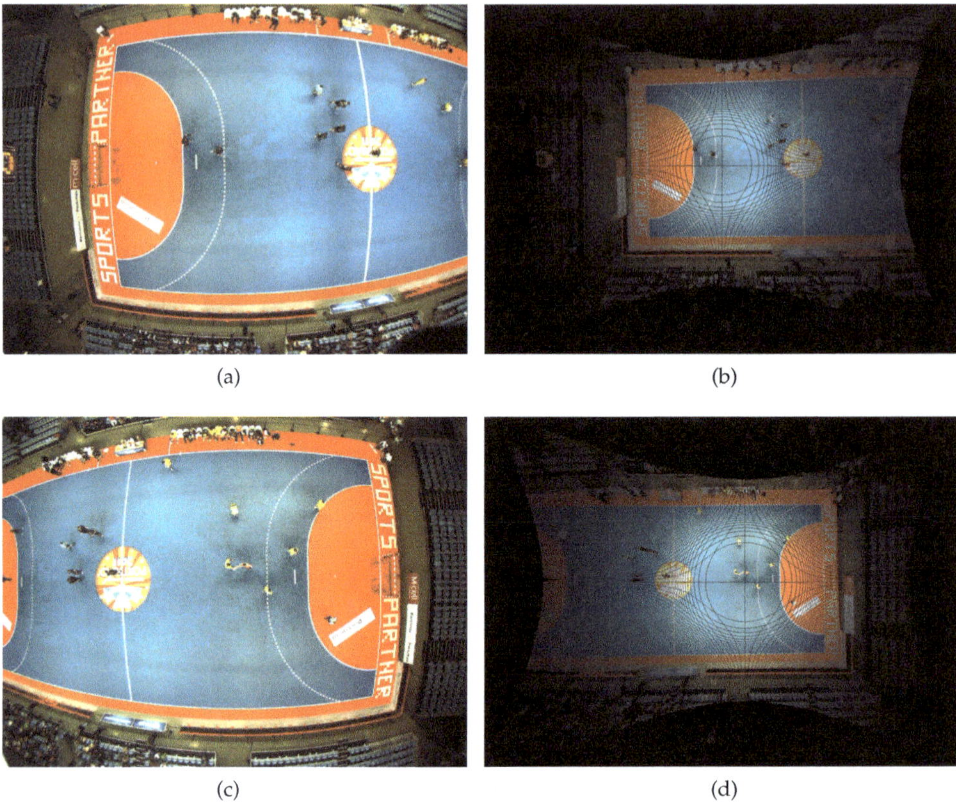

Fig. 3. (a) and (b) left image before and after removing the barrel effect distortion. (c) and (d) right image before and after removing the barrel effect.

Where A is the state model matrix and assumes the form of Eq.11 (where Δt is the time between frames - in this case corresponds to $\frac{1}{30}$ seconds), H is the observation model matrix (Eq.12) and the random variables w_k and v_k represent the process and measurement noise. The usage of real world coordinates allows for a transparent tracking between the two video streams.

$$A = \begin{bmatrix} 1 & 0 & \Delta t & 0 \\ 0 & 1 & 0 & \Delta t \\ 0 & 0 & 1 & 0 \\ 0 & 0 & 0 & 1 \end{bmatrix} \tag{11}$$

$$H = \begin{bmatrix} 1 & 0 & 0 & 0 \\ 0 & 1 & 0 & 0 \end{bmatrix} \tag{12}$$

Whenever the user indicates a player (using the mouse), a new Kalman filter is added to the vector with the real world position of the player and a default velocity of 0 m/s. Afterwards, the players' locations on the subsequent frames are predicted using Eq.9. The area around

the predicted measure is searched according to the process explained on 3.2.3 to generate a measure (z_k) to updated the estimate.
In addition, and since the players' velocity is not constant through out the game, at each frame each player velocity is updated.
By predicting the position of the players on the subsequent frames it is possible to reduce the computational cost because only a few regions of the entire image are search for players.

3.4 Generation of human meaningful video

This chapter concerns primarily with the generation of a single complete image video stream that is of the utmost importance for the possible human end-users of our system, for example a sports' scientist, educator or coach.
Generating a single, high quality, "undistorted", video stream is a complex task due to the usage of complex optical systems (wide angle zoom lenses) and the need for accuracy of the system that involves dealing with 2 sets of intrinsic and extrinsic camera parameters for the so called pinhole models of the cameras. High accuracy mapping of the pixels of the images onto real world is needed, with the added difficulty of covering large real world areas. This is an even greater task given that high resolution and high frame rate are of interest, thus producing large amounts of data (2 cameras @ 1024 x 768 resolution, RGB colour depth @ 30fps). By using advanced camera calibration techniques, the two different sets of parameters were found, thus allowing image to world accurate mapping (on separate images) - as seen in Fig. 3.
In order to produce the "undistorted" Human Meaningful Video stream, the first task is to optimize the algorithm used. Firstly a pair of "static", off-line created, Look Up Tables (LUTs) are created to map real world pixels into their origin in the original images. Mapping non-overlapped image areas is not complex if all data is available. Additional processing is necessary for the overlapped parts of the image, in order to get a human meaningful image (without neither repeated nor cut objects). LUTs are very useful because they exchange complex mathematical operations with memory storage for the repeated computations which is very interesting in terms of general performance and parallel computing in particular.
One of the issues to be solved is to show an interesting view of the overlapped portions of the images shown in Fig. 1. This is an issue because near the centre line of the handball field, where images overlap, the same player is seen tilted in opposite directions by both cameras (as shown in Fig. 4). The strategy is to have complete objects always from one of the cameras - this is always possible because the largest object is inferior in image size to the size of the overlap in the images (see Fig. 4). It is not computationally immediate, though, because objects and background have to be identified and reconstructed.

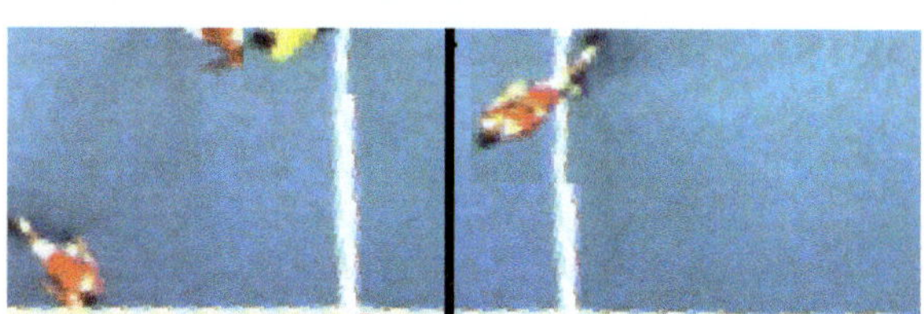

Fig. 4. Same player over time, transitioning from camera to the other.

The implementation was built in C++, under windows and Microsoft Visual C++, using previously mentioned frameworks as OpenMP (OpenMP, 2011) and Compute Unified Device Architecture (CUDA) (NVIDIA, 2011). An example of the processed image is shown in Fig. 5.

Fig. 5. Single frame from the "undistorted" human meaningful video.

4. Results

4.1 Player detection and tracking

In order to validate the approach, the system was mounted at the public sports hall of Portimão to film the games of the portuguese SuperCup. The video footages collected validated the engineering solution since we were able to cover the entire field with good resolution and a good overlapped zone as shown on Fig. 1.

Initially two distinct teams (team A and team B - examples of both teams can be seen on Fig. 6(b)) were calibrated as explained on section 3.2.1. The original seeds were selected by clicking on the players of both teams which resulted on the initial colour subspaces illustrated on Fig. 6(a).

After 1200 frames and a time between expansions (t_{expans}) of 30 frames (which corresponds to 34 expansions), it is possible to verify that the teams' colour subspaces have updated and grown around the initial seeds resulting on the new colour subspaces of Fig. 7.

Table 3 provides an overview of the colour spaces dimensions during the auto calibration process. The @F represents the frame number (remember that the auto expansion occurs at every 30th frame).

The table clearly illustrates that from the initial colour subspaces (Fig. 6(a)) to the final (Fig. 7) the number of colour triplets that are seeds (C_S) increases so that the colour subspaces can adapt to the colour conditions of that team through the entire field and through time.

In addition, if the colour subspace is more condensed, triplets that resemble the colour (C_L) and are colour (C_F) tend to exist in higher number (Team B) than when the colour seeds are more spread in the entire colour space (Team A).

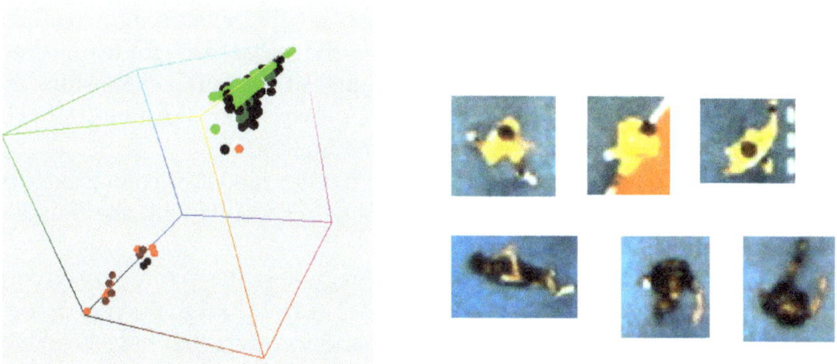

(a) (b)

Fig. 6. (a) Initial colour subspaces. Green dots correspond to team A and red dots to team B. Lighter dots are seed colours (C_S), intermediate are team colour (C_F) and the darkest resemble the team colour (C_L). (b) Examples of players from team A (up) and team B (down).

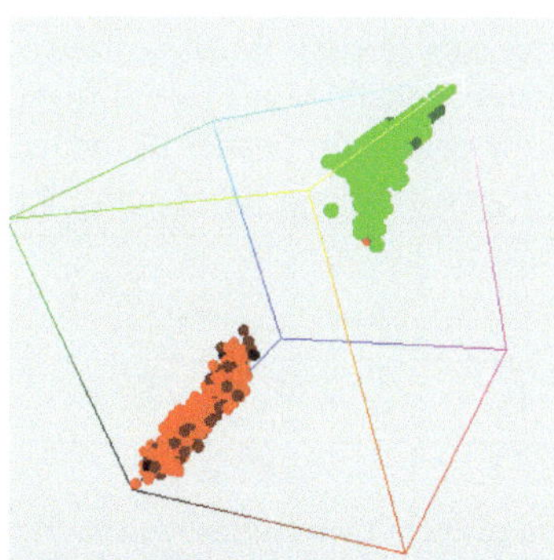

Fig. 7. Final colour subspaces after 34 auto expansions (team A in green and team B in red).

Team	B	@F1	@F5	@F31	@F35	@F61	@F301	@F421	@F691	@F811	@F1021
A	C_L	34	28	12	10	10	10	3	1	4	0
	C_F	6	6	6	6	8	4	9	11	3	7
	C_S	39	39	44	44	51	81	92	105	109	112
B	C_L	2	2	1	0	1	2	4	7	7	3
	C_F	6	6	6	6	6	11	18	42	44	43
	C_S	8	8	12	12	12	25	39	103	116	149

Table 3. Evolution of the number of colour triples that belong to each team and the respective belonging degree. The expansion is performed at every 30th frame ($t_{expans} = 30$)

It is also possible to verify the effect of colour persistence, namely the number of colour triplets that are C_L decreases on Team A from frame 1 to frame 5 (the persistence of these colours is of 3.75 ($p_{S_c} = \frac{1}{8} t_{expans}$)) also they tend to have a more erratic behaviour because they belong to the periphery of the colour subspace and therefore do not appear so often, and triplets with C_F decrease on Team A from frame 61 to frame 301 and on Team B from frame 811 to frame 1021.

However, it is possible to verify that the initial seed choice will influence how well the colour subspace adapts to the environment conditions. In fact, the initial seeds for team A resulted in a faster adaptation: the colour subspace growth is higher at the initial expansions and tends to stabilize, while for team B there is still a reasonable growth even at frame 1021. The initial seed choice is a well known problem of region growing methods.
Comparing the results with and without the Fuzzy inspired model of colour expansion it is possible to verify that the players' detection achieves better results with the mutable colour subspaces as depicted on Fig. 8.
These images show the players' detection at frame 762, where the green crosses correspond to players detected from team A and the purple crosses correspond to players detected from team B. The green and red highlighted pixels correspond to pixels that have been labelled as belonging to one of the teams (green correspond to team A and red to team B).
Analysing the two images it is possible to verify that using the Fuzzy inspired auto expansion model all fielders from both teams were detected (Fig. 8(b)), while using the initial colour subspaces (Fig. 6(a)) the system is unable to detect four players of team B (Fig. 8(a)). In addition, the detected area of the players is higher with the Fuzzy model which allows to have a better measure of the player center of mass.
Moreover, the detection rate increases along with the colour subspaces update as illustrated on Table 4. This increase rate is more visible on players from team B, since its initial colour subspace did not reflect so well the colour properties of the team.

Team	player	1-100	101-200	201-300	301-400	401-500	501-600	601-700	701-800
A	1	1.00	1.00	1.00	1.00	1.00	1.00	1.00	1.00
	2	1.00	1.00	1.00	0.95	1.00	1.00	1.00	1.00
	3	1.00	1.00	1.00	0.96	1.00	1.00	1.00	1.00
B	1	0.00	0.27	0.00	0.70	0.63	1.00	0.93	1.00
	2	0.03	0.35	0.33	0.49	0.92	0.90	0.90	1.00
	3	0.13	0.43	0.24	0.56	1	1	0.98	0.99

Table 4. Player detection of three players from teams A and B from frame 1 until frame 800.

The usage of Kalman filters to perform the tracking allows not only to make the processing time shorter, but also to minimize the miss-detections because the predictive stage (Eq. 10) determines the next position of the player (taking into account the model of the player movement). This position is used on the next frame to search for the blob and in case of not finding a measure the value is used as the position of the player. The following table (Table 5) shows how the tracking of players from team B is improved using the Kalman filter.

Player	1-100	101-200	201-300	301-400	401-500	501-600	601-700	701-800
1	0.00	0.79	0.15	1.00	1.00	1.00	0.95	1.00
2	0.16	0.83	0.57	0.90	1.00	0.98	0.93	1.00
3	0.74	0.86	0.75	1.00	1.00	1.00	0.99	1.00

Table 5. Detection improvement using the Kalman filter of the three players of team B from Table 4.

Additionally, by performing the tracking in real world coordinates it is possible to track the players between cameras in a transparent way. Figure 9 illustrates players from team B crossing the middle line and being identified initially by the left camera and afterwards by the right camera, without the need for user intervention.

(a)

(b)

Fig. 8. Results of the player detection at frame 762: (a) without colour auto expansion and (b) with colour auto expansion.

4.2 Generation of human meaningful video

In order to ascertain the importance of parallel processing in this application, the same algorithm was ported to the OpenMP and CUDA frameworks, yielding the results shown in Table 6. Naturally, parallel processing is heavily dependent on hardware and results are shown for two laptop solutions, only one of each is able to run the proprietary CUDA toolkit for GPU processing.

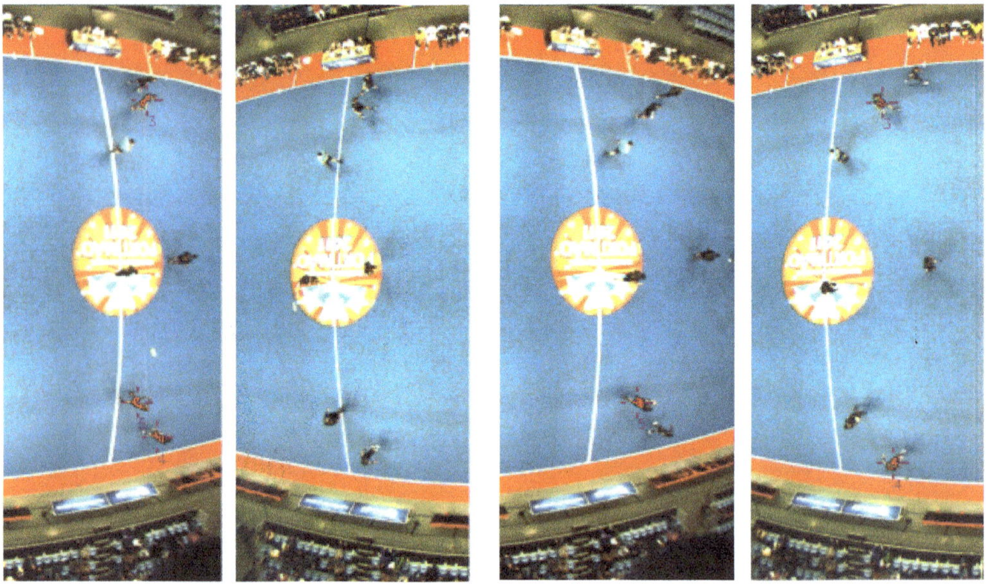

(a) (b)

Fig. 9. Tracking players between cameras. (a) Players are being tracked by the left camera. (b) Some players start to be tracked by the right camera.

PC (Year) / Processor / Video Processor / O. S. / CPUs (GPUs)	OMP× CUDA×	OMP✓ CUDA×	OMP× CUDA✓	OMP✓ CUDA✓
Asus V6J (2006) Intel Core Duo no CUDA capabilities Windows 7 2@1.87GHz (none)	145.1+42.1 = 187.2 ms (5.3 fps)	137.9+34.5 = 172.4 ms (5.8 fps)	-	-
Toshiba Tecra S11 (2011) Intel core i7-640M NVidia NVS2100M 512MB Windows 7 4@2.8+GHz (16@0.5+GHz)	91.1+69.7 = 160.8 ms (6.2 fps)	40.7+28.6 = 69.3 ms (14.42 fps)	91.1+22.2 = 113.3 ms (8.83 fps)	40.7+21.5 = 62.5 ms (16.02 fps)

Table 6. Parallel computing execution times: the parcels (a+b) are: a - other algorithms; b - undistorting and correctly joining incoming images. The × indicates NO and the ✓ indicates WITH.

By studying Table 6 and Figure10, it can be found that a single better CPU improves performance marginally - a much better CPU running with a clock at least 1.49 times faster yields about 17% performance gain. This is probably due to the large amount of data being manipulated in this application - the limitations are in the memory area, not pure single CPU power. The same results also demonstrate the advantage of parallel computing with joint usage of OpenMP (OMP) and CUDA, with a performance gain little over 2.6 times (in the same computer). Even larger gains were initially expected but the complexity of the algorithm on the selection of the objects to show on the overlapped portion of the initial images, limited

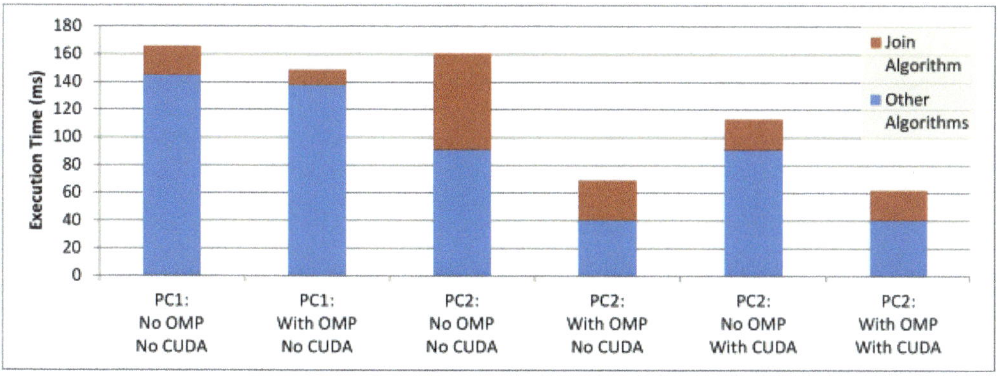

Fig. 10. Execution times for several parallel computing techniques (see Table 6).

considerably the overall performance. As recent processors offer high frequency dual/quad (or more) cores, the interest of using 16 much slower processors is also to be considered - but using the GPU processors of the video card will free up the main CPU for other tasks.

The previous study is related to producing a single frame of the human meaningful video stream with parallel computing. Future work includes using distributed computing to producing the stream, for example, using 2 computers to produce alternating frames is expected to produce excellent performance gains (almost halving processing time). This expectation is justified as little dependencies exist among consecutive frames and, as such, parallel processing is most effective and the "slow" transmission over network would not introduce significant performance loss, when compared to the expected benefits of the double computational power available for processing. A similar approach (several frames at once) would also be possible in a single computer but would likely not be as interesting as most available resources are used fully most of the time. As mentioned earlier, actual performance gains are very much hardware dependent.

In order to really have a meaningful video for humans, the found objects (handball players, etc) have to be marked onto the image, a task that was not parallelized.

5. Conclusions

This chapter presented a visual automatic system for detecting and tracking players in indoor games. The main objectives were to implement a system that could be adaptative and take into consideration the light changes (and subsequent colour change) that frequently occurs in sports pavilions (for example due to windows, clouds, sun orientation, ...). The ultimate goal is to further improve information for sports agents and sportive quality.

Players are identified by color segmentation. The used techniques include foreground detection and classification of team colours using a Fuzzy inspired categorization model that allows a single colour to belong to both teams simultaneously. The colour subspaces have no particular shape and grow or shrink over time in order to take into account spatial and temporal changes of the recognized colours.

Player tracking is improved by making use of a Kalman Filter per player and the resulting information is organized and shown in a single undistorted image view of the entire field,

adequate for human studies. Although the system is multi-camera, the usage of parallel processing allows efficient generation of this final image.
The proposed methodology was validated with a real game footage filmed at an important Portuguese handball championship. Results show that, using simple features such as colour combined with a powerful tool for tracking (Kalman Filter), it is possible to detect and track players throughout the game area with very limited user intervention - initial colour calibration by user and little more. The usage of adaptative colour subspaces generated by a Fuzzy inspired methodology allows to better define the teams' colour properties during the game and increasing detection rates.

5.1 Future work

Future work includes exploring methodologies to automatically define the time between expansions along the game according to the colour subspaces dynamics and also the detection rate and incorporate more information on the Kalman Filter model in order to make it more robust to long merging and occlusion situations and also enable both the detection and the tracking with the benefits of parallel processing.
Additionally, interesting game sequences are also intended to be automatically detected (taking into consideration the players positions) and the generated video tagged accordingly. This video is to be made "active" (searchable, "jumpable", ...) to be integrated into a human interface navigation tool to allow for an easy usage of the extracted information by the final end users.

6. Acknowledgments

We would like to thank Fundação Calouste Gulbenkian by the support given through a PhD scholarship with ref. 104410.

7. References

Alahi, A., Boursier, Y., Jacques, L. & Vandergheynst, P. (2009). Sport players detection and tracking with a mixed network of planar and omnidirectional cameras, *Distributed Smart Cameras, 2009. ICDSC 2009. Third ACM/IEEE International Conference on*, IEEE, pp. 1–8.

APIDIS (2008). Autonomous Production of Images based on Distributed and Intelligent Sensing.
URL: *http://www.apidis.org/index.htm*

Barron, J. & Thacker, N. (2005). Tutorial: Computing 2D and 3D optical flow, *Tina Memo Internal* (2004-12).

Canny, J. (1986). A computational approach to edge detection, *IEEE Trans. Pattern Anal. Mach. Intell.* 8: 679–698.

Chapman, B., Jost, G. & Van Der Pas, R. (2007). *Using OpenMP: portable shared memory parallel programming*, Vol. 10, The MIT Press.
URL: *http://books.google.com/books?hl=en&lr=&id=MeFLQSKmaJYC&oi=fnd&pg=PR7&dq=Using+OpenMP,+Portable+Shared+Memory+Parallel+Programming&ots=5zOPjR26VC&sig=R3WOZRMwGX1tAw-pcR46NuId3xw*

Cheng, H., Jiang, X., Sun, Y. & Wang, J. (2001). Color image segmentation: advances and prospects, *Pattern recognition* 34(12): 2259–2281.

Delannay, D., Danhier, N. & De Vleeschouwer, C. (2009). Detection and recognition of sports (wo)men from multiple views, *Distributed Smart Cameras, 2009. ICDSC 2009. Third ACM/IEEE International Conference on*, IEEE, pp. 1–7.

Deng, Y. & Manjunath, B. (2001). Unsupervised segmentation of color-texture regions in images and video, *IEEE Transactions on Pattern Analysis and Machine Intelligence* 23(8): 800–810.

Franks, I. M. & Nagelkerke, P. (1988). The Use of Computer Interactive Video in Sport Analysis, *Ergonomics* 31(11): 1593–1603.

Franks, I., Willison, G. E. & Goodman, D. (1987). Analysing a team sport with the aid of computers, *Canadian Journal of Sport Sciences* 12(2): 120–125.

Grimson, W., Stauffer, C., Romano, R. & Lee, L. (1998). Using adaptive tracking to classify and monitor activities in a site, *Computer Vision and Pattern Recognition, 1998. Proceedings. 1998 IEEE Computer Society Conference on*, IEEE, pp. 22–29.

Group, K. (2011a). The khronos group inc.
URL: *http://www.khronos.org/*

Group, K. (2011b). Opencl - the open standard for parallel programming of heterogeneous systems.
URL: *http://www.khronos.org/opencl/*

Halfhill, T. (2008). Parallel Processing with CUDA Nvidia's High-Performance Computing Platform Uses Massive Multithreading, *Microprocessor Report* 22: 1–8.
URL: *http://scholar.google.com/scholar?hl=en&btnG=Search&q=intitle:Parallel+Processing+With+CUDA:+Nvidia's+High-Performance+Computing+Platform+Uses+Massive+Multithreading#0*

Heikkila, J. & Silvén, O. (1999). A real-time system for monitoring of cyclists and pedestrians, *Visual Surveillance, 1999. Second IEEE Workshop on,(VS'99)*, IEEE, pp. 74–81.

Hu, M., Chang, M., Wu, J. & Chi, L. (2011). Robust Camera Calibration and Player Tracking in Broadcast Basketball Video, *Multimedia, IEEE Transactions on* 13(2): 266–279.

Kalman, R. & Others (1960). A new approach to linear filtering and prediction problems, *Journal of basic Engineering* 82(1): 35–45.

Kasiri-Bidhendi, S. & Safabakhsh, R. (2009). Effective tracking of the players and ball in indoor soccer games in the presence of occlusion, *Computer Conference, 2009. CSICC 2009. 14th International CSI*, IEEE, pp. 524–529.

Koprinska, I. & Carrato, S. (2001). Temporal video segmentation: A survey, *Signal processing: Image communication* 16(5): 477–500.

Kristan, M., Perš, J., Perše, M. & Kovačič, S. (2009). Closed-world tracking of multiple interacting targets for indoor-sports applications, *Computer Vision and Image Understanding* 113(5): 598–611.

Monier, E., Wilhelm, P. & Ruckert, U. (2009). Template matching based tracking of players in indoor team sports, *2009 Third ACM/IEEE International Conference on Distributed Smart Cameras (ICDSC)* pp. 1–6.

Munkres, J. (1957). Algorithms for the Assignment and Transportation Problems, *Journal of the Society for Industrial and Applied Mathematics* 5(1): 32–38.

Needham, C. & Boyle, R. (2001). Tracking multiple sports players through occlusion, congestion and scale, *British Machine Vision Conference*, BMVA, pp. 93–1022.

NVIDIA (2011). Cuda zone.
URL: *http://www.nvidia.com/object/cuda _home _new.html*

OpenMP (2011). OpenMP.org.
URL: *http://openmp.org/wp/*

Roberts, L. G. (1963). *Machine Perception of Three-Dimensional Solids*, Outstanding Dissertations in the Computer Sciences, Garland Publishing, New York.

Santiago, C., Sousa, A. & Reis, L. (2011). Real time colour based player tracking in indoor sports, *Computational Vision and Medical Image Processing* 19: 17–35.

Sobel, I. E. (1970). *Camera models and machine perception*, PhD thesis, Stanford, CA, USA. AAI7102831.

Stelzer, A., P. K. & Fischer, A. (2004). Concept and Application of LPM - A Novel 3-D Local Position Measurement System, *IEEE Transactions on Microwave Theory Techniques* 52: 2664–69.

Welch, G. & Bishop, G. (2002). An introduction to the Kalman filter, *Technical report*, Department of Computer Science University of North Carolina at Chapel Hill, EUA, North Carolina.

6

New IM Torque Control Scheme with Improved Efficiency and Implicit Rotor Flux Tracking

Bojan Grčar, Peter Cafuta and Gorazd Štumberger
University of Maribor, Faculty of Electrical Engineering and Computer Science
Slovenia

1. Introduction

The induction machine serves as a workhorse in the majority of industrial and commercial applications requiring speed or torque controlled drives. Control design that satisfies a high number of different performance objectives is becoming an increasingly important and challenging issue. Although the research community has proposed several control structures for this purpose, only two major schemes have been accepted by the industry: the well established field-oriented control (FOC) and the more recent direct torque control (DTC).

The popularity and wide applicability of FOC arises from its relatively simple, and decoupled (rotor flux linkage versus torque) structure (Blaschke, 1971). A deeper understanding of FOC performance has been achieved by control theorists, including feedback linearization (Marino, 1999), (Bodson & Chiasson, 1998), (Chiasson, 1998), passivity (Ortega, 1998), and flatness (Martin & Rouchon, 1996). Formal stability proofs, together with guidelines for controller parameter settings, are now available.

The main characteristic of standard FOC, namely invariance of the magnitude of the rotor flux linkage vector, enables a globally stable solution for the non-holonomic (double) integrator problem, which in essential describes the IM rotor dynamics (Brockett, 1996), (Grcar, 2011). Since the FOC concept requires implicit system inversion (mapping of the oriented reference current vector into voltage vector), an inner current loop is needed. An outer loop for speed and/or torque control is usually designed for the reduced model, assuming that high gain current controllers achieve perfect tracking of the current command. Recent results improve the FOC capability and efficiency by explicit rotor field tracking (Peresada, 2003), (Chakraborty & Hori, 2003). It should be pointed out that in all various modifications of FOC some kind of estimator (open-loop) or observer (closed-loop) is required at least for necessary coordinate transformations. Proving rigorously overall global stability based on the estimated variables while considering the rotor flux tracking, current limits and parameter variations is a difficult task and is still an ongoing challenge.

On the other hand, DTC (Takagashi & Noguchi, 1986), (Depenbrock, 1986), introduces a different control philosophy based on stator flux linkage rotation. This concept is voltage-based and operates without explicit current controllers (Ortega, 2000), (Attaianese, 1999). Some authors claim that, by introducing stator quantities into the torque control, substantial reduction in parameter sensitivity is achieved. Control signal (stator voltage vector) is generated in accordance with the finite number of possible voltage-source inverter

(VSI) states, resulting in a complex nonlinear hybrid feedback system. The switching logic can be expressed in the form of algebraic inequalities to enable both stability analysis and determination of feasible operation areas. Some additional performance criteria (minimal losses, best tracking) might be simultaneously satisfied by selecting alternative switching patterns.

In this chapter, we examine a control structure that does not fit in either of the two classes although the proposed control uses a cascaded structure similar to FOC and vector rotation concept along with assumption of known machine torque similar to DTC. Our design is based on an IM model in a reference frame aligned with the stator current vector. Although it is well known that structural machine properties are invariant under different bijective geometric transformations, the choice of the reference frame have indeed some relevant implications in physical interpretation of the control design, especially in practice. The choice of stator current reference frame is justified due to the following facts:

- Dynamic inversion of the rotor dynamics enables the unique determination of maximal torque-per-amp ratio ($T_{el} / \|i_s\|$) equilibrium for all required torques.
- All machine fluxes (stator, air-gap, rotor) have the same torque-producing component, orthogonal to the stator current vector.
- The suggested reference frame enables the determination of a safe operation region delimited by the maximal torque-per-amp ratio thus offering the possibility for implicit rotor flux linkage changes through the manipulation of the stator current vector only.
- The stator current vector is a measured quantity, therefore the corresponding coordinate transformation is parameter invariant.
- In the implementation of the proposed control, only one of the orthogonal flux linkage components is necessary to obtain reliable machine torque estimate. The influence of the parameter uncertainty could be therefore reduced.

We propose that current vector magnitude and its relative rotation speed are changed simultaneously in accordance with the reference torque command so that maximal torque per ampere ratio is achieved for any feasible steady state (assuming perfect knowledge of parameters). Rotor flux linkage vector, not directly used in the proposed torque control scheme, is therefore allowed to change freely in accordance with the actual rotor dynamics. This concept results in a single input-two output system instead of the two input-two output system encountered in other schemes. During transients, the rotor flux linkage vector can be forced to remain close to the maximal torque-per-amp ratio bounds inside the sector of safe operation. This important property is achieved by adjusting amplitude and frequency modulation of the stator current vector. In addition, no singularities restrict the operation at zero state (zero torque, zero flux linkage, zero mechanical speed) or during torque reversal. Assuming that signal conditioning and estimation of the machine torque are solved adequately, almost proportional torque responses are obtained for smooth or even step references except in the obvious case when the machine rotor flux linkage starts near zero state. The problem of static "field weakening" and more demanding flux tracking is therefore solved concurrently for all feasible operating conditions since the machine operates with a minimal rotor flux linkage magnitude needed to generate the required torque. Usually un-modelled saturation effects could be substantially reduced since proposed reference frame is not affected by these effects and the rotor flux linkage is not directly used in the proposed

control scheme. The upper bounds of reachable torque and mechanical speed are given by the current and voltage constraints. Extension towards speed control can be achieved in the standard way, with an additional (PI) control loop. Proposed feedback structure is relatively simple, easy to implement on the standard hardware, and is mostly based on physical considerations.
The organization of the chapter is as follows. First, the IM model in fixed stator $\alpha - \beta$ reference frame is transformed into the reference frame aligned with the stator current vector. In Section 2, a second order nonlinear reduced model, derived from the wide adopted assumption of high gain current controllers, is introduced. Partial dynamical inversion and maximal torque-per-amp ratio equilibrium conditions are presented. We also introduce, for particular initial conditions, a set of guidelines to design our torque controller. In Section 3, the main result introducing several versions of open- or closed-loop torque controller for nominal and perturbed parameter case is given. The inner current loop designed in rotor $\gamma - \delta$ reference frame with almost perfect tracking capability and sufficient robustness is presented in Section 5. Adopting well justified time separation of the stator and rotor dynamics the time-varying linear current controllers based on the internal model principle are introduced. In Section 6, we include different experimental results illustrating the potential and performance characteristics of the proposed IM torque control. In Appendix I detailed stability analysis of the feedback system is presented while in Appendix II the description of the experimental set-up along with motor data and controller parameters are given.

2. IM model in stator current reference frame

Under standard modelling assumptions (Krause, 1986) for linear magnetics and by choosing the stator current vector and the rotor flux linkage vector as state variables, an $\alpha - \beta$ model in a fixed reference frame for the two pole machine is obtained in the form of

$$\begin{bmatrix} \dot{i}_{\alpha\beta} \\ \dot{\psi}_{\alpha\beta} \\ \dot{\omega}_m \end{bmatrix} = \begin{bmatrix} -\tau_\sigma^{-1} i_{\alpha\beta} + \frac{M}{L_\sigma L_r \tau_r}(\mathrm{J}^T \omega_m \tau_r + 1)\psi_{\alpha\beta} + \frac{1}{L_\sigma} u_{\alpha\beta} \\ -\tau_r^{-1}(\mathrm{J}^T \omega_m \tau_r + 1)\psi_{\alpha\beta} + \tau_r^{-1} M i_{\alpha\beta} \\ \frac{M}{J L_r} i_{\alpha\beta}^T \mathrm{J} \psi_{\alpha\beta} - T_L / J \end{bmatrix} \tag{1}$$

Introducing a new set of state variables $z = [\|i_s\|, \theta_i, \psi_I, \psi_\perp, \omega_m]^T$, along with the nonlinear state transformation

$$z = T(x) = \begin{bmatrix} \sqrt{i_\alpha^2 + i_\beta^2} \\ \tan^{-1}(\frac{i_\beta}{i_\alpha}) \\ \cos(\theta_i)\psi_\alpha + \sin(\theta_i)\psi_\beta \\ -\sin(\theta_i)\psi_\alpha + \cos(\theta_i)\psi_\beta \\ \omega_m \end{bmatrix} \tag{2}$$

and with the new input vector

$$\begin{bmatrix} u_I \\ u_\perp \end{bmatrix} = \begin{bmatrix} \cos(\theta_i)u_\alpha + \sin(\theta_i)u_\beta \\ -\sin(\theta_i)u_\alpha + \cos(\theta_i)u_\beta \end{bmatrix} \tag{3}$$

the transformed model ($\dot{z} = \frac{\partial T(x)}{\partial x}\dot{x}$) in the stator current vector reference frame is obtained in the form of

$$\begin{bmatrix} \|\dot{i}_s\| \\ \dot{\theta}_i \\ \dot{\psi}_I \\ \dot{\psi}_\perp \\ \dot{\omega}_m \end{bmatrix} = \begin{bmatrix} -\tau_\sigma^{-1}\|i_s\| + \frac{M}{L_\sigma L_r \tau_r}\psi_I + \omega_m \frac{M}{L_\sigma L_r}\psi_\perp + \frac{1}{L_\sigma}u_I \\ -\omega_m \frac{M}{L_\sigma L_r}\frac{\psi_I}{\|i_s\|} + \frac{M}{L_\sigma L_r \tau_r}\frac{\psi_\perp}{\|i_s\|} + \frac{1}{L_\sigma \|i_s\|}u_\perp \\ -\tau_r^{-1}\psi_I + (\omega_i - \omega_m)\psi_\perp + M\tau_r^{-1}\|i_s\| \\ -\tau_r^{-1}\psi_\perp - (\omega_i - \omega_m)\psi_I \\ -\frac{1}{J}(\frac{M}{L_r}\psi_\perp\|i_s\| - T_L) \end{bmatrix} \quad (4)$$

Assuming that, with some appropriate current controllers, the tracking problem of $\|i_s\|^\star, \omega_i^\star$ can be solved, the reduced second order IM model representing the rotor dynamics is derived as

$$\begin{bmatrix} \dot{\psi}_I \\ \dot{\psi}_\perp \end{bmatrix} = \begin{bmatrix} -\tau_r^{-1}\psi_I + \omega_r\psi_\perp + M\tau_r^{-1}\|i_s\| \\ -\tau_r^{-1}\psi_\perp - \omega_r\psi_I \end{bmatrix} \quad (5)$$

along with the algebraic output equation for the generated electrical torque

$$T_{el} = -\frac{M}{L_r}\psi_\perp\|i_s\| \quad (6)$$

In (5) we introduced relative speed ω_r as the difference between the rotation speed of the stator current vector ω_i and the rotor speed ω_m. In the following, the rotor speed is simply interpreted as a bounded time-varying parameter, and the dynamics of the mechanical part is therefore omitted from further analysis. Considering the reduced model (5), along with the output equation (6) it can be seen that unique inverse mapping $T_{el} \rightarrow \psi_I, \psi_\perp, \|i_s\|, \omega_r$ does not exist. Additional conditions must therefore be introduced to the obtain required equilibrium. By manipulating only the first control input $\|i_s\|$, assuming that $\omega_r = 0$, only the rotor flux linkage component ψ_I will be changed. The second input ω_r must therefore also be changed in order to establish necessary $\psi_\perp$ and thus generate the required machine torque. Since current vector rotation is required for all transient and steady-states (except at $T_{el} = 0$ and $\omega_m = 0$), the corresponding equilibrium conditions must be determined. Steady-state characteristics describing relations between machine torque, stator current vector magnitude and relative speed are given in Fig. 1. Consequently, different control strategies could be

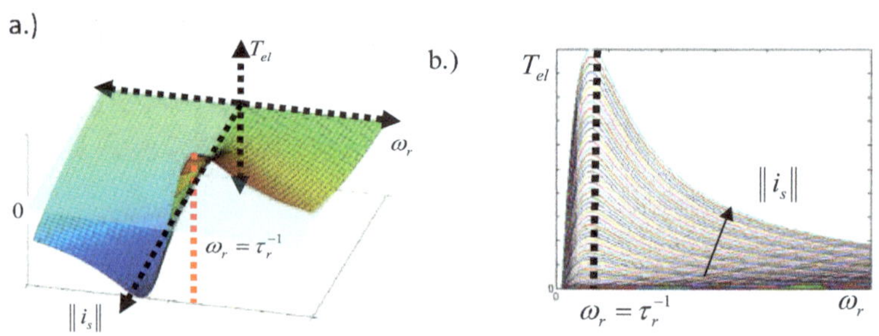

Fig. 1. Steady-state relations between electrical torque T_{el}, stator current vector magnitude $\|i_s\|$ and relative speed ω_r.

employed in the torque generation. Keeping the current vector magnitude constant at some nominal value, high torque dynamics on the cost of reduced torque-per-amp ratio is obtained. On the other hand, if we keep the second input ω_r constant, the efficiency could be improved, but only poor dynamics would be achieved due to the slow dynamics in the rotor flux linkage magnitude. Thus, the real control challenge lies in the question of how to manipulate both control inputs simultaneously to achieve acceptable torque tracking and global stability, as well as the maximal torque-per-amp ratio in steady-state.

3. Partial dynamic inversion

Considering the reduced model (5) and assuming that ω_r is a constant parameter, equation (5) can be written in linear state-space form

$$\dot{\psi}_r = \begin{bmatrix} -\tau_r^{-1} & \omega_r \\ -\omega_r & -\tau_r^{-1} \end{bmatrix} \psi_r + \begin{bmatrix} \tau_r^{-1}M \\ 0 \end{bmatrix} ||i_s|| \tag{7}$$

where $\psi_r = [\psi_I, \psi_\perp]^T$. Calculating the corresponding transfer functions

$$\begin{aligned} G_1(s) &= \frac{\psi_I}{||i_s||} = \frac{(s\tau_r + 1)M}{s^2\tau_r^2 + 2s\tau_r + 1 + \omega_r^2\tau_r^2} \\ G_2(s) &= \frac{\psi_\perp}{||i_s||} = -\frac{\omega_r\tau_r M}{s^2\tau_r^2 + 2s\tau_r + 1 + \omega_r^2\tau_r^2} \end{aligned} \tag{8}$$

we can easily conclude that maximal steady-state gain between $||i_s||$ and $\psi_\perp$ is obtained if the following dynamic condition

$$\omega_r = \tau_r^{-1} \tag{9}$$

is satisfied. Note that introduced relative rotation speed ω_r equals the slip speed in the steady-state. Since the quantities $\psi_\perp$, $||i_s||$ are torque producing, this implies that maximal torque per ampere ratio is obtained at any steady-state resulting from the equilibrium condition (9). From steady-state analysis It further follows that

$$\begin{bmatrix} \psi_I \\ \psi_\perp \end{bmatrix} = \begin{bmatrix} \frac{M||i_s||}{1+\omega_r^2\tau_r^2} \\ -\frac{\omega_r\tau_r M||i_s||}{1+\omega_r^2\tau_r^2} \end{bmatrix} \tag{10}$$

Additional characteristic feature is derived from (10) on condition that (9) is satisfied

$$\psi_I = |\psi_\perp| \tag{11}$$

where ψ_I is restricted to positive values. Furthermore, we can calculate the steady-state torque producing flux linkage vector component and the corresponding current vector magnitude for a given torque command $T_{el}^\star$

$$\begin{aligned} \psi_\perp^\star &= -\text{sign}(T_{el}^\star)\sqrt{\frac{|T_{el}^\star|L_r}{2}} \\ ||i_s||^\star &= \frac{2}{M}|\psi_\perp^\star| \end{aligned} \tag{12}$$

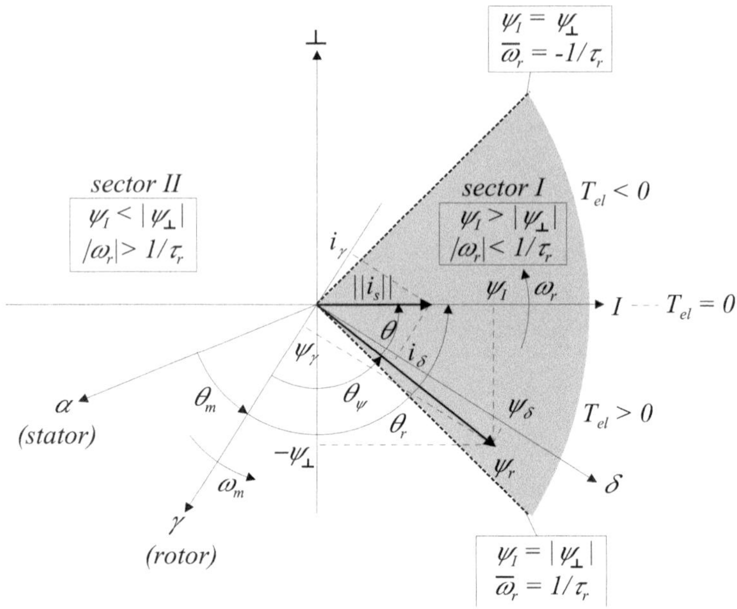

Fig. 2. Desired operating sector I and reference frames.

In Fig. 2 a graphical interpretation of the desired operating sector and the relevant reference frames are given. The trajectories of the rotor flux linkage vector inside the $\pi/2$ wide sector I always ensure torque responses without stall by bounding ω_r. However, only steady-states on the boundary of sector I additionally fulfill the maximal torque-per-amp ratio requirement. Operation in sector II without bounding ω_r is avoided due to oscillatory torque responses and danger of stall.

The control design challenge is to force the flux linkage trajectories to remain predominantly inside sector I during transients and in steady-states to lie as close as possible to the maximal torque-per-amp line. A careful study of (5) allows us to conclude that the rotor flux linkage will remain, for zero initial rotor flux linkage, inside sector I provided the following conditions are satisfied

$$\begin{aligned} &\psi_I \geq |\psi_\perp|; \ \forall T_{el}^\star \\ &||i_s|| = f(\omega_r) \leq ||i_s||\text{max} \\ &|\omega_r| \leq \tau_r^{-1} \quad \text{in steady-state} \\ &\dot{\psi}_\perp \leq 0; \ T_{el}^\star > 0 \ \text{ or} \\ &\dot{\psi}_\perp \geq 0; \ T_{el}^\star < 0 \ \text{ or} \\ &\psi_\perp = 0; \ T_{el}^\star = 0 \end{aligned} \tag{13}$$

The analysis of different flux trajectory families resulting from (5), existence of unique equilibrium (9), (11) and (12) along with the introduced conditions (13), motivated us to design the torque controller introduced in the next Section.

4. Torque controller design

The proposed control implies that machine torque is known. Several estimation and measuring techniques are available. Advanced, robust estimation methods use voltage sensors and stator equations (Verghese & Sanders, 1988), (Vas, 1998) and (Briz, 2002) instead of more parameter sensitive rotor estimators. Note that only in the reference frame that is proposed estimation of any (stator, rotor or air-gap) orthogonal flux linkage projection, with the respect to the measured stator current vector magnitude, is sufficient for machine torque calculation. In some drives with large IM, air-gap flux is actually measured (Plounkett, 1979). In this particular case, it is a straightforward task to calculate machine torque based on the air-gap flux linkage projection orthogonal to the measured stator current vector. This is always a dynamic process, and the first order dynamics, with the time constant τ_f will be introduced to model this measurement. A standard singular perturbation argument involving estimation time constant ($\tau_f \to 0$) recovers the algebraic expression for the actual machine torque. In the stability analysis, both situations will therefore be discussed; in the first part the machine torque measurement will be assumed, and in in the second part the simple rotor flux linkage estimator will be introduced.

Indirect open-loop controller

In order to simultaneously satisfy requirements for high dynamic performance and maximal torque-per-amp ratio, the following indirect open-loop controller is proposed first, assuming nominal parameter case

$$\begin{aligned} \|i_s\| &= \frac{L_r}{M}\frac{|T_{el}|^\star}{|\hat{\psi}_\perp|} && ;0 \leq \|i_s\| \leq I_{\text{max}} \\ \omega_r &= \frac{T_{el}^\star R_r}{\hat{\psi}_\perp^2 2} && ;|\omega_r| \leq |\omega_{\text{max}}| \end{aligned} \tag{14}$$

where estimate $\hat{\psi}_\perp$ is obtained from the nominal rotor model (5). First control input $\|i_s\|$ is obtained from the equation for the machine torque (6) while the second input ω_r is derived from the instantaneous power equilibrium $\|i_s\|^2 R_r/2 = T_{el}\omega_r$ that is valid along the boundary of sector I. For nominal parameter case the control (14) assures the best possible performance, forcing the rotor flux trajectory to move along the the boundary of sector I for all required torques (except for zero initial condition in rotor flux linkage vector). The growth of the machine torque starting from zero initial condition in the rotor flux linkage vector depends on the maximal available stator current magnitude I_{max} and maximal rotation speed ω_{max}. After transients, ω_r converges to τ_r^{-1} satisfying equilibrium conditions (11) and (12). Assuming nominal parameters simple rotor flux linkage estimator is introduced

$$\begin{aligned} \dot{\hat{\psi}}_I &= -\frac{1}{\tau_r}\hat{\psi}_I + \omega_r\hat{\psi}_\perp + \frac{M}{\tau_r}\|i_s\| && ;\hat{\psi}_I(0) = 0 \\ \dot{\hat{\psi}}_\perp &= -\frac{1}{\tau_r}\hat{\psi}_\perp - \omega_r\hat{\psi}_I && ;\hat{\psi}_\perp(0) = 0 \end{aligned} \tag{15}$$

Defining the estimation errors

$$e_I = \psi_I - \hat{\psi}_I$$
$$e_\perp = \psi_\perp - \hat{\psi}_\perp$$

Straightforward computations show that estimation errors satisfy the equations

$$\begin{aligned} \dot{e}_I &= -\tfrac{1}{\tau_r} e_I + \omega_r e_\perp \\ \dot{e}_\perp &= -\tfrac{1}{\tau_r} e_\perp - \omega_r e_I \end{aligned} \tag{16}$$

It is easy to show that (16) is globally exponentially stable with Lyapunov function $V_o = \frac{1}{2}e_I^2 + \frac{1}{2}e_\perp^2$, and an estimate of the machine torque is obtained as

$$\hat{T}_{el} = -\frac{M}{L_r}\hat{\psi}_\perp \|i_s\| \tag{17}$$

If the rotor time constant τ_r^{-1} is considered as uncertain parameter due to variations in the rotor resistance, estimate $\hat{\tau}_r$ is used in the perturbed estimator

$$\begin{aligned} \dot{\hat{\psi}}_I &= -\tfrac{1}{\hat{\tau}_r} \hat{\psi}_I + \omega_r \hat{\psi}_\perp + \tfrac{M}{\hat{\tau}_r}\|i_s\| \qquad ; \hat{\psi}_I(0) = 0 \\ \dot{\hat{\psi}}_\perp &= -\tfrac{1}{\hat{\tau}_r} \hat{\psi}_\perp - \omega_r \hat{\psi}_I \qquad ; \hat{\psi}_\perp(0) = 0 \end{aligned} \tag{18}$$

Consequently the open-loop control (14) will shift the equilibrium point away from the boundary of sector I, additionally steady state error in the estimated machine torque will be observed

$$\begin{aligned} |\overline{\hat{\psi}}_\perp| = \overline{\hat{\psi}}_I &= \sqrt{\tfrac{|T_{el}^\star| L_r}{2}} \\ \overline{\psi}_I &= \tfrac{L_r}{1+\tau_r^2/\hat{\tau}_r^2} \tfrac{|T_{el}^\star|}{|\hat{\psi}_\perp|} \\ |\overline{\psi}_\perp| &= \tfrac{\tau_r}{\hat{\tau}_r} \overline{\psi}_I \\ \overline{\hat{T}}_{el} &= 2\tfrac{\tau_r \hat{\tau}_r}{\tau_r^2 + \hat{\tau}_r^2} T_{el}^\star \end{aligned} \tag{19}$$

Direct open-loop controller

To reduce the influence of parameter perturbation we propose a more efficient procedure to estimate torque producing flux component. Instead of using rotor model based perturbed estimator (15), consider the stator voltage equation in the stationary reference frame

$$\hat{\psi}_s = \int_0^t (u_s - R_s i_s) d\tau \qquad ; \hat{\psi}_s(0) = 0 \tag{20}$$

where $u_s = [u_\alpha, u_\beta]^\mathrm{T}, i_s = [i_\alpha, i_\beta]^\mathrm{T}, \psi_s = [\psi_\alpha, \psi_\beta]^\mathrm{T}$ and R_s is the stator resistance. Expressing stator current and flux linkage vectors in the polar coordinates $\|i_s\| = \sqrt{i_\alpha^2 + i_\beta^2}, \varphi_i = \tan^{-1}(i_\beta / i_\alpha)$ and $\|\hat{\psi}_s\| = \sqrt{\hat{\psi}_\alpha^2 + \hat{\psi}_\beta^2}, \varphi_\psi = \tan^{-1}(\hat{\psi}_\beta / \hat{\psi}_\alpha)$, torque producing flux linkage component $\hat{\psi}_{eff}$ that is orthogonal to the stator current vector is obtained by using a simple transformation

$$\hat{\psi}_{eff} = \|\hat{\psi}_s\| \sin(\varphi_i - \varphi_\psi) \tag{21}$$

from where also the torque estimate can be calculated as $\hat{T}_{el} = -\|i_s\|\hat{\psi}_{eff}$. Estimation of the ψ_{eff} requires knowledge of the stator voltage vector u_s. If the corresponding voltage sensor is available, stator voltage is measured. In the opposite case, the reference voltages obtained from the current controllers could be used instead. A direct version of the open-loop torque controller (14) could therefore be rewritten as

$$\begin{aligned} \|i_s\| &= \frac{|T_{el}|^\star}{|\hat{\psi}_{eff}|} && ;0 \le \|i_s\| \le I\max \\ \omega_r &= \frac{T_{el}^\star R_r}{\hat{\psi}_{eff}^2 2} && ;|\omega_r| \le |\omega\max| \end{aligned} \tag{22}$$

Note that the method of using torque producing flux linkage components $\widehat{\psi}_\perp$ or $\widehat{\psi}_{eff}$ for the torque control is a obvious consequence of selected reference frame aligned with the stator current vector.

Indirect closed-loop controller

To assure ultimate steady-sate accuracy in the estimated machine torque a feedback version of the open-loop controller (14) based on the torque error $\widetilde{T}_{el} = T_{el}^\star - \widehat{T}_{el}$ is proposed in the following form

$$\begin{aligned} \|i_s\| &= \frac{L_r}{M}\frac{|T_{el}|^\star}{|\hat{\psi}_\perp|} && ;0 \le \|i_s\| \le I\max \\ \dot{v} &= k_i \widetilde{T}_{el} && ;k_i > 0, v(0) = 0, |v| \le \hat{\tau}_r^{-1} \\ \omega_r &= v + k_p \widetilde{T}_{el} && ;k_p > 0, |\omega_r| \le |\omega\max| \end{aligned} \tag{23}$$

Calculating the fifth order feedback system based on rotor model (5), perturbed estimator (18) and control law (23) results in

$$\begin{aligned} \dot{\psi}_I &= -\tau_r^{-1}\psi_I + (v + k_p\widetilde{T}_{el})\psi_\perp + L_r\tau_r^{-1}\frac{|T_{el}^\star|}{|\hat{\psi}_\perp|} \\ \dot{\psi}_\perp &= -\tau_r^{-1}\psi_\perp - (v + k_p\widetilde{T}_{el})\psi_I \\ \dot{\hat{\psi}}_I &= -\hat{\tau}_r^{-1}\hat{\psi}_I + (v + k_p\tilde{T}_{el})\hat{\psi}_\perp + L_r\hat{\tau}_r^{-1}\frac{|T_{el}^\star|}{|\hat{\psi}_\perp|} \\ \dot{\hat{\psi}}_\perp &= -\hat{\tau}_r^{-1}\hat{\psi}_\perp - (v + k_p\widetilde{T}_{el})\hat{\psi}_I \\ \dot{v} &= k_i\widetilde{T}_{el} \end{aligned} \tag{24}$$

with unique equilibrium point

$$\begin{aligned} \overline{\psi}_I &= \sqrt{\frac{\hat{\tau}_r^3|T_{el}^\star|L_r}{\tau_r\hat{\tau}_r^2 + \tau_r^3}} \\ |\overline{\psi}_\perp| &= \frac{\tau_r}{\hat{\tau}_r}\overline{\psi}_I \\ |\overline{\hat{\psi}}_\perp| = \overline{\hat{\psi}}_I &= \sqrt{\frac{|T_{el}^\star|L_r}{2}} \\ \overline{\omega}_r = \overline{v} &= \hat{\tau}_r^{-1} \\ \overline{\hat{T}}_{el} &= T_{el}^\star \end{aligned} \tag{25}$$

Consequently, the actual rotor flux linkage vector will be forced to move slightly away from boundary into sector I provided $\widehat{\tau}_r^{-1} \le \tau_r^{-1}$. Linearizing feedback system (24) around equilibrium point (25), we observe that rotor and estimator dynamics are governed by the stable eigenvalues $\lambda_{1,2} = -\tau_r^{-1} \pm j\hat{\tau}_r^{-1}$ and $\lambda_{3,4} = -\hat{\tau}_r^{-1} \pm j\hat{\tau}_r^{-1}$. Since the term $|T_{el}^\star|/|\hat{\psi}_\perp|$ is bounded by controller construction the feedback system (24) is locally stable. the restriction of local stability could eventually be relaxed by introducing Lyapunov function candidate $V_c = 1/2(\psi_I^2 + \psi_\perp^2 + \hat{\psi}_I^2 + \hat{\psi}_\perp^2 + v^2)$. Calculating the time derivative of V_c along the solution of (24) gives

$$\dot{V}_c = -\tau_r^{-1}(\psi_I^2 + \psi_\perp^2 - \frac{\psi_I L_r |T_{el}^\star|}{|\hat{\psi}_\perp|}) - \hat{\tau}_r^{-1}(\hat{\psi}_I^2 - \hat{\psi}_\perp^2 + \frac{\hat{\psi}_I L_r |T_{el}^\star|}{|\hat{\psi}_\perp|}) \tag{26}$$

As ψ_I and $\hat{\psi}_I$ are positive semi-definite due to chosen reference frame, the derivative of V_c is obviously negative definite.

Full information controller

Assuming that the electro-mechanical torque (6) is measured, the following dynamic controller is proposed[1]

$$\begin{aligned} \omega_r &= \text{sat}_{[-k_4, k_4]}\left(k_p \widetilde{T}_{el} + v\right) \\ \| i_s \| &= -2\,\tfrac{\hat{\tau}_r}{M}\,\psi_\perp^\star\,\omega_r \\ \dot{\widetilde{T}}_{el} &= \tau_f^{-1}\left(T_{el}^\star - \widetilde{T}_{el} - T_{el}\right) \\ \dot{v} &= k_i\,\widetilde{T}_{el}, \end{aligned} \tag{27}$$

where k_p, k_i, $\tau_f \ll \tau_r$ and k_4 are positive design constants, $\hat{\tau}_r$ is an estimate of the rotor time constant and $\widetilde{T}_{el}$ is an auxiliary state.
The closed–loop dynamics yields

$$\begin{aligned} \dot{\psi}_I &= -\tfrac{1}{\tau_r}\,\psi_I + (k_p\,\widetilde{T}_{el} + v)\,\psi_\perp - 2\tfrac{\hat{\tau}_r}{\tau_r}\,\psi_\perp^\star\,(k_p\,\widetilde{T}_{el} + v) \\ \dot{\psi}_\perp &= -\tfrac{1}{\tau_r}\,\psi_\perp - (k_p\,\widetilde{T}_{el} + v)\,\psi_I \\ \dot{\widetilde{T}}_{el} &= \tau_f^{-1}\left[T_{el}^\star - \widetilde{T}_{el} - 2\,\tfrac{\hat{\tau}_r}{L_r}\,\psi_\perp^\star\,\psi_\perp\,(k_p\,\widetilde{T}_{el} + v)\right] \\ \dot{v} &= k_i\,\widetilde{T}_{el} \end{aligned} \tag{28}$$

thus, the closed–loop equilibria are the solutions of the following equations

$$\begin{aligned} 0 &= -\tfrac{1}{\tau_r}\,\psi_{Ie} + v_e\,\psi_{\perp e} - 2\tfrac{\hat{\tau}_r}{\tau_r}\,\psi_{\perp e}\,v_e \\ 0 &= -\tfrac{1}{\tau_r}\,\psi_{\perp e} - v_e\,\psi_{Ie} \\ 0 &= \tau_f^{-1}\left[T_{el}^\star - 2\,\tfrac{\hat{\tau}_r}{L_r}\,\psi_\perp^\star\,\psi_{\perp e}\,v_e\right] \end{aligned} \tag{29}$$

From the first two equations of (29) we obtain ψ_{Ie} and $\psi_{\perp e}$. Now, replacing ψ_{Ie}, $\psi_{\perp e}$ and $\psi_\perp^\star$ into the third relation and considering $|T_{el}^\star| = \text{sign}(T_{el}^\star)\,T_{el}^\star$, we obtain the following cubic polynomial with respect to v_e

$$f(v_e) = 1 + {v_e}^2\,\tau_r^2 - 2\,\text{sign}(T_{el}^\star)\,\tau_r\,\hat{\tau}_r^2\,v_e^3 = 0 \tag{30}$$

Hence, the number of equilibrium points of (28) is determined by the number of real roots of the polynomial (30). Basic computations show that for $T_{el}^\star > 0$ ($T_{el}^\star < 0$), $f(v_e)$ has a minimum

[1] With $\text{sat}_{[a,b]}(u) = \begin{cases} u \text{ if } a \le u \le b \\ a \text{ if } u < a \\ b \text{ if } u > b \end{cases}$.

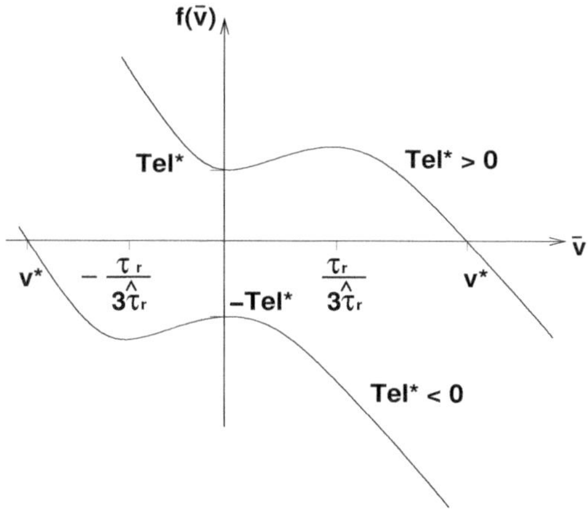

Fig. 3. Equilibrium points of $f(v_e)$.

(maximum) at $v_{e_1} = 0$ and a maximum (minimum) at $v_{e_2} = \frac{1}{3}\frac{\tau_r}{\hat{\tau}_r^2}$ $(v_{e_2} = -\frac{1}{3}\frac{\tau_r}{\hat{\tau}_r^2})$, see Fig. 3; moreover, $f(v_{e_2}) > f(v_{e_1}) > 0$ $(f(v_{e_2}) < f(v_{e_1}) < 0)$. Thus, it can be concluded that (30) has only one real root so that the closed-loop dynamics (28) has a unique equilibrium point given by

$$\begin{bmatrix} \psi_{Ie} \\ \psi_{\perp e} \\ \widetilde{T}_{el} \\ v_e \end{bmatrix} = \begin{bmatrix} -\frac{2\,\hat{\tau}_r\,v^\star\,\psi_\perp^\star}{1+v^{\star 2}\,\tau_r^2} \\ \frac{2\,\tau_r\,\hat{\tau}_r\,v^{\star 2}\,\psi_\perp^\star}{1+v^{\star 2}\,\tau_r^2} \\ 0 \\ v^\star \end{bmatrix} \tag{31}$$

with $v^\star$ the unique real root of (30). It is important to point out that, for small values of $\hat{\tau}_r$, with respect to τ_r the value of $v^\star$ increases and the risk of stalling also increases since $\omega_r^\star = v^\star$. In order to avoid this risk, we need to choose $\hat{\tau}_r$ larger than τ_r, thus keeping $v^\star$ small, (see Fig. 3). A detailed stability analysis is given in App.I.

5. Current control

The task of the inner current controllers is demanding since reference tracking, disturbance suppression and voltage drop compensation of the VSI must be achieved simultaneously. Note that inverse mapping between the current vector magnitude $\|i_s\|$ and the voltage u_I is characterized by perturbed first-order dynamics and that inverse mapping between ω_i and voltage $u_\perp$ is purely algebraic and nonlinear. Before introducing the current controllers a physical interpretation of the nominal mapping $\|i_s\|, \omega_i \to u_I, u_\perp$ is given.

Analyzing the stator equations of (4) in polar form

$$\begin{aligned} \dot{\|i_s\|} &= -\tau_\sigma^{-1}\|i_s\| + \tfrac{M}{L_\sigma L_r \tau_r}\psi_I + \omega_m \tfrac{M}{L_\sigma L_r}\psi_\perp + \tfrac{1}{L_\sigma}u_I \\ \dot{\theta}_i = \omega_i &= -\omega_m \tfrac{M}{L_\sigma L_r}\tfrac{\psi_I}{\|i_s\|} + \tfrac{M}{L_\sigma L_r \tau_r}\tfrac{\psi_\perp}{\|i_s\|} + \tfrac{1}{L_\sigma \|i_s\|}u_\perp \end{aligned} \tag{32}$$

and assuming synchronous operation first, when stator current vector and rotor flux linkage vector are aligned and characterized by the conditions $\omega_r = 0 \rightarrow T_{el} = 0, \psi_\perp = 0$ and $\omega_m \neq 0, \|i_s\| \neq 0 \rightarrow \omega_i = \omega_m$, $\psi_I \neq 0$, the required steady state voltage vector can be expressed as

$$\begin{aligned} u_I &= (R_\sigma - \tfrac{M^2}{L_r\tau_r})\|i_s\| \\ u_\perp &= (L_\sigma + \tfrac{M^2}{L_r})\|i_s\|\omega_m \end{aligned} \tag{33}$$

where the relation $\psi_I = M\|i_s\|$ is valid at synchronous operation. The control voltage u_I therefore influences only the flux linkage magnitude while $u_\perp$ is needed to enable synchronous rotation of the current vector. On the other hand, if the machine rotor is locked ($\omega_m = 0 \rightarrow \omega_i = \omega_r$) and the maximmum torque-per-amp operation is achieved ($\omega_r = \tau_r^{-1}$), the corresponding control voltage components are obtained in the form

$$\begin{aligned} u_I &= (R_\sigma - \tfrac{M^2}{2L_r\tau_r})\|i_s\| \\ u_\perp &= (L_\sigma + \mathrm{sign}(T_{el})\tfrac{M^2}{2L_r})\|i_s\|\omega_r \end{aligned} \tag{34}$$

where the relations $\psi_I = \|i_s\|M/2$ and $|\psi_\perp| = \psi_I$ were considered. In general, when the machine is rotated at a certain mechanical speed and an arbitrary electrical torque is generated, simultaneously satisfying maximum torque-per-amp requirement, the control voltage cannot be expressed simply as a superposition of (33) and (34) since flux linkage vector changes its magnitude and relative position between any steady state operation point. Control voltage vector, assuming operation at maximal torque-per-amp ratio $\omega_r = \tau_r^{-1}$ and considering that $\omega_i = \omega_m + \omega_r$, can therefore be calculated as

$$\begin{aligned} u_I &= (R_\sigma + (\mathrm{sign}(T_{el})\omega_m - \omega_r)\tfrac{M^2}{2L_r})\|i_s\| \\ u_\perp &= (L_\sigma + \tfrac{M^2}{2L_r})\|i_s\|\omega_m + (L_\sigma + \mathrm{sign}(T_{el})\tfrac{M^2}{2L_r})\|i_s\|\omega_r \end{aligned} \tag{35}$$

Note that both voltage components in (35) are expressed with known or measured quantities and that u_I predominantly changes rotor flux linkage magnitude and that $u_\perp$ influences the angle between the stator current vector and rotor flux linkage vector. Both voltages also define the steady state lower bound for u_I and upper bound for $u_\perp$ if stable operation inside sector I is supposed to be achieved. It is clear that the control voltage component $u_\perp$ is significant with respect to transient torque response, torque-per-amp ratio and stability. Poorly damped oscillatory torque responses are obtained if the control voltage $u_\perp$ is too high. In an extreme case, instability could even due to stall, as the angle between the stator current vector and the rotor flux linkage vector approaches $\pm\pi/2$. In the opposite case, when the rotation speed is too small, the angle between the stator current and the rotor flux linkage vectors is also small. The machine is therefore forced to operate with an unnecessarily large rotor flux linkage. Poor torque-per-amp ratio is the obvious consequence although the required torque is generated. It should be pointed out that the inverse mapping $\|i_s\|, \omega_i \rightarrow u_I, u_\perp$ is quite sensitive in practical implementation due to the partially algebraic plant nature, influence of the signal noise, parameter uncertainty and VSI operation. To avoid algebraic loops and obtain symmetrical current control structure with simple tuning rules and sufficient robustness, the stator equations of (4) are transformed into the rotor reference frame.

Current controllers in rotor reference frame

The task of current controller in the rotor reference frame is the mapping of reference currents $i_\gamma^\star$ and $i_\delta^\star$ into machine voltages u_γ and u_δ. The reference currents $i_\gamma^\star$ and $i_\delta^\star$ are obtained from torque controller output as:

$$\begin{aligned} \theta_r &= \textstyle\int_0^t \omega_r(\tau)d\tau + \vartheta_f \quad ;\theta_r(0) = 0 \\ i_\gamma^\star &= \|i_s\| \cos(\theta_r) \\ i_\delta^\star &= \|i_s\| \sin(\theta_r) \end{aligned} \tag{36}$$

where feed-forward angle θ_f can be set as:

$$\theta_f = -\text{sign}(T_{el}^\star)\frac{\pi}{4} \tag{37}$$

The angle θ_f offers an addition degree of freedom in the torque control providing phase modulation of the stator current vector, while amplitude and frequency modulation are provided by the torque controller. This additional input can be used for fast (almost instantaneous) angle changes between the stator current vector and the rotor flux linkage vector in the case of sign changes of the reference torque (active breaking). Stator equation in rotor reference frame is obtained in the vector form as:

$$\begin{aligned} \dot{i}_{\gamma,\delta} &= -\tau_\sigma^{-1}(\mathrm{J}\omega_m\tau_\sigma + 1)i_{\gamma,\delta} + \tfrac{M}{L_\sigma L_r \tau_r}(\mathrm{J}^T\omega_m\tau_r + 1)\psi_{\gamma,\delta} \\ &+ \tfrac{1}{L_\sigma}u_{\gamma,\delta} = (A + \mathrm{J}\omega_m)i_{\gamma,\delta} + Bu_{\gamma,\delta} + d \end{aligned} \tag{38}$$

where the influence of the rotor flux linkage is captured in the disturbance d. Based on the well justified time separation of the stator and rotor dynamics bounded disturbance d could be neglected since that stator current dynamic is predominantly driven by the eigenvalues of the time variant system matrix $\lambda_{1,2}(A) = -\tau_\sigma^{-1} \pm j\omega_m$. For the cross-coupling effects between the current components $i_{\gamma,\delta}$ due to the motion induced voltages $L_\sigma \mathrm{J}\omega_m i_{\gamma,\delta}$, time separation is not justified since this voltage can change as fast as the stator current. To obtain a satisfactory tracking performance, these terms must be considered in control design. Simple feed-forward compensation with opposite sign provides only moderate performance due to uncertainty in the estimation of machine leakage inductance L_σ. Introducing the current error $\widetilde{i}_{\gamma,\delta} = i_{\gamma,\delta}^\star - i_{\gamma,\delta}$ the PI controller using internal model principle is therefore proposed in the following form:

$$u_{\gamma,\delta} = k_{pi}\widetilde{i}_{\gamma,\delta} + k_{pi}k_{ii}\int_0^t (\frac{\mathrm{J}\omega_m}{k_{ii}} + 1)\widetilde{i}_{\gamma,\delta}d\tau \tag{39}$$

where $k_{pi}, k_{ii} > 0$ are free design parameters. Dynamic of the augmented system is therefore given as:

$$\begin{aligned} \dot{i}_{\gamma,\delta} &= A(\omega_m)i_{\gamma,\delta} + Bk_{pi}\widetilde{i}_{\gamma,\delta} + Bz \\ \dot{z} &= (k_{pi}k_{ii} + k_{pi}\mathrm{J}\omega_m)\widetilde{i}_{\gamma,\delta} \qquad ;z(0) = 0 \end{aligned} \tag{40}$$

By choosing $k_{ii} = \tau_\sigma^{-1}$ and setting $i^\star_{\gamma,\delta} = 0$ the closed loop system matrix is obtained in the form:

$$A_{cl}(\omega_m) = \begin{bmatrix} -\tau_\sigma^{-1} - \frac{k_{pi}}{L_\sigma} & \omega_m & \frac{1}{L_\sigma} & 0 \\ -\omega_m & -\tau_\sigma^{-1} - \frac{k_{pi}}{L_\sigma} & 0 & \frac{1}{L_\sigma} \\ -\frac{k_{pi}}{\tau_\sigma} & k_{pi}\omega_m & 0 & 0 \\ -k_{pi}\omega_m & -\frac{k_{pi}}{\tau_\sigma} & 0 & 0 \end{bmatrix} \tag{41}$$

For any constant ω_m corresponding closed-loop eigenvalues are:

$$\begin{aligned} \lambda_{1,2} &= -k_{pi}/L_\sigma \\ \lambda_{3,4} &= -\tau_\sigma^{-1} \pm j\omega_m \end{aligned} \tag{42}$$

The eigenvalues $\lambda_{1,2}$ define the dynamics of the stator current, while $\lambda_{3,4}$ are related with the dynamics of the auxiliary variable z.

6. Experimental results

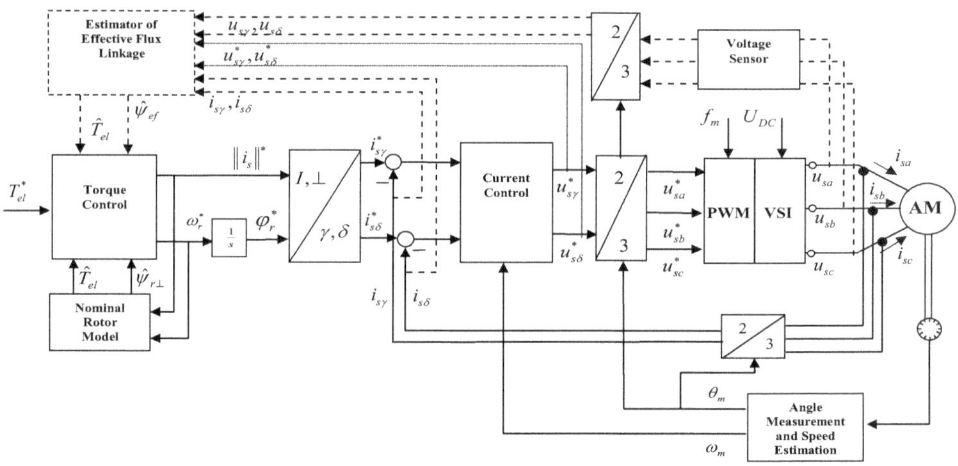

Fig. 4. Overall IM control scheme including torque controller, current controller, and estimator based on nominal rotor model. The dotted lines indicate advanced observers based on measured machine currents and voltages.

The experimental setup, motor data, along with the parameters of the torque and current controllers, are given in the Appendix. To avoid the danger of escaping, the DC motor was connected to the shaft. Breaking torque proportional to mechanical speed was generated ($t_l = k\omega_m$) so that the effect of the linear friction was simulated. In all experiments the most simple rotor flux estimator and torque calculator based on (15) and (17) were used

$$\begin{aligned} \dot{\widehat{\psi}}_I &= -\widehat{\tau}_r^{-1}\widehat{\psi}_I + \omega_r\widehat{\psi}_\perp + M\widehat{\tau}_r^{-1}\|i_s\| \\ \dot{\widehat{\psi}}_\perp &= -\widehat{\tau}_r^{-1}\widehat{\psi}_\perp - \omega_r\widehat{\psi}_I \\ \widehat{T}_{el} &= -\frac{M}{L_r}\widehat{\psi}_\perp\|i_s\| \end{aligned} \tag{43}$$

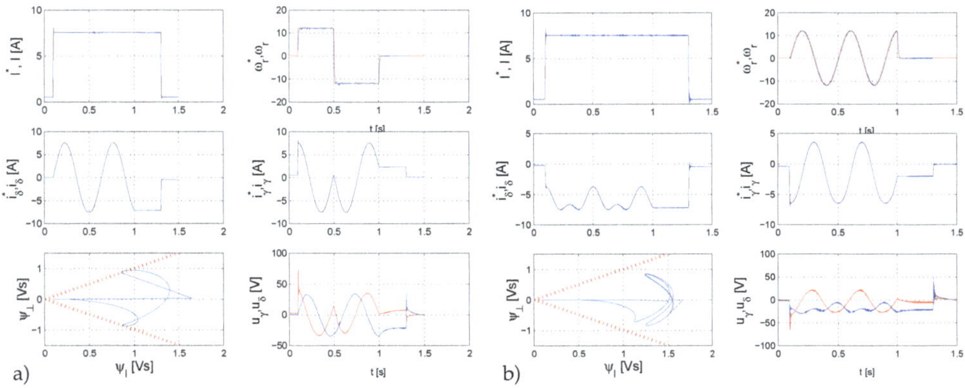

Fig. 5. Current control step response (a) and current control response to sinusoidal $\omega_r^\star$ (b).

where the actual stator current module $\|i_s\|$ and the relative speed ω_r were measured, and the nominal value of $\widehat{\tau}_r$ was used. The problem of all rotor based estimators is the uncertain rotor time constant $\widehat{\tau}_r$, completely neglecting the saturation effects influencing M and L_r. If the estimate of actual τ_r is wrong, than the steady-state error between estimated and actual machine torque occurs. Just to calibrate and verify steady-state torque and flux estimates at $\omega_m = 0 \rightarrow \omega_i = \omega_r$ (locked rotor), the torque sensor HBM-T20WN (10 Nm) was mounted between the IM and DC motor. By using torque sensor output T_{elm}, these estimates were corrected off-line by adjusting $\widehat{\tau}_r$

$$\begin{aligned} \widehat{\psi}_\perp &= -\frac{T_{elm} L_r}{M\|i_s\|} \\ \dot{\widehat{\psi}}_I &= -\widehat{\tau}_r^{-1}\widehat{\psi}_I - \frac{T_{elm} L_r}{M\|i_s\|}\omega_r + M\widehat{\tau}_r^{-1}\|i_s\| \qquad ;\widehat{\psi}_I(0) = 0 \end{aligned} \tag{44}$$

in such a way that $\widehat{T}_{el} \approx T_{elm}$ was achieved in steady-state. It should be pointed out that the estimated flux linkage component $\widehat{\psi}_I$ was used only to evaluate the transient performance and torque-per-amp ratio while the estimate $\widehat{\psi}_\perp$ was needed in torque calculation. The overall control scheme is shown in Fig. 4. The first two experiments show the performance of the current controller. Step changes in both reference current magnitude $\|i_s\|^\star$ and relative speed $\omega_r^\star$ are presented in Fig. 5a. Note that in all presented diagrams the current vector magnitudes $\|i_s\|^\star$, $\|i_s\|$ are denoted as $I^\star$, I) and machine torques $T_{el}^\star$, $\widehat{T}_{el}$ as $t_{el}^\star$, t_{el}. From the estimated rotor flux linkage trajectory it can be seen that the maximal torque-per-amp line is reached for each steady state as long as the condition $\omega_r \approx \tau_r^{-1}$ is satisfied. During transients, the flux linkage trajectory changes inside the desired operating sector I.

In Fig. 5b an experiment is shown where current vector magnitude was kept constant, while rotation speed ω_r was changed as a sinusoidal function with the magnitude between $\pm\tau_r^{-1}$. Since current magnitude is constant, the rotor flux linkage component ψ_I increases as ω_r decreases. Perfect current tracking could be observed in both experiments since reference and actual currents actually overlap.

The next few experiments show the responses of the torque controlled machine using nominal estimate of τ_r. Only positive torque reference was used during the first experiment in Fig. 6a, more demanding tracking task is presented in Fig. 6b, and torque reversal is shown in Fig. 7a.

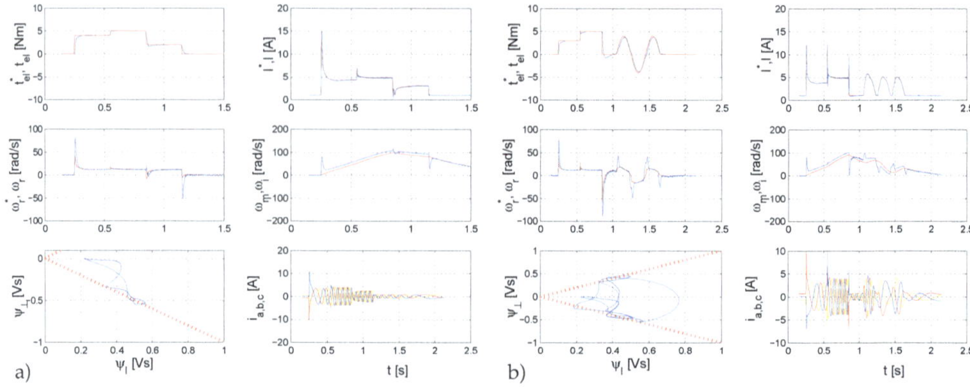

Fig. 6. Tracking of the positive torque reference (a) and tracking of the general torque reference (b).

The torque was built up completely after $\approx$ 50ms, starting practically from zero field; almost perfect tracking performance of the generated torque can be observed once a sufficient rotor field is established. The growth of ψ_I is slightly faster than growth of $\psi_\perp$, and the flux linkage trajectory is therefore forced to change inside the desired sector I. As the magnitude of the flux linkage vector components equalizes, the maximal torque-per-amp ratio operation is reached in all steady-states. In Fig. 7a the flux linkage trajectory also moves in sector II for a short period of time. This effect could be prevented by using "reset integrator" in torque controller. In Fig. 7b, an experiment is shown where short torque pulses were generated representing the high dynamic capability of the control scheme.

The next two experiments concern the perturbed rotor time constant case. The experiment in Fig. 8a was performed with $\hat{\tau}_r = 3\,\tau_r$, while the experiment in Fig. 8b was performed with $\hat{\tau}_r = 0.7\,\tau_r$. Note that in the first case the flux linkage trajectory is forced to move deeper in sector I, while in the second case the flux linkage trajectory moves also outside sector I. Decreasing the $\widehat{\tau}_r$ further would therefore threat the stability while increasing $\widehat{\tau}_r$ would just

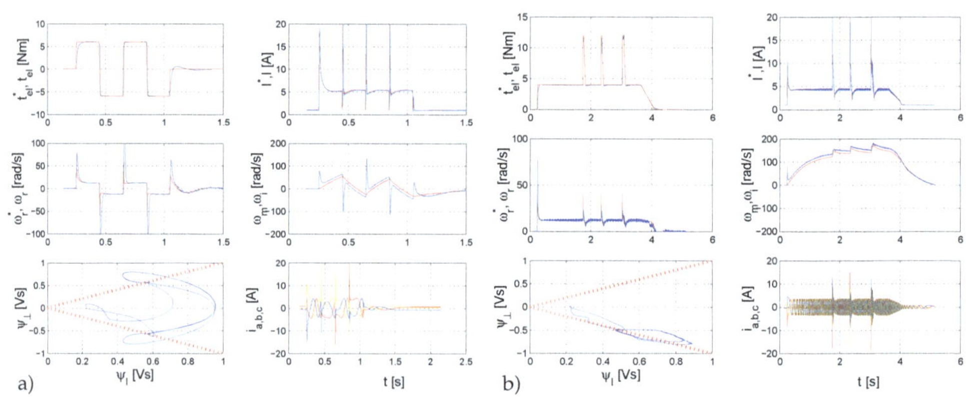

Fig. 7. Torque reversal (a) and generation of the torque pulses (b).

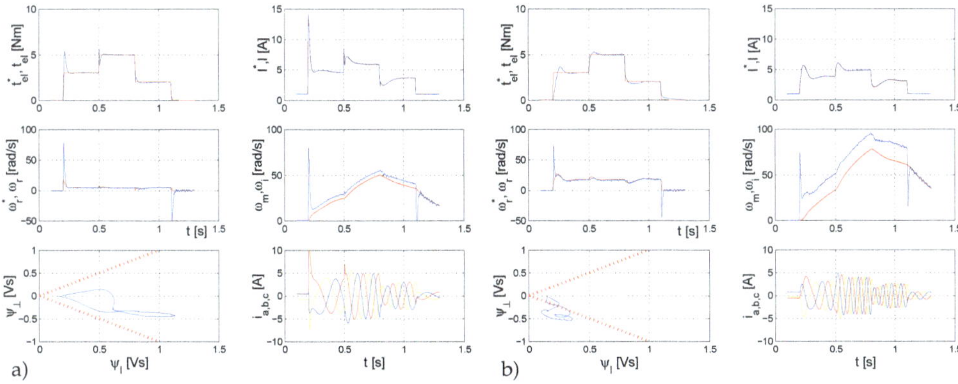

Fig. 8. Torque control with perturbed rotor time constant: $\hat{\tau}_r = 3\,\tau_r$ (a) and $\hat{\tau}_r = 0.7\,\tau_r$ (b).

increase the current vector magnitude and, consequently, the rotor field. In both cases stability was preserved, and the required torque was generated, however the maximal torque-per-amp operation was not reached.

The experiment in Fig. 9 shows that the proposed controller is able to handle operation at a higher rotor speed (approximately three times of the nominal speed). Experiments actually show responses similar to those in field weakening mode. Although the required torque is reduced, mechanical speed increases and reaches about three times the nominal speed. In all experiments could be observed how the stator current rotation speed changes relative to the rotor speed. During all steady-states, this relative speed equals the slip speed.

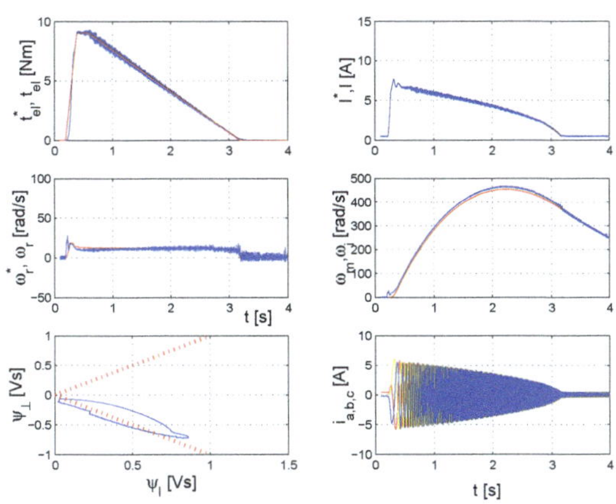

Fig. 9. Operation above nominal rotor speed.

7. Conclusion

The chapter presents a control scheme for torque control of IM based on a stator current vector reference frame. The overall design is motivated by physical interpretations of typical transient and steady-state IM phenomena rather than by concepts from abstract control system theory. Nevertheless, the derived IM control based on introduced partial dynamic inversion enables an acceptable tracking performance, asymptotic stability for all physically feasible initial conditions, robustness with respect to rotor time constant perturbations and implicitly optimized torque-per-amp ratio. These properties were achieved while avoiding the FOC concept of a decoupled control based on rotor flux vector orientation. The key requirement in our solution is that all rotor flux linkage vector trajectories are implicitly restricted to the desired operating sector I and that, in steady-state, the trajectories end as close as possible to the maximal torque-per-amp line, simultaneously fulfilling basic control objectives. The proposed control assures implicit changes in the rotor flux vector without separate control actions. In addition, the proposed control is less sensitive with respect to saturation effects compared to other control schemes, whose reference frames are attached to the estimated machine fields. The implementation of the proposed control is possible on standard industrial hardware, assuming that the machine torque is calculated based on the flux estimate and measured current.

8. Appendix

Appendix I:

In order to analyze stability of the closed–loop dynamics, (28) can be written as

$$\dot{x} = A\,x + g(x) \tag{45}$$

where

$$x = \begin{bmatrix} \psi_I - \psi_{Ie} \\ \psi_\perp - \psi_{\perp e} \\ \widetilde{T}_{el} \\ \omega_r - v^* \end{bmatrix}, \quad g(x) = \begin{bmatrix} \widetilde{\psi}_\perp \\ -\widetilde{\psi}_I \\ -\frac{2\tau_r\,\psi_{\perp e}}{M\,\tau_f}\,\widetilde{\psi}_\perp \\ -\frac{2k_p\tau_r\,\psi_{\perp e}}{M\,\tau_f}\,\widetilde{\psi}_\perp \end{bmatrix} \widetilde{\omega}_r$$

$$A = \begin{bmatrix} -\frac{1}{\tau_r} & v^* & 0 & -\psi_{\perp e} \\ -v^* & -\frac{1}{\tau_r} & 0 & \frac{\psi_{\perp e}}{\tau_r\,v^*} \\ 0 & -\frac{2\tau_r\,v^*\,\psi_{\perp e}}{M\,\tau_f} & -\frac{1}{\tau_f} & -\frac{2\tau_r\,\psi_{\perp e}^2}{M\,\tau_f} \\ 0 & -\frac{2k_p\tau_r\,v^*\,\psi_{\perp e}}{M\,\tau_f} & k_i - \frac{k_p}{\tau_f} & -\frac{2k_p\tau_r\,\psi_{\perp e}^2}{M\,\tau_f} \end{bmatrix}$$

Straightforward computations show that the linear part of (45) is exponentially stable provided that $\tau_f < 1$ and $k_p \geq k_i\,\tau_f$. Stability of the linear part of (45) implies that there exist a positive definite matrix P such that

$$P\,A + A^\top P = -I$$

with I the identity matrix. Note that the time derivative of $V = x^\top P\,x$ along (45) gives

$$\dot{V} = -x^\top x + 2\,\widetilde{\omega}_r\,x^\top P\,G\,x$$

where

$$G = \begin{bmatrix} 0 & 1 & 0\,0 \\ -1 & 0 & 0\,0 \\ 0 & -\frac{2\tau_r\,\psi_{\perp e}}{M\,\tau_f} & 0\,0 \\ 0 & -\frac{2k_p\,\tau_r\,\psi_{\perp e}}{M\,\tau_f} & 0\,0 \end{bmatrix}$$

therefore, we have

$$\dot{V} \leq -\|x\|^2 + 2\,\bar{k}_4\,\|PG\|\,\|x\|^2$$

for all $|\widetilde{\omega}_r| < \bar{k}_4$ with $\bar{k}_4 = k_4 + |v^\star|$ and $\dot{V}$ is negative definite provided

$$\bar{k}_4 < \frac{1}{2\,\|PG\|} \tag{46}$$

Thus, it can be concluded that the dynamics (5) in closed–loop with the controller (27) is exponentially stable in the domain

$$D = \left\{ x \in R^4 \mid \sqrt{x_1^2 + x_2^2 + x_3^2 + \bar{k}_4^2} \leq r \right\}$$

An estimate of the region of attraction in the plane $\psi_I - \psi_\perp$ can be obtained as follows. Consider the positive definite function $V_1 = \frac{1}{2}\widetilde{\psi}_I^2 + \frac{1}{2}\widetilde{\psi}_\perp^2$ whose time derivative along (45) is given by

$$\dot{V}_1 = -\frac{1}{\tau_r}(\widetilde{\psi}_I^2 + \widetilde{\psi}_\perp^2) - \psi_\perp^\star\,\widetilde{\omega}_r\,\widetilde{\psi}_I + \frac{\psi_\perp^\star}{\tau_r\,v^\star}\widetilde{\omega}_r\,\widetilde{\psi}_\perp$$

from (46) we have

$$\dot{V}_1 \leq -\frac{1}{\tau_r}\|\widetilde{\psi}\|^2 + \frac{\bar{k}_4\,|\psi_\perp^\star|}{\tau_r\,v^*}\sqrt{1 + \tau_r^2\,v^{\star 2}}\|\widetilde{\psi}\|$$

where $\widetilde{\psi} = [\,\widetilde{\psi}_I\;\widetilde{\psi}_\perp\,]$. Thus, $\dot{V}_1$ is negative definite provided

$$\|\widetilde{\psi}\| > \frac{\bar{k}_4\,|\psi_\perp^\star|}{v^*}\sqrt{1 + \tau_r^2\,v^{\star 2}}$$

and an estimate of the domain of attraction $\Omega_c$is given as

$$\Omega_c = \{V_1(\widetilde{\psi}) = c\}$$

with $c > \frac{\bar{k}_4^2\,|\psi_\perp^{\star 2}|}{v^{\star 2}}(1 + \tau_r^2\,v^{\star 2})$.

Remark. *It should be stressed, that in steady state, the controller output provides useful information for estimation of the rotor time constant. Note that, in steady state, we have access to $v^\star$ so that the only unknown in (30) is the rotor time constant, which can be computed as*

$$\tau_r = \begin{cases} \frac{\hat{\tau}_r^2\,v^{\star 2} + \sqrt{\hat{\tau}_r^4\,v^{\star 4} - 1}}{v^\star}, & \text{for } T_{el}^\star > 0 \\ \frac{\hat{\tau}_r^2\,v^{\star 2} - \sqrt{\hat{\tau}_r^4\,v^{\star 4} - 1}}{v^\star}, & \text{for } T_{el}^\star < 0 \end{cases} \tag{47}$$

Remark. *Figure 3 shoes that efficiency depends on the accuracy of the rotor time constant as any mismatch in the estimated rotor time constant $\hat{\tau}_r$ affects the magnitude of $v^\star$. Thus, recalling that $\omega_r^\star = v^\star$, the rotor dynamics (5) may not operate at the maximal torque per ampere ratio defined by (9); however stable operation is preserved.*

Stability analysis using estimated machine torque.

Simple computations show that by replacing the machine torque (6) with its estimated value (17) in (27) gives a closed–loop dynamics described by equations (45) perturbed by an additive term which is bounded by $\kappa\,|e_\perp|$ for a positive constant κ and it is exponentially decaying to zero. Consider now the Lyapunov function V and note that

$$\dot{V} \leq -\|x\|^2 + 2\,\bar{k}_4\,\|PG\|\,\|x\|^2 + \kappa\,\|P\|\,\|x\|\,|e_\perp| \leq \kappa\,\|P\|\,\|x\|\,|e_\perp| \tag{48}$$

for $\bar{k}_4$ satisfying (46). THE equation (48) implies that, along the trajectories of the closed–loop system with estimated electro-mechanical torque, V is bounded. As a result, $\|x\|$ is bounded and, converges to zero, since $e_\perp$ exponentially decays to zero.

Appendix II:

Beside the tested IM the experimental setup consisted of: a DSPACE DS1103 PPC Controller board, a host PC with installed development environment, an incremental encoder Iskra TELA TGR 10 with 2500 pulses per revolution, current sensors LEM LT 100-P, Semikron inverter (modules SKH1 22, SKM 50GB 123D and SKD 51/14, up to 800 V at the dc bus and currents up to 30 A RMS) and DC motor. The DSPACE DS 1103 PPC Controller board consists of: the IBM PowerPC 604e, the slave digital signal processor (DSP) TMS320F240, interfaces for incremental encoders, and AD and DA converters. Although ω_i can be obtained directly by (noisy) derivation of the current angle vector θ_i a second order observer as proposed in (Harnefors & Nee, 2000) was used. During the tests, data acquisition, transformations, and control were executed on the PowerPC, while the slave DSP was used for vector modulation running at 4 kHz. All experiments were performed with the sampling time of 250 μs. The program codes for the PowerPC and for the slave DSP were developed with Real Time Interface and Simulink.

Stator resistance	$R_s = 1.976\Omega$
Rotor resistance	$R_r = 2.91\Omega$
Mutual inductance	$M = 0.223$H
Stator inductance	$L_s = 0,2335$ H
Rotor inductance	$L_r = 0.2335$ H
Stator leakage inductance	$L_{s\sigma} = 0.0105$ H
Rotor leakage inductance	$L_{r\sigma} = 0.0105$ H
Number of pole pairs	$p = 2$
Nominal torque	$T_{el} = 10$ Nm

Table 1. Nominal motor data

k_p	15
k_i	800
τ_f	0.005s
Current bounds	$0.5 \leq \|\|i_s\|\| \leq 20$A
k_{pi}	20
k_{ii}	225.5

Table 2. Controller parameters

9. List of symbols

$i_{\alpha,\beta}, u_{\alpha,\beta}, \psi_{\alpha,\beta}$	components of stator current, stator voltage and rotor flux linkage vectors in stationary reference frame
$u_{I,\perp}, \psi_{I,\perp}$	parallel $(.)_I$ and orthogonal $(.)_\perp$ components of stator voltage and rotor flux linkage vectors
$i_{\gamma,\delta}, u_{\gamma,\delta}, \psi_{\gamma,\delta}$	components of stator current, stator voltage and rotor flux linkage vectors in rotor reference frame
$\|\|.\|\|$	norm L_2 of vectors or matrices
ω_m	rotor speed (mechanical)
ω_i	stator current vector rotation speed
$\omega_r = \omega_i - \omega_m$	relative rotation speed of stator current vector
θ_i	absolute stator current vector angle
θ_r	relative stator current vector angle
θ_f	feed-forward angle
M	mutual inductance
R_r, L_r	rotor resistance and self-inductance
$\tau_r = L_r/R_r$	rotor time constant
T_{el}, T_L	machine torque, load torque
J	drive inertia
R_s, L_s	stator resistance and self-inductance
$\tau_\sigma = L_\sigma/R_\sigma$	leakage time constant
$L_\sigma = L_s - (M)^2/L_r$	leakage inductance
$R_\sigma = R_s + (M/L_r)^2 R_r$	leakage resistance
$\mathrm{J} = [0, -1; 1, 0]$	2×2 rotation matrix
k_p, k_i, k_4	torque controller parameters
τ_f	torque estimator time constant
v	output from integral control action
k_{pi}, k_{ii}	current controller parameters
$(.)^\star, \widehat{(.)}, \widetilde{(.)}, \overline{(.)}$	reference values, estimated values, errors, steady state values
$(.)_e$	equilibrium values
$e_I, e_\perp$	rotor flux linkage estimation errors

10. References

Attaianese C., Nardi V., Perfetto A. & Tomasso G. (1999). Vectorial Torque Control: A Novel Approach to Torque and Flux Control of Induction Motor Drives, *IEEE Transactions on Industry Applications*, Vol. 35, No. 6, pp. 1399–1405.

Blaschke F. (1971). Das Prinzip der Feldorientierung, die Grundlage für die Transvektor-Regelung von Drehfeldmaschinen, *Siemens Zeitschrift*, Vol. 10, No. 45.

Bodson M. & Chiasson J. (1998). Differential-Geometric Methods for Control of Electric Motors, *International Journal of Robust and Nonlinear Control*, No. 8, pp. 927–952.

Briz F., Degner M.W., Diez A. & Lorenz R.D. (2002). Static and Dynamic Behavior of Saturation-Induced Saliencies and Their Effect on Carrier-Signal-Based Sensorless AC Drives, *IEEE Transactions on Industry Applications*, Vol. 38, No. 3, pp. 670–678.

Brockett R. (1996). Characteristic Phenomena and Model Problems in Nonlinear Control, *IFAC, 13th Triennial World Congres, San Francisco*, 2b II, pp. 257–262.

Grčar B., Cafuta P., Štumberger G., Stankovič A., Hofer A. (2011). Non-holonomy in Induction Torque Control, *IEEE Transactions on control System Technology*, Vol. 19, No. 2, pp. 367–375.

Chakraborty C., Hori Y. (2003). Fast Efficiency Optimization Techniques for the Indirect Vector-Controlled Induction Motor Drives, *IEEE Transactions on Industry Applications*, Vol. 39, No. 4, pp. 1070–1076.

Chiasson J. (1998). A New Approach to Dynamic Feedback Linearization Control of an Induction Motor, *IEEE Transactions on Automatic Control*, Vol. 43, No. 3, pp. 391–396.

Depenbrock M. (1986). Direct Self-Control (DSC) of Inverter-Fed Induction Machine, *IEEE Transactions on Industry Applications*, Vol. IA-22, No. 5, pp. 820–827.

Harnefors L., Nee H.P. (2000). A General Algorithm for Speed and Position Estimation of AC Motors, *IEEE Transactions on Indusrtial Electronics*, Vol. 47, No. 1, pp. 77–83.

Krause P.C. (1986). *Analysis of Electric Machinery*, McGraw-Hill.

Marino R., Peresada S. & Tomei P. (1999). Global Adaptive Output Feedback Control of Induction Motors with Uncertain Rotor Resistance, *IEEE Transactions on Automatic Control*, Vol. 44, No. 5, pp. 967–980.

Martin P., Rouchon P. (1996). Flatness and Sampling Control of Induction Motor, *IFAC, 13th Triennial World Congres, San Francisco*, 2b 27 2, pp. 389–394.

Ortega R., Loria A., Nicklasson P.J. & Siera-Ramirez H. (1998). *Passivity-based Control of Euler-Lagrange Systems*, Springer-Verlag, Berlin.

Ortega R., Barabanov N., Escobar G. (2000). Direct Torque Control of Induction Motors: Stability Analysis and Performance Improvement, *IEEE Transactions on Automatic Control*, Vol. 46, No. 8, pp. 967–980.

Peresada S., Tilli A., Tonielli A. (2003). Theoretical and Experimental Comparison of Indirect Field-Oriented Controllers for Induction Motors, *IEEE Transactions on Power Electronics*, Vol. 18, No. 1, pp. 151–163.

Plounkett A.B., D'Atre J.D., Lipo T.A. (1979). Synchronous Control of a Static AC Induction Motor Drive, *IEEE Transactions on Indusrty Applications*, Vol. 5, No. 4, pp. 430–437.

Takagashi I., Noguchi T. (1986). A New Quick-Response and High-Efficiency Control Strategy of an Induction Motor, *IEEE Transactions on Industry Applications*, Vol. IA-22, No. 5, pp. 820–827.

Vas P. (1998). *Sensorless Vector and Direct Torque Control*, Oxford University Press, Oxford.

Verghese G.C., Sanders S.R. (1988) Observers for Flux Estimation in Induction Machines, *IEEE Transactions on Industrial Electronics*, Vol. 35, No. 1, pp. 85–93.

7

Logistics Services and Intelligent Security Control for Transport Companies

José F. Díez-Higuera, Francisco J. Díaz-Pernas, Miriam Antón-Rodríguez, David González-Ortega and Mario Martínez-Zarzuela
Department of Signal Theory, Communications and Telematics Engineering Telecommunications Engineering School, Valladolid University, Valladolid, Spain

1. Introduction

In recent years, both growth of urban areas and trade expansion have increased freight traffic on the public roads. This increase in traffic affects transport companies, which suffers long delays in delivery of goods. Therefore, they experience degradation in service quality while increasing operating costs. Intelligent transportation systems have shown their potential to improve significantly the management and the operation of existing transportation systems (Siwek, 1998). Transport companies can benefit from using this technology to improve the quality of their services and to optimize the use of their resources. In order to increase the safety of their workers, these systems may also include the monitoring of the driver fatigue level, an important aspect considering that driver fatigue causes 60% of fatal accidents in which trucks are involved (AWAKE Consortium, 2004).
The developed system meets these requirements by means of a corporative web site application where the whole telematics infrastructure of the service provider is hosted. The system has a fully customizable friendly-user interface based on Joomla!, which integrates a real-time geographic information system, in addition to other functions such as both customer management and dispatching scheduling for different user profiles (administrators, operators, managers, ...) in a company and different company profiles (backbone, retail, multimodal, etc.). Besides the usual real-time route guidance and communication functions, the on-board system incorporates an intelligent security control, which monitors driver fatigue and alertness level in real time. This functionality is accomplished by a computer vision system based on both face tracking and gesture recognition techniques. Collected data are processed to avoid accidents caused by driver distractions or sleepiness.

1.1 Outsourcing

One of the most complex problems that companies with transport and/or delivery activities must address nowadays is to have suitable information to improve some of their tasks (operational efficiency and management, customer services, outputs, and other factors) by managing and monitoring the different operational variables that transport operations need. Outsourcing is the trend solution to minimize costs with no loss of efficiency. On the other

hand, not only is providing logistic assistance for drivers in freight transportation operations important and but also providing a safer driving is essential.
In 1980, logistics represented 17.9% of US Gross Domestic Product (GDP). Today, it is 7.7%. The cost of logistics in the US was $1.1 trillion in 2009, which is 7.7% of (GDP), according to CSCMP's 21st Annual "State of Logistics Report®" (CSCMP, 2009). By comparison, estimated logistics costs represent from 15 to 16% of China's GDP and from 11 to 13% of India's GDP. Logistics costs in Europe are significantly lower due to a combination of less area and long-established transportation infrastructure, including rail, rivers, and highways. On average, logistics costs represent 7.15% of European GDP.
According to a study conducted by Capgemini (Langley & Capgemini, 2010), one of the leading suppliers in the world of consulting, technology, and outsourcing services, a significant number of shippers are increasing their use of third-party logistics (3PL). Overall, 65% of shippers report an increase in their use of outsourced logistics services, and 78% of 3PL shippers agree that this is what they are noting from their customers. Regionally, 57% of North America shippers have increased that use, as well as 65% of European shippers, 81% of Asia-Pacific, and 69% of Latin American shippers. Based on these results, it seems that the predominant direction is to move toward increased use of outsourced logistics services, confirming findings also reported in previous studies.
As should be expected, the innovation currently occurring in the marketplace is focused on the more complex and integrated services. The outsourcing of supply chain management is coming true and also requires a more advanced use of technology. Of course, this concerns not just selling tools but rather delivering services. The provider must offer economies of scale for relatively scarce skills or for activities that are only needed incidentally by clients (Giloth & van dort, 2011).

1.2 Resources optimization

Many researches on routing problems have appeared in the literature. Comprehensive and detailed explanations of theoretical models and solutions of them are given by Toth & Vigo (Toth & Vigo, 2002). In Kokubugata (Kokubugata & Kawashima, 2008) a variety of routing problems are introduced and followed by the explanation of features of routing problems. Likewise, there are many commercial solutions that provide fleet management and route optimization, such as EnGenX, Trackroad, Routesmart, GeoBase, SubeX, Telogis Route, GMV, E-Drive Technology, etc.
However, most of the relevant groups in this area, simply take into account the use of static data (distance, road types, ...), which means that, regardless of the used optimization method used, the route calculated in special circumstances (traffic jams, snow, ...) can not only be optimal, but a bad choice. Moreover there is no system that optimizes the assignment of cargo, vehicle, and driver for a particular route.

1.3 Driver alertness monitoring

Many vision-based methods have been proposed for eye tracking and eyelid closure detection. They can be based on active infrared or passive illumination. With infrared illumination, pupils can be detected by a simple thresholding of the difference between the dark and the bright pupil images (Ji et al., 2004) and factors such as the brightness, size of the pupils, and the external illumination can influence performance. Approaches using standard cameras and passive illumination in cluttered scenes have also been presented.

D'Orazio et al. (D'Orazio et al., 2007) proposed a neural classifier to recognize the eyes in the image selecting two candidates regions that might contain the eyes by using iris geometrical information and symmetry. Smith et al. (Smith et al., 2003) presented a system to detect eye blinking and eye closure based on color statistics. Królak proposes a system which uses two active contours, one from each eye, for eye blink detection from previous eye tracking (Królak & Strumillo, 2008).

Statistics show that between 10% and 20% of all the traffic accidents in Europe are due to drivers with a reduced vigilance level caused by fatigue (AWAKE Consortium, 2004). These figures show the importance that driver alertness monitoring applications can have to decrease the number of traffic accidents. Fatigue measurement is a difficult problem as there are few direct measures and most of them are measures of the outcomes of the fatigue rather than of fatigue itself. An important physiological measure that has been studied to detect fatigue is eye motion. Several eye motions were used to measure fatigue such as blink rate, blink duration, long closure rate, blink amplitude, saccade rate, and peak saccade velocity. PERCLOS measure is the percentage of eyelid closure over the time and reflects slow eyelid closures rather than blinks (Dinges & Grace, 1998). A PERCLOS drowsiness metric was established as the proportion of time in a minute that the eyes are at least 80 percent closed and it is a reliable and widely used driver performance and bio-behavioral measure (Wang et al., 2006).

1.4 System relevant features of the system

With the proposed system, transport companies can increase productivity of their drivers, optimize roadmaps, promote communication among all members of the company, and reduce the cost of transport operations. Each company has its own intranet, accessible via a corporate portal fully configurable by the user, and supported by Joomla!. Among the applications available on the intranet, in addition to the usual ones, there is a proprietary Enterprise Resources Planning (ERP), a Resources Optimization Module, and an Intelligent Security Control.

The system minimizes the impact that the deployment, management, and maintenance of the central computer system have on the transport company. It outsources most of this workload to a third party logistics. Thus, the transport company can manage its resources more conveniently and transparently, eliminating the need for a large investment in computer equipment and the requirement to train or hire an expert operator. In turn, the company providing the service can meet the needs of multiple carriers with a lower overall cost, providing a specific "Know How" value to the quality of service.

Each transport company may customize its contract selecting the services that it wants to outsource. Likewise, shippers may elect, at the instant of requesting the load, the associated services that they want to use. The installation of software on the transport company is not necessary because all management can be done from the telematics platform. In addition, the system is easily scalable, so the company can meet the progressive increase in demand.

A resources optimization module makes route calculations take into account all data that can affect driving (works, accidents, planned events, weather, traffic ...), in addition to the points of distribution, number of available vehicles, etc.. Furthermore, regarding transport backbone, the system automatically selects the best suited truck-load-driver triad for a particular case.

The system incorporates a new concept in fleet management systems: the driver's safety. Real-time monitoring of driver's face detects whether the driver's vigilance level drops below a certain threshold to control his level of fatigue. In the current system this task is performed on a laptop that sends data to the mobile terminal. The data are sent to the central control along with other information about the vehicle and freight, so that it may carry out monitoring and preventive control of the state of the driver. The vehicle on-board equipment incorporates a safety control system for the driver, which periodically sends data to the central control to prevent accidents due to fatigue or distraction of the driver.
The rest of this chapter is organized as follows: the infrastructure of the proposed system is described in Section 2. The main modules included in the framework are described in Sections 3 to 6: kernel, Intelligent Security Control, Resources Optimization module, and on-board equipment.

2. System description

This section describes a complete multi-management system that incorporates advanced information technology and communications. The developed system is a telematics platform of logistics services, which allows companies to manage their mobile resources effectively, control them and get all the benefits of having timely and accessible information. The system is based on the use of computer science, and mobile and satellite telecommunications. These characteristics enable to convert a commercial vehicle fleet into a mobile intranet and to integrate mobile resources in the computing infrastructure of the company.

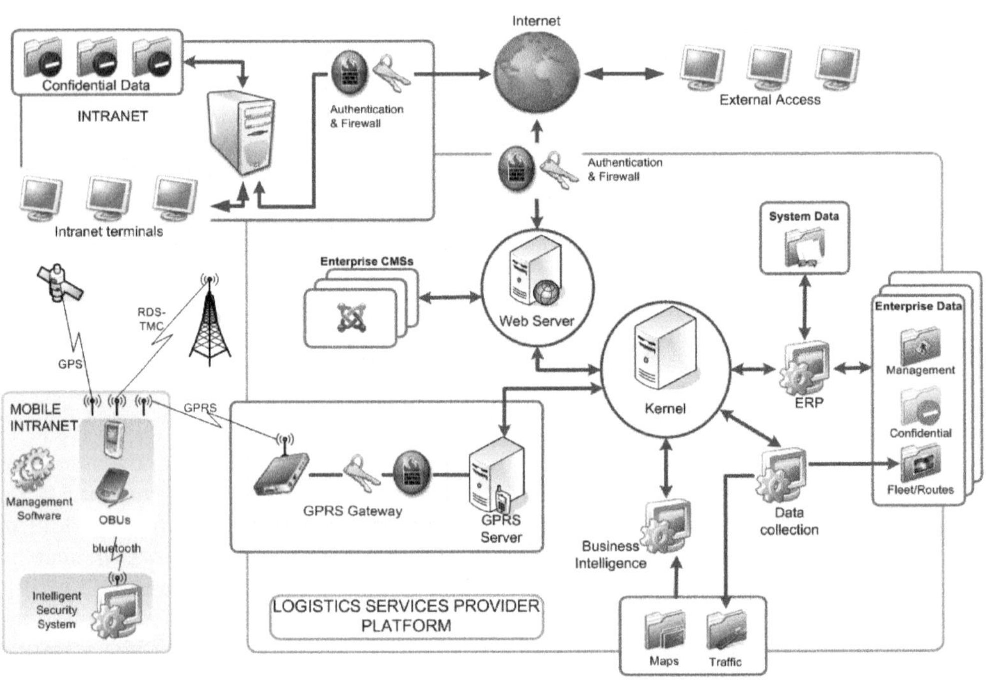

Fig. 1. Scheme for the telematics platform of logistics services.

The general scheme of the system in figure 1 shows the "classic" intranet of the company, the new Mobile Intranet, and different subsystems comprising the system. This scheme also shows the relationship between the different modules and the way information flows between them. Basically, the system consists of three major modules, namely Kernel, Intelligent Security Control, and Resources Optimization Module.
The system is designed modularly to facilitate the addition or substitution of specific software modules responsible for carrying out system tasks. Thereby, the system is scalable and it is feasible to adapt its configuration to the specific requirements of the wide variety of transportation companies.
The following sections describe the major elements of the proposed system, with special emphasis on the modules that distinguish our system from others of similar characteristics: the kernel, the Resources Optimization module, and the Intelligent Safety Control.

3. Kernel

In the developed system, the flow of information among the various components is carried out through the kernel, whose main functions are to provide access to databases for the different modules when they need it, and to establish a channel for information flow between them. This last feature allows the information contained in the databases to be centralized and managed by a single process (the kernel) that performs operations on data concurrently, thus ensuring data consistency and replica control.
The system modules are connected in such a way that each module can be executed on separate machines connected to the machine running the kernel. The system tasks can either be both distributed in a network of computers or executed on a single machine. The Central Control is responsible for proper communication and proper transfer of information among the subsystems of the platform. The kernel exchanges information and controls the communications, the Web server, and the Data Acquisition and Management services.
Since one of the roles played by this module is to carry the information flow among the rest of the modules of the system, sometimes its behavior will either be client or server. Therefore it has two channels for communication, one to serve information and other to receive it. The channel of communication between the modules of the system is established using sockets, a specific communication protocol, and client/server architecture.

3.1 Kernel development

The kernel of the system has been programmed completely in Java, using the NetBeans IDE suite. This kernel has been developed so that it is easily scalable and editable, due to a good structure and good use of object-oriented programming. Thus, adding new functionality (or changing an existing one) involves the modification of a few lines of code in one or a few files, each corresponding to a particular object.
As regards access to databases, unnecessary resource consumption has been minimized. For this purpose, all information on the database for a particular company, which is stored in the system database, will be stored in a cache bin each time they are accessed for the first time and will remain in this bin until a certain time has elapsed since the last access. This way, multiple simultaneous queries from a client will only impact on the database system the first time they are accessed, avoiding redundant queries. To implement this feature, we have opted for the use of both threads running in parallel to the rest of the program, which

does not increase the memory needed to run the program, and dynamic tables which allow a quick and easy access.

3.2 Communications

The main method of communication between the kernel and the outside is through XML messages, a procedure that is already being widely used in B2B communication. Java libraries and JAXB tools have been used to facilitate the handling of this format. These tools allow fitting up objects with XML messages. Thus, any addition or modification on the structure of the XML is very easy to implement, because it would be equivalent to including an additional variable in the class of the specified object. Figure 2 shows the diagram of communications among the kernel and the rest of the components of the system: internal modules, external modules, optimization module, and vehicle on-board terminal.

The kernel maintains connections with databases permanently, as well as with clients who connect with it. Therefore, to avoid problems of persistence, a thread in parallel to the core is permanently executed. This thread is responsible for detecting when someone wants to exit the system, so that it is responsible for closing one by one all the connections opened at the same time, ensuring that the system is closed only when all the data that the system has to process are ready.

A communication procedure based on XML messages has been developed so that external modules can communicate with the kernel. To achieve it, the kernel creates a thread that is permanently waiting for incoming connections. When a connection is detected, it is accepted and sent to a connection manager, which will create a new thread for the new connection. The kernel goes back to the listening mode while the new connection is processed.

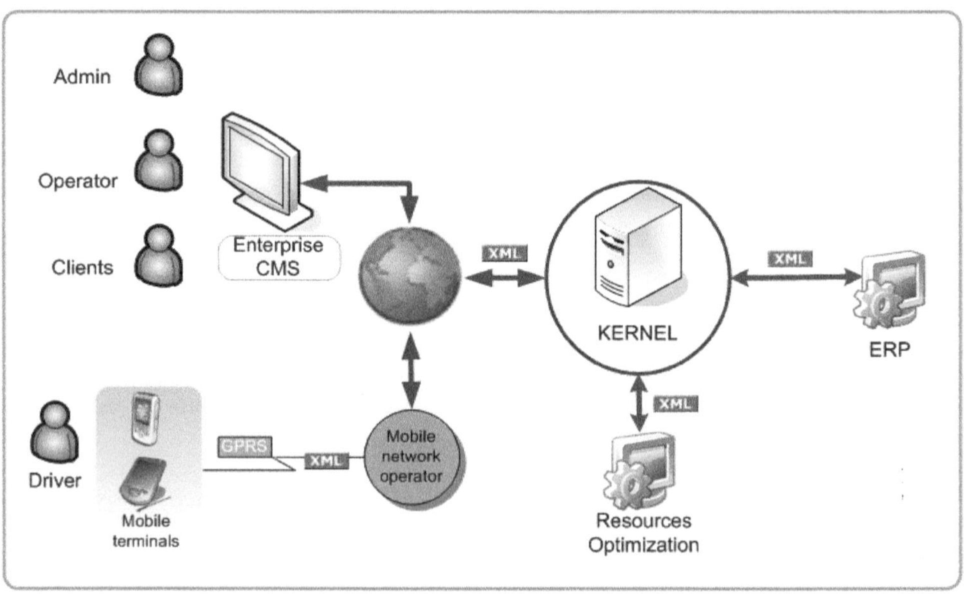

Fig. 2. Diagram of communications among the kernel and the rest of the components of the system.

Then, the kernel processes the received XML message. To do this, the kernel has a "parser" or interpreter of XML messages, which transforms the received message in a series of particular objects, so that they can be accessed easily from any system method. However, these objects are encapsulated within a specific object for requests which is in charge of returning the fields requested by any part or function of the code so that if the format of XML messages or the format of communication are changed, only the inside of the objects, and not the rest of the system, has to be changed.

Once the message has been interpreted, it is redirected to the internal module indicated in the header of the specific request, or to the link with an external module if necessary. These modules are responsible for generating a response based on the received information.

For the answer, there is an object similar to the object used in request, which generates the same XML data from Java objects.

Once the response is complete, it is returned to the kernel thread that was processing the connection, which is responsible for returning through the socket connection to the external module. After sending the message in XML, the kernel closes the connection to the module, and ends the thread that was processing.

On the side of the client or external module, the system is similar: when a module requires certain information or needs to add or update fields in the database of a particular company, it generates a specific XML request, creates a connection to the Kernel, and sends the XML message. Subsequently, the module is waiting for a response and, once received, interprets or processed it, if necessary.

3.2.1 Communication between the kernel and the optimization module

Optimization module opens a TCP/IP socket server, which is listening to client connection requests. The kernel initiates communication through an authentication message when it requires a service of resource optimization. After a validation message from the server, the kernel generates a request message to calculate routes that are sent via socket to the optimization module. This message contains the source and destination of all cargo and vehicles involved in the process of optimization. It also indicates the identification of drivers available for each vehicle and the date and time of pickup/delivery of each load.

When the optimization module receives the routing request message, it checks the format and content. If the validation is successful the application starts the processes of simplification, preoptimization, and tabu search with the received parameters. These processes are described in the section dedicated to the optimization of resources. Upon completion of these processes, the server generates a message containing the calculated route for each load, assigned vehicle and driver, as well as identifiers of the main crossing points (and the associated time) that are crossed by the vehicle between the source and destination. Message with the route and the transmission completion message are sent to the kernel, which terminates the communication after reporting the correct reception.

3.2.2 Communication between the kernel and the mobile terminal

This module implements the communications with the Mobile Intranet or GPRS network, consisting of all commercial vehicles equipped with a GPRS device.

The network of mobile terminals (also called Mobile Intranet or Fleet Area Network) communicates with the company's kernel through two channels: the first, through the Internet, and the second through a multi-channel GPRS modem that is used to send data in

real time to mobile terminals. The choice between the two channels or a hybrid solution depends largely on the comparison of connection and installation costs.
The first channel is used for the regular transmission of data from the vehicle to the kernel. To achieve it, a plug-in is installed on the PDA, which allows establishing a fast and encrypted connection with the Web. When the connection with the PDA is established, the XML message is sent periodically throughout the mobile service provider to the system's Web site. Then the kernel sends this information to the corresponding module which will process the service request.
The second channel communicates with the GPRS modem system, through the Mobile network operator. Its main task is to send messages to modify or update data, and above all, emergency messages, although it can also be used for the transmission of positional information when the other channel is not available.

3.2.3 Communication between the vehicle on-board terminal and the intelligent security control

Currently the intelligent security control is implemented on a laptop that communicates with the mobile terminal through a client socket that connects to the server socket of the mobile terminal when it is ready to send information.
The application communicates with the server core using WSDL (Web Services Description Language) language that describes the public interface to the Web service, i.e., the methods that the device application can use.
It is based on XML and describes the form of communication, i.e., the protocol requirements and the message formats required to interact with the services listed in its catalogue. The client program (the application) connects to this Web service and reads the WSDL to know what functions are available on the server and the class of data that can exchange between them.
WSDL has been used in combination with SOAP 1.1, XML Schema, and HTTP GET/POST. The client program connects to the Web service using the SOAP (Simple Object Access Protocol) protocol to make the call to each of the listed functions. Special data types are included in the XML Schema format.
At the present time, a client-server connection based on bluetooth is being developed. This connection will replace the current connection to prevent the driver from having to manually connect on-board mobile devices. This bluetooth connection also allows a much easier migration when the intelligent control of security is integrated into the on-board computer, or into a specific device.

3.3 ERP module

The main features of the ERP module are:
It allows the user to enter both deliveries (road map) and destinations of a vehicle and to specify the information associated with each or them (customer, sales order, invoice, freight or product description, name of driver, etc...). These schedules (sequences of deliveries) can be displayed to control their specific performance.
Client management allows the user both to geo-reference and to enter information from their clients, taking into account the basis of a Geographical Information System (GIS). Geo-references can be structured according to different categories and levels. It has ease of search and updates the entered information (addresses, categories, etc.).

The stored and processed information (by Individual Taxpayer Identification Number (ITIN), truck code, etc.) related to events associated with the vehicles and their planning (e.g. location, mean speed, vehicles stops, deviations from the planned route, visiting customers checks, etc.) is used to measure performance and draw conclusions that will improve the overall management.

Planning and vehicle fleet operations can be managed and displayed by an on-line dynamic control: location of customers and vehicles, pending and completed delivery routes, checkpoints passing, speed and dwell time of vehicles, state of driver fatigue, and other data that the client can set as part of their own particular configuration.

3.4 Access to the system

The system can define different user roles (administrator, operator, customer, driver), and control access levels to the various tools and information services.

The customer (freight forwarder) accesses the system through the web site of the company. Once the access has been validated, the freight forwarder has a menu with the basic freight hiring service, in addition to those services that he wants to hire for a specific load. For example, he can request online information about the status of his freight. The system generates a web page with the information about the freight status, the estimated time of arrival, as well as a map with the planned route, current route, and the one followed by the vehicle.

The fleet operator accesses the system through the web site of the company. Only register operators can access the system, and different operators can have different privileges. According to those privileges, an operator can access the data from all currently active services, accept orders from new costumers, request locations of their trucks, etc.

The drivers have a mobile terminal that establishes a bi-directional communication via GPRS to the central system. On the one hand, the system allows updating or modifying the geographic data through the collection of traffic data sent by the vehicles, as well as the RDS-TMC information provided by the Traffic Department. Those data together with the indices of driver fatigue are sent periodically to the central system. On the other hand, the central system can send alerts, incidents, new assignments, or routes to the mobile terminal.

3.5 System interface

In the latest version of the developed system, all the users access the system through the website of the company. Once the company has hired the provider services, the system generates a Web portal based on the CMS Joomla! This configuration allows the company to customize its website, offering information that it considers useful, and in turn, managing registration and access to the portal.

This portal is also used as an access point for both fleet operators and potential customers. This configuration has another advantage: the carrier does not need to install anything on their computers. Even the resources of the customized web portal can be hosted on a server of the provider.

The interface is permanently connected with the kernel located in the service provider through the Internet. This will establish a constant flow of information, since the nucleus is responsible for transmitting requests, from both operators and customers, to the central system, and in the opposite direction, receives and presents the responses generated by the kernel to such requests.

3.5.1 Agents

Potential agents that can use this interface can be classified as: the system administrator, webmaster of the company, traffic agent of the company (or fleet operator), and freight forwarders. Here are the characteristics associated with each one.

The system administrator is responsible for managing the hiring and maintenance of administrative data and resources hired by each company. For these tasks the administrator uses the provider's own web portal, which can connect directly to the kernel just as the rest of the portal. This agent is the only one who has access to the administrative database of the companies.

The webmaster manages the portal web site of the company. He is responsible for maintaining the data of registered users and for updating commercial content that the company can offer its clients through the portal. Another function of the webmaster is to register the traffic agents of the company, and validate the users who register. He doesn't have access to the central system. This role can be performed by an employee of the company. Otherwise this provision can be included in the portfolio of services under contract with the supplier if the company prefers it. Anyway, it is not necessary for the company to install anything on its computers.

The traffic agents are directly recorded by the web administrator of the company. This register allows them to access all the services hired by the company. Obviously, the menu at the interface of the traffic agent differs substantially from the Web administrator menu: one manages the portal Joomla! and the other manages the fleet. Once the company has registered an agent, he can access the system and have all the information from the business and take action on it. He can access the data of all the currently active services, accept the hiring of new clients, request location and status of their trucks, send instructions to the driver to change the service, and so on.

Freight forwarders can register through the website of the company, or can contact the company to apply for registration as a customer. In the latter case, the webmaster makes the registration on the website and the traffic agent recorded them in the company database. Obviously, each company has its own database of customers. The administrative databases of the companies are independent, which ensures the confidentiality of the data.

Once registered, the client can hire the services of the company through the Internet. The client can access the company Web site, log into his account and through a form, and enter the necessary data about the load to request a transport service. If the load is accepted, data is recorded in the corresponding database.

Once the loads are hired, the customer can request information about the state of their loads. The system responds, through the Web connection, by sending the next information: a map of the location of the truck with its load and an estimated time of arrival.

3.5.2 Interface description

The system's web interface has been developed entirely in PHP and HTML, and is ready to function as an integrated component within the content management system "Joomla". However, it has chosen to give greater flexibility to the interface, so the business logic of the interface can run on any web server with PHP support.

As the kernel, the interface has been developed with an object-oriented architecture, so you can make changes or include new features without changing any code. Thus, there are objects only responsible for generating requests and processing responses, others that are in

charge of interpreting the parameters of an answer, and another, more complex, which is responsible for displaying all the necessary information in HTML format.
The design has been chosen to look simple but effective, based on intensive use of tables (playing with the sizes and edges), and the help of some JavaScript code to give certain dynamism to the content. Due to the intensive use of tables, we have developed from scratch a module (object) that is responsible for creating virtual tables in the system memory, with a large variety of options and features, so that the code for the object generator of HTML is more understandable and scalable.
Subsequently, a function of this module encrypts all the information in HTML code which will be integrated into the web interface. This module also allows nesting tables and inserting a table within another without being in HTML. Thus, modifying a particular field is much easier and more intuitive than if you modify the HTML directly.
The external view of the interface (what the user sees) is similar to most of the fleet management systems available, so this chapter does not include screenshots.

4. Intelligent security control module: Driver alertness monitoring

This module is aimed to monitor the driver alertness level of each fleet vehicle with a view to both sending the fatigue level to the central control and preventing the driver when the alertness level is too reduced so that driving can be considered dangerous.
We developed a robust real-time module to detect the eye state of the driver with a consumer-grade computer, an inexpensive Universal Serial Bus camera, and passive illumination. While previous approaches have proposed many different methods and features to fulfill eye and eyelid closure detection, our module focuses on the calculation of multiple features of different nature in the eye region to distinguish three eye states, open, nearly close, and close, in the eye regions. The use of three eye states is necessary to calculate PERCLOS as in (Dinges & Grace, 1998) accurately. We accomplished the discrimination among these states of a non-rigid object with high variability such as the eye with a multiple-feature scheme. The module is based on Multi-Layer Perceptron classifiers, which are used to measure PERCLOS, runs at 40 fps and achieved an overall accuracy of 97% with videos with different users, environments, and illumination.

4.1 Development of the driver alertness monitoring module

We aimed to monitor the driver alertness with the PERCLOS measure. PERCLOS is meaningful in highway driving when the driver face is mainly in frontal position with respect to a camera placed on the car dashboard and there is low head motion. For the face detection, we use the frontal face detector based on the AdaBoost algorithm and with Haar-like features available in the framework of the Intel Open Source Computer Vision Library (OpenCV, 2009). This detector has a great performance with frontal and near frontal faces.
Rectangular pair of eyes region location is determined using face and head anthropometry measures from the output given by the face detection. As the face detection is mainly based on the eyes, they can be accurately located. Assuming that eyelid closure is simultaneous in both eyes, the eye state has to be determined in each frame of a video sequence to compute the PERCLOS measure. Therefore, we had to discriminate among open and close eye state and also the nearly close state, which is characteristic of a high level of drowsiness of the driver. In this state, the proportion of visible iris is below 20% of its total height. Due to the

large variability of the eye appearance and dynamics, our approach was not to extract an individual feature or a small set of features as individual features are influenced by the fact that a feature with a very wide class of invariance loses the power to discriminate among other differences. On the contrary, we have extracted many features of different nature to achieve robustness to illumination, presence or absence of structural components such as glasses, and facial expression.

The extracted features were: 1-D grayscale histograms, 2-D color histograms, horizontal and vertical projections, and the entire rectangular eye images. Regarding grayscale images, an eye can be characterized by the intensity of two regions, one corresponds to the iris and the other to the sclera so 1-D grayscale histograms are extracted. 2-D histograms of the chromaticity components (HS) of the HSV color space were extracted to discriminate between an open and a close eye with some illumination invariance as skin colors have a certain invariance regarding chromaticity components (Sigal et al., 2004). Horizontal and vertical projection functions of a grayscale image I(x,y) in intervals [y1,y2] and [x1,x2] are expressed in Eq. (1) and (2), respectively. Zhou & Geng (Zhou & Geng, 2004) proposed the horizontal and vertical generalized projection functions. We calculated the four mentioned projection functions.

$$IPF_h(y) = \sum_{x=x_1}^{x_2} I(x,y) \tag{1}$$

$$IPF_v(x) = \sum_{y=y_1}^{y_2} I(x,y) \tag{2}$$

Once we have the information of an eye image captured in a feature set, there are two possibilities from endowing them with meaning: make a unilateral interpretation from the feature set or compare the feature set with some elements on the basis of a similarity function. As proposed in (Bradski, 1998), the complexity of the eye detection problem makes it necessary the use of a similarity function to achieve the desired accuracy in analyzing eyelid closure with non-strictly frontal face, in motion, and with illumination changes. The elements to compare the feature set of an eye image are the templates of open eye, nearly close eye and close eye.

4.2 Similarity functions

With the information obtained from the eye regions, a big number of similarity measures between the eye regions and the three templates were extracted. Our approach is to compute a set of different similarity measures so that the best measures be selected later. First, we obtained four histogram-based similarity measures: correlation, χ^2, intersection and Bhattacharyya distance.

The following four template matching methods to measure the similarity are applied to the entire eye regions and the projection functions of the regions: correlation matching, correlation coefficient matching, and the normalized versions of the two matching methods. Unlike the histogram-based measures, these methods compute the spatial distribution of pixels. The correlation matching method multiplicatively matches the template T against the

image I to obtain the matching result. The correlation coefficient matching method matches a template relative to its mean against the image relative to its mean. The normalized version of the correlation and the correlation coefficient matching methods are obtained dividing them by the same normalization coefficient. They are useful as they can help reduce the effects of lighting differences between the template and the image.
Each similarity measure between an eye region and a template represents an eye region feature.

4.3 Feature selection and classifiers

After computing a big number of features, dimensionality reduction is necessary. Many features are not suitable to classify eye images because noisy and redundant inputs can have a bad influence on the classification performance (Devijver & Kittler, 1982). Dimensionality reduction approaches can be classified in feature extraction and feature selection algorithms. Unlike feature extraction, feature selection allows to make inferences on how input variables affect the model results. We have compared two selection algorithms: the $J_5(\xi)$ criterion (Devijver & Kittler, 1982) and the SFS (sequential forward selection) (Jain & Duin, 2000).
We used classifiers widely used in data classification and pattern recognition: MLP. MLP is a type of Artificial Neural Network (ANN). ANNs, which are inspired by biological neural networks, are composed of neural-like units connected together through input and output paths which have adjustable weights (Haykin, 2008). The MLP is an ANN which has been very successful in a variety of applications, producing results that are at least competitive and often exceed other existing approaches.

4.4 Driver alertness monitoring module performance

We selected two MLP-based classifiers from many experiments for the eye state monitoring module: one for the pair of eyes and the other for the individual eyes. We used 12 videos to test the module: six videos from the ZJU Eyeblink database and six videos from our private database, all with 320×240 pixels, with a face in frontal or near frontal position and a wide range of conditions: translation and scale changes, some people wear glasses and different illumination conditions. Two videos from our private database were recorded with a camera on the dashboard of a car while a user is driving in real conditions. The average overall accuracy is 97% for the 12 videos. Video frames were different from the training images, although some people in the videos from our private database, apart from the videos recorded inside a car, were included in the training set. The system performance is also high in videos in which the people do not appear in the training set. In-car videos have a lower performance due to the poor illumination conditions, unlike the rest of the videos, which are indoor videos with better illumination conditions. In spite of that, overall accuracy percentages of the in-car videos allow a reliable PERCLOS computation to monitor the driver alertness because most of the errors are produced when the eyes are in a transition state, in which the eyes are open but not completely.
Processing times were taken with an Intel Core 2 Duo processor at 3 GHz and 4 GB RAM, with the software implemented in Visual C++. Each frame takes 25ms on average, out of which 14 ms are taken for the face detection and the remaining time is mainly devoted to the eye region feature calculation and eye state classification. AdaBoost-based face detection is applied to image subwindows varying in position and size from a minimum size. We

adapted this minimum size depending on the face size in the previous frame, with width and height two-thirds the face width and height in the previous frame. These processing times give rise to a frame rate of 40 frames per second (fps) so that real-time driver vigilance monitoring is achieved. Frame rate was limited by the 30 fps given by the two cameras used in the experiments: Logitech Quick Cam Zoom and Philips SPC315NC.

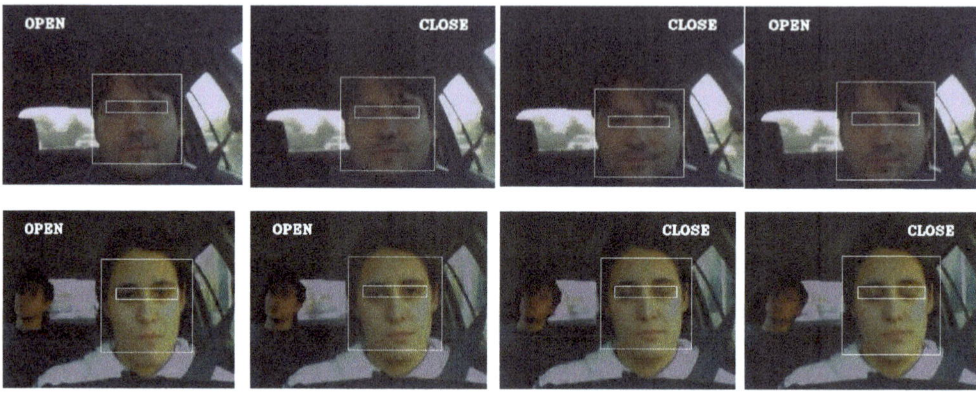

Fig. 3. Output of the driver alertness monitoring system with two video sequences.

Figure 3 shows frames of two video sequences in real driving conditions. Face detection and eye region location are drawn in the frames together with the output of our eye state monitoring module. Our module is used to monitor the driver alertness through not only the PERCLOS measure but also the non-detection of the face or the close eye state detection in a series of consecutive frames. There is one threshold for each of the three measures: $TH_{PERCLOS}$ is the maximum value of PERCLOS allowed for secure driving; TH_{face} is the maximum time without detecting the face in consecutive frames allowed for secure driving; TH_{eyes} is the maximum time detecting close eye state in consecutive frames for secure driving. These thresholds can be adapted to each driver and the level of alertness required for driving (high, medium, or low). At the time when any threshold value is overcome, alarm state will be triggered. The driver will have to react and return to a secure driving state to abandon the alarm state.

5. Resources optimization

In a transport company the drivers' management strategy is a fundamental aspect to achieve an optimal allocation of resources. The factors that influence this strategy are diverse, and the most relevant are: the limitation of the maximum driving time imposed by law for drivers, the possibility of incorporating two or more drivers within the same vehicle, the availability of drivers which restricts the number of vehicles that can be on the road, and the own cost of hiring or maintenance of drivers.

The information on loads and trucks was used in an early version of this module. In the current version, the information about the company's drivers is incorporated. This new addition resulted in a very significant decrease in the performance of the previous optimization module. This module was implemented using evolutionary programming, and

these changes made slower genetic operators be slower. On the other hand, the increase in the number of parameters leads to a more complex evaluation function.
This fact led to the need to design and implement a new algorithm that could adapt to new conditions, while maintaining a high level of performance.
The main factor that determines the choice of the most appropriate methodology to solve the problem is the very large number of possible cases that need to be evaluated during the process of resource allocation when the management of drivers on the transport routes is included. This factor has a direct impact on the processing time of the algorithm, which will increase with the size of the company's fleet.
In order to ensure a rapid implementation, the resolution is divided into two phases: problem simplification and resource allocation optimization.

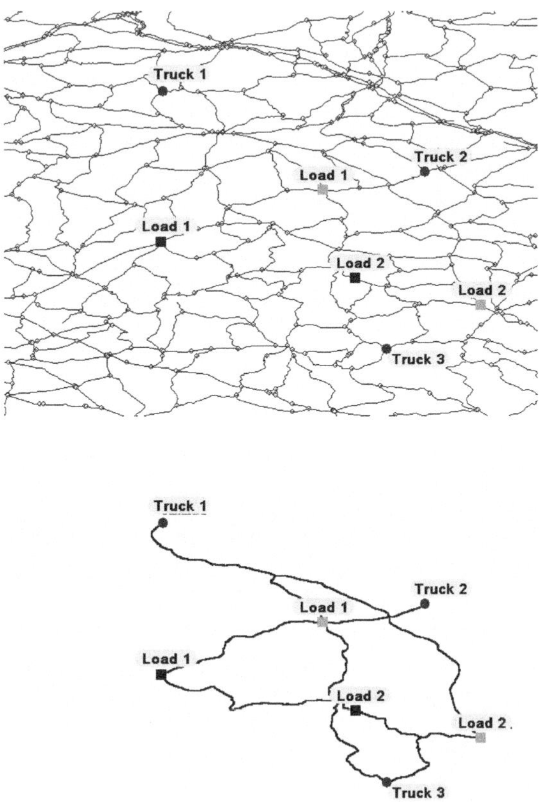

Fig. 4. Result of the simplification process.

Simplification process consists in considering only the routes involving a shorter route between trucks and cargo, leaving out the rest of sections of the road network (see figure 4). The algorithm takes into account the updated information about the road status, the traffic density, and the distance between the origin point and target point. These routes are classified into three categories: routes that link the origin with the destination of each load, routes that lead the trucks to each load, and routes from the target of each load to the origin of other loads or to the current location of the transport vehicles. Achieving the best route in each of these categories refers to a problem of solving the Shortest Path Problem (SPP) (Helgason et al., 1993). The method chosen for implementation is Dijkstra's algorithm with double buckets (Zhan & Noon, 1996) because its proven speed and its perfect adaptation to the class of graph generated by real networks of roads, characterized by their low level of connectivity.

This solution must also fulfill that the selected vehicle is suitable for transporting the assigned load and that the delivery time does not exceed the specified delivery time. There are several algorithms that allow to find this solution, and which provide an acceptable computing time. For the proposed system, we have chosen the TabuSearch algorithm (Glover & Laguna, 1997). TabuSearch is a mathematical optimization method, which belongs to the class of trajectory based techniques. TabuSearch enhances the performance of a local search method by using memory structures that describe the visited solutions: once a potential solution has been determined, it is marked as "taboo" so that the algorithm does not visit that possibility repeatedly.

Its most important feature is the fast convergence when the solution is approaching to the optimal solution. Its computational cost is low when assessing very few solutions in comparison with other methods; generally it does not exceed 5 or 10%. The fact that the found solution may not be the best, but a local maximum, constitutes one of its disadvantages. However, the application of different techniques can reduce this possibility to null or very small margins.

Depth-First Search & Breadth-First Search algorithms (Cormen et al., 2009) were also tested. These methods were dismissed due to its low performance in problems with a very large number of possible solutions. The response time of genetic algorithms and evolutionary programming does not grow exponentially when the data of the problem increase as it happens with searches in breadth and depth. However, its performance drops considerably when the used genetic operators execution slows down and when the degree of the evaluation function increases.

The algorithms we have developed for the stages of simplification and optimization of resources are integrated into the system through the kernel. The kernel is responsible for providing both the communication with the database that contains information about the company's resources and communication with other components of the system.

The performance tests were performed on a PC equipped with processor Intel Pentium 4 2.5 GHz with a Windows XP operating system and 256 MB of RAM. The Geographic information used in the tests includes the main towns and roads of the Spanish road network, which is a graph with 4574 arcs and 3636 nodes.

The evaluation of the simplification process implemented using Dijkstra's algorithm with double buckets was carried out considering routes from 100 origin nodes to a set of 300 target nodes, both chosen randomly. This result applied to a transportation company with a fleet of 50 trucks and 200 hired loads, allows calculating optimal routes in 11.5 seconds. In the light of these results it is clear that Dijkstra's algorithm with double buckets is a valid option for this process.

To achieve an effective implementation of TabuSearch for the resources allocation to the previously computed optimal routes, the following aspects were considered during its development:

- Each solution consists of a set of routes. A vehicle, a driver, and a series of charges are allocated to each route. The variable to optimize is the total travel time for all the routes.
- Criteria for validation of routes are fixed. These criteria are based on the number of transported cargo, on the compliance with delivery times, and on the restrictions arising from the drivers and vehicles.
- Identification of neighboring solutions is determined by a double replacement of vehicles between routes. This procedure maintains a reduced range of tested solutions that ensures the achievement of the absolute maximum in the 100% of the tested cases.
- The starting point to begin the optimization process is to assign to each load the nearest vehicle. This is possible thanks to the ordered list of closest vehicles prepared by Dijkstra's algorithm in the simplification phase. This way the algorithm is considerably faster because it is highly likely that the initial solution is already very close to the optimum. In fact, some vehicles are already definitively assigned to a load in this point.

The tests performed in the same conditions and with the same cartographic information that the tests performed in the simplification process, produced times of execution of the algorithm below 0.5 seconds for the specific case of 15 vehicles, 15 drivers, and 10 loads.

6. On-board equipment

The on-board equipment consists of two devices: intelligent security control and fleet management module.

On the one hand, the security module performs a monitoring in real time of the driver fatigue status, critical to workers whose working life is spent on the road. It is currently implemented on a laptop.

On the other hand, drivers have a mobile terminal that graphically shows both the position of the vehicle and the calculated route. This terminal receives information about the state of the driver from the laptop and sends it to the kernel along with information about the position of the vehicle.

The following sections describe the main features of the mobile terminal application.

6.1 Mobile terminal features

Each mobile terminal includes a plug-in that enables bidirectional communication between the vehicle and the base via GPRS, and obtains the geographical position of the vehicle by reading GPS signals. It also has a graphical interface that allows the driver to see the route to follow and receive update messages or warnings about incidents on the road or load, or in situations where the vehicle deviates unusually from the established route.

Each driver has a unique identifier (own code of the mobile device). When a driver runs the application on his mobile device, the application connects to the kernel to authenticate him. If authentication is successful, it returns the assigned route o routes. For each destination, the driver is provided with the best route to follow, previously calculated in the corresponding module, as well as detailed data for each delivery/pickup point (cargo, address, city, telephone contact, etc.).

The application allows real-time tracking of each vehicle by sending the coordinates of the current position to the database system. When the vehicle arrives at a destination, the driver can record the date and time of arrival at the point of delivery, pick up the recipient's signature and record the cause in the event that delivery has not been possible. The driver can use the application to inform the system about incidents (accidents, breakdowns, traffic jams) produced along the route. Besides, the driver can access the company intranet for information or to report something, or he can communicate with the kernel to request the calculation or recalculation of a route.

The application uses Google Maps for showing the optimal routes. Because of its low usage of resources, Google Maps does not require much space in the device memory, and the driver can display the full path to follow. The following section describes this feature in more detail.

6.2 Driver Interface

The most important function of the driver interface is to show the physical maps for each of optimal route, and inform the user about the points he has to reach along his route. To achieve this, it was initially decided to use the Google Maps API, which allows developers to integrate maps free in any Web application with JavaScript.

Fig. 5. Result of integration of the Google Static Maps API on the PDA.

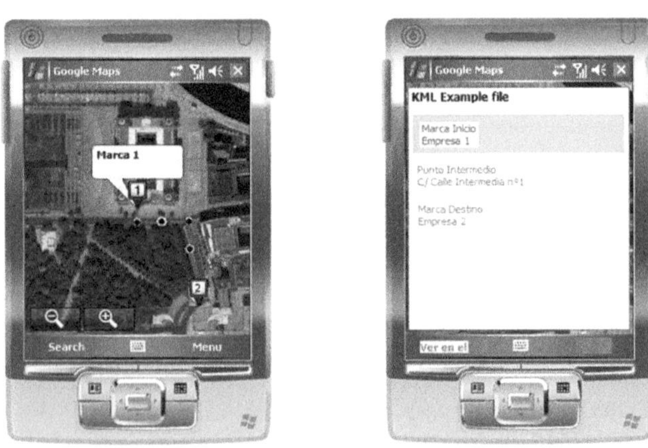

Fig. 6. Integration of Google Maps with route data. By clicking on each of the marks, the application shows more detailed information.

Figure 5 shows the result of integration of the Google Static Maps API on the PDA. It shows this type of map for each pair of destinations associated with each driver. In this map the driver can find origin marks, target marks, and the optimal route generated by the optimization module. All these data are sent in the URL of the HTTP request.
Later this solution was found to be inadequate, since, being a static image, the user cannot zoom in to view the streets to cross, for example. Moreover, if it is not possible to write any "window of information" or any mark, the driver cannot know the geographic data of the intermediate points of the route.
For this reason we chose the following solution, which is more appropriate. Google Maps provides an executable file (compatible with Windows Mobile 5.0) that can be downloaded and installed on any mobile device. Having completed this step, we designed a C# application that runs Google Maps in the mobile terminal and loads the file that contains information about marks and intermediate points in the optimal routes. The driver can perform zooms and any movement on the loaded maps and can access all the information by clicking on each map element (see figure 6).
The files that are loaded are described in KML (Keyhole Markup Language) language. KML is XML-based and is used to represent geographic data in 3D. It was developed especially for Google Earth, though their version 2.0 is also supported by Google Maps. Each file specifies a title, a description of the place to represent, its coordinates (latitude, longitude, and altitude if desired) and some additional information.

7. Conclusions

The proposed system is not only directly relevant in the automation of the processes involved in managing commercial fleets, but it also presents a business opportunity to large communications companies. This whole project results in a strategic weapon for companies, which provides management services and an essential tool for transport companies, strengthens their market position, and makes them more competitive and efficient in their processes.
The main innovations of the proposed system are focused on the design of the platform of logistics services, the optimization of the process of allocation of resources and routes, and the monitoring of the physical condition of the driver in real time.
We have successfully implemented the core modules of the multi-manager platform. We have also established a robust and scalable communications protocol based on XML message exchange, and have designed and implemented a database system, using the MySQL engine.
The developed kernel is fully modular, well-structured, and very stable against bugs, doing a high control of exceptions to prevent it from collapsing in anomalous situations. Moreover, due to the use of multiple threads, we have achieved a more flexible kernel that makes a better use of system resources, ensures greater integrity of concurrent access to the database, and minimizes the accesses to expedite the top requests.
The Web interface has been designed through sober but functional design patterns, and shows a fairly accessible and professional look. In addition, we have implemented all sections of the interface with the same physical appearance to make it easier to use, and to allow their integration into the other components of Joomla! to be as transparent as possible.
With regard to the communication protocol, we have chosen an implementation based on XML messages. This option will have the greatest future growth and is the most

understandable by humans. This way, it is easier to understand the operation of the communications protocol and add or modify some aspect to this protocol. It is not necessary to generate new code in the system modules to interpret and generate these messages, since there are XML parsers for the majority of the programming languages.

The module of optimization of resources allows optimizing the allocation of loads, vehicles, and drivers, and the dynamic and adaptive calculation of distribution routes taking into account the special incidents along the route that may influence the traffic.

To achieve greater precision in the route optimization we have to feed the module, in addition to the data supplied by the traffic authorities, with data submitted by the vehicles of the companies that are using this platform. Unfortunately, we cannot have this valuable information until the system is in production.

On the other hand, the incorporation of faster processors in the future will allow processing an increasing volume of traffic and cartographic data and calculating and updating the path in real time with more precision.

The system includes a module to monitor the driver alertness level of each fleet vehicle. Although the module runs in a standard laptop computer and a Universal Serial Bus camera, a PDA cannot be used to run it in real time. Experimental results showed the robustness of the module to changes of driver appearance, illumination, and environment. The module can be used to compute the PERCLOS measure, which showed the clearest relation with driver performance compared to a number of other potential drowsiness detection devices in literature. Besides the PERCLOS measure, the module obtains the consecutive time of non-face and close eye state detection for robust monitoring driver vigilance. The massive use of such in-car security modules would have a big impact on the number of traffic accidents, decreasing their huge economical and personal cost.

Finally, the integration of all these components has been successful and has generated a stable and robust final system as laboratory simulations and field tests have shown.

8. Acknowledgment

This work has been partially supported by the Goverment of Castilla y León (Spain), under the project VA171A11-2.

9. References

[AWAKE Consortium, 2004] Awake Consortium (IST 2000-28062), (2004). *AWAKE - System for effective assessment of driver vigilance and warning according to traffic risk estimation,* 2001-2004. 15.06.2011, Available in http://www.awake-eu.org.

[Bradski, 1998]Bradski, G.R. (1998). *Real time face and object tracking as a component of a perceptual user interface,* Proc. of the IEEE Workshop on Appl. of Computer Vision, (1998), pp. 214-219.

Cormen et al., 2009 Cormen T.H., Leiserson, C.E, Rivest R.L. and Stein, C. (2009). *Introduction to Algorithms,* Third Edition. MIT Press, Inc., Cambridge, MA.

CSCMP, 2009 CSCMP - Council of Supply Chain Management Professionals. *State of Logistics Report. Fast Facts on the Global Supply Chain,* 21.06.2011, Available in http://cscmp.org/press/fastfacts.asp.

[D'Orazio et al., 2007] D'Orazio, T., Leo, M., Guaragnella, C. and Distante, A. (2007). *A visual approach for driver inattention detection*, Pattern Recognition, vol. 40, no. 8, (2007), pp. 2341-2355.

[Devijver & Kittler, 1982] Devijver, P.A. and Kittler, J. (1982). *Pattern Recognition: A Statistical Approach*, Prentice Hall.

[Dinges & Grace, 1998] Dinges, D.F. and Grace, R. (1998). *PERCLOS: A valid psychophysiological measure of alertness as assessed by psychomotor vigilance*, U.S. Department of Transportation, Federal Highway Administration, Office of Motor Carriers, Publication number FHWA-MCRT-98-006.

Giloth & van dort, 2011 Giloth, J. and van Dort. (2011). *Supply chain business process outsourcing*. EVO Logistics Yearbook, 2011 edition.

[Glover & Laguna, 1997] Glover F., M. Laguna, *Tabu Search*, Kluwer, Boston, 1997.

[Haykin, 2008] Haykin, S. (2008). Neural networks and learning machines, Prentice Hall, 3rd ed., 2008.

Helgason et al., 1993 Helgason, R.V., Kennington, J.L. and Stewart, B.D. (1993). *The one-to-one shortest-path problem: An empirical analysis with the two-tree Dijkstra algorithm*. Computational Optimization and Applications, 1, (1993), 47-75.

[Jain & Duin, 2000] Jain, A.K., Duin, R.P.W. and Mao, J. (2000). *Statistical pattern recognition: a review*, IEEE Trans. Pattern Analysis Machine Intelligent., vol. 22, no. 1, (2000), pp. 4-37.

[Ji et al., 2004] Ji, Q., Zhu, Z. and Lan, P. (2004). *Real-time nonintrusive monitoring and prediction of driver fatigue*, IEEE Trans. Vehicular Technology, vol. 53, no. 4, (2004), pp. 1052-1068.

[Kokubugata & Kawashima, 2008] Kokubugata, H. and Kawashima, H. (2008). *Application of Simulated Annealing to Routing Problems in City Logistics. In Simulated Annealing*, Book edited by: Cher Ming Tan, ISBN 978-953-7619-07-7, (February 2008), pp. 420, , I-Tech Education and Publishing, Vienna, Austria.

[Królak & Strumillo, 2008] Królak, A. and Strumillo, P. (2008). *Vision-based eye blink monitoring system for human-computer interfacing"*, Proc. of the Conf. on Human Systems Interactions, (2008), pp. 994-998.

Langley & Capgemini, 2010 Langley Jr., C.J. & Capgemini (2010). *Third-party logistics. The State of Logistics Outsourcing*. 15.08.2011, Available from http://www.3plstudy.com/downloads/download-the-2010-3pl-study/

[OpenCV, 2009] Open Computer Vision Library.(2009), 24.07.2011, Available in http://sourceforge.net/projects/opencvlibrary.

[Sigal et al., 2004] Sigal, L., Sclaroff, S. and Athitsos, V. (2004). *Skin color-based video segmentation under time-varying illumination*, IEEE Trans. Pattern Analysis Machine Intell., vol. 26, no. 7, (2008).

[Siwek, 1998] Siwek, S. J. & Associates. (1998). *Transportation Planning and ITS: Putting the Pieces Together*. Prepared for FHWA, U.S. DOT, Washington, D.C.: 1998.

[Smith et al., 2003] Smith, P., Shah, M. and da Vitoria Lobo, N. (2003). *Determining driver visual attention with one camera*, IEEE Transactions on Intelligent Transport System, vol. 4, no. 4, (2003), pp. 205-218.

[Toth & Vigo, 2002] Toth, P. & Vigo, D. (Eds.) (2002). *The Vehicle Routing Problem*, SIAM, Philadelphia.

[Wang et al., 2006] Wang, Q., Yang, J., Ren, M. and Zheng, Y. (2006). *Driver fatigue detection: a survey*, Proceedings of the World Congress on Intelligent Control and Automation, (2006), pp. 8587-8591.

Zhan & Noon, 1996 Zhan, F. B., & Noon, C.E. (1996). *Shortest Path Algorithms: An Evaluation Using Real Road Networks*, Transportation Science.

[Zhou & Geng, 2004] Zhou, Z.H. and Geng, X. (2004). *Projection functions for eye detection*, Pattern Recognit., vol. 37, no. 5, (2004), pp. 1049-1056.

8

Implementation of Control Design Methods into Matlab Environment

Radek Matušů and Roman Prokop
Department of Automation and Control Engineering
Faculty of Applied Informatics, Tomas Bata University in Zlín
Czech Republic

1. Introduction

Computer-aided tools for analysis and synthesis of control systems are widely employed by many users from a range of researchers, control engineers or students. The reason is obvious. Such toolboxes represent comfortable and effective way of dealing with an array of complex control problems, sometimes even without deeper knowledge of the specific method. For example, Control System Toolbox, Robust Control Toolbox or Polynomial Toolbox (PolyX, 2011) for Matlab belong among the most popular ones in the control field.

The main aim of this chapter is to present two simple and freely downloadable Matlab programs which allow user-friendly work for two selected specific control design issues by means of Graphical User Interface (GUI).

First of the packages (Matušů, 2010; Matušů & Prokop, 2011a) is focused on algebraic design of continuous-time controllers under assumption of interval plants. The program takes advantage of Matlab + Simulink + Polynomial Toolbox (PolyX, 2011) environment and it represents an easy but effective and user-friendly way to control synthesis, robust stability analysis and simulation.

The second of the presented programs (Matušů & Prokop, 2010, 2011b, 2011c) deals with control of time-delay systems using three various modifications of Smith predictor. The software implementation includes the modification for unstable and integrating processes, PI-PD modification for systems with long dead time, and modification applying control design by Coefficient Diagram Method (CDM).

The described software products, which can be used both for research and educational purposes, are freely available on the Internet (Matušů & Prokop, 2011a, 2011b). Their application potential is going to be illustrated on several control examples.

The chapter is organized as follows. The Section 2 focuses on algebraic design of controllers for interval plants. It is divided into three partial subsections dealing with brief outline of basic theoretical background, description of the developed program itself and demonstration of its capabilities by means of an illustrative example, respectively. Analogically, the Section 3 has the very same structure but it presents the control of time-delay systems using three modifications of Smith predictor. Finally, Section 4 offers some conclusion remarks.

The partial versions of this work have been already presented in (Matušů & Prokop, 2010, 2011c; Matušů 2010).

2. Algebraic design of controllers for interval systems

2.1 Theoretical background

Problems of analysis and synthesis of control systems under uncertainty have attracted the attention of researchers and engineers for decades. The origin of uncertainty in constructed mathematical model of an industrial process can be seen in the effort to consider the process as an linear time invariant system, in spite of the fact, that the real behaviour is oftentimes different, much more complicated, and moreover because, strictly speaking, the physical parameters are never exactly known, possibly they can vary according to operating conditions. A possible and very popular approach to uncertain description supposes the known structure but uncertain knowledge of actual physical parameters of the controlled system. Their possible values are usually bounded by intervals and many of such plants can be assumed as so-called interval systems. The simple example of the interval system is represented by transfer function (1).

The common problem is to design a cheap controller with simple structure and fixed parameters which guarantees stability and often also required control behaviour for all possible values of the uncertain parameters. A potential solution to this task relies on the application of continuous-time robust linear controllers designed via the general solutions of Diophantine equations in the ring of proper and (Hurwitz-)stable rational functions (R_{PS}). This technique is based on the ideas of Vidyasagar (1985) and Kučera (1993) and it is proposed and analyzed e.g. in (Prokop & Corriou, 1997). The algebraic method brings a single scalar parameter $m > 0$ which influences the dynamics and robustness of the closed control loop. The controllers are designed for nominal systems while the robust stability can be tested through some standard instruments, for example the value set concept in combination with the zero exclusion condition and the overbounding method together with the Kharitonov theorem (Barmish, 1994).

The well known closed-loop control configurations with one degree of freedom (1DOF) or two degrees of freedom (2DOF) are supposed. In both cases, the loop is assumed to contain an interval controlled plant and a fixed controller. The regulator is designed for a nominal plant and consequently it is applied to plant under uncertainty while the robust stability is tested using some standard tools.

A fractional approach to continuous-time control design (Vidyasagar, 1985; Kučera, 1993) is based on general solutions of Diophantine equations in R_{PS}. The set of stabilizing controllers is given by known Youla-Kučera parameterization and the choice of the appropriate controller according to user requirements (asymptotic tracking, disturbance rejection and attenuation) consists in utilization of divisibility conditions in the specified ring.

One of advantages of this algebraic synthesis lies in the existence of single scalar tuning parameter $m > 0$ which can serve for additional influencing the final closed-loop control behaviour. The very topical question is how to choose m to gain the appropriate controller. The papers (Matušů & Prokop, 2008, 2011d) outline a possible technique for selection of m based on user-defined nominal control behaviour in the form of the first under/overshoot size. An alternative method consists in minimization of sensitivity function (Vidyasagar, 1985; Kučera, 1993) using H_∞ norm. In this instance, such m which tunes the "most robust" controller towards changes in controlled system (or in closed loop transfer function) is found. The simplest nevertheless in practice often sufficient solution is to select this parameter more or less "randomly" or on the basis of "engineering feeling" and

subsequently test the regulation by means of simulation. Even an inexperienced user is usually able to find a suitable m after several steps.
Much more details about the synthesis method and specific controller design and tuning can be found e.g. in (Prokop & Corriou, 1997; (Matušů & Prokop, 2008, 2011d).
Once the nominally stabilizing controller is designed, one need to verify the stability of the closed control loop with this fixed controller and an interval plant. The robust stability can be investigated, for example, using the theory of polytopes (closed-loop characteristic polynomial has an affine linear uncertainty structure) or via the overbounding method in combination with the classical Kharitonov theorem. An excellent overview of this field provides e.g. books (Barmish, 1994; Bhattacharyya et al., 1995).

2.2 Description of the program

This section introduces a simple user-friendly program (Matušů, 2010; Matušů & Prokop, 2011a) for synthesis and simulation of control systems under assumption that controlled plants are affected by interval uncertainty. It incorporates selected controller design algorithms and tools for robust stability analysis as they have been adumbrated hereinbefore. The developed software tool takes advantage of functions and GUI of MATLAB (tested on various versions from 6.5.1 - R13SP1 to 7.9 - R2009b) and also benefits of simulation environment SIMULINK and support of the Polynomial Toolbox 2.5 (PolyX, 2011). The program can be downloaded from web page (Matušů & Prokop, 2011a). It must be decompressed and launched in the Matlab via "start.m" file. Then, it is very intuitive and easy to use. The main menu window of the product is shown in Fig. 1 which is accompanied by the following concise description of program possibilities according to numbered items:

1. The definition of nominal system (with fixed parameters) which is used for controller design.
2. The definition of perturbed system (with interval parameters) which is used for simulation of control and robust stability tests.
3. Size of perturbations (expressed in percentage).
4. The choice of strategy for controller tuning. The first eventuality allows to define an arbitrary value of tuning parameter $m > 0$ while the second one minimizes the sensitivity function and searches for the "most robust" regulator to given nominal plant.
5. The selection of one from two basic closed control loop configurations - 1DOF or 2DOF control system.
6. The option of desired properties of the controller - either asymptotic tracking of reference signal or simultaneous tracking and disturbance rejection.
7. Adjustments of basic simulation parameters such as simulation time, reference signal, load disturbance and controller saturation.
8. Possibility of harmonic disturbance setting (in the output of the controlled plant).
9. Possibility of ramp disturbance setting (in the output of the controlled plant).
10. The selection of simulation results which should be displayed. The important item is "Number of partial intervals for simulation" defining how many intervals is each uncertain parameter in controlled system divided into. In other words, this number increased by one expresses the quantity of "sampled" values in individual uncertain coefficients. The aim is to create some "representative set of systems" (RSS) used for simulation process. However, be careful, the higher numbers noticeably increase computational time.
11. The buttons for start of simulation and exit from the program.

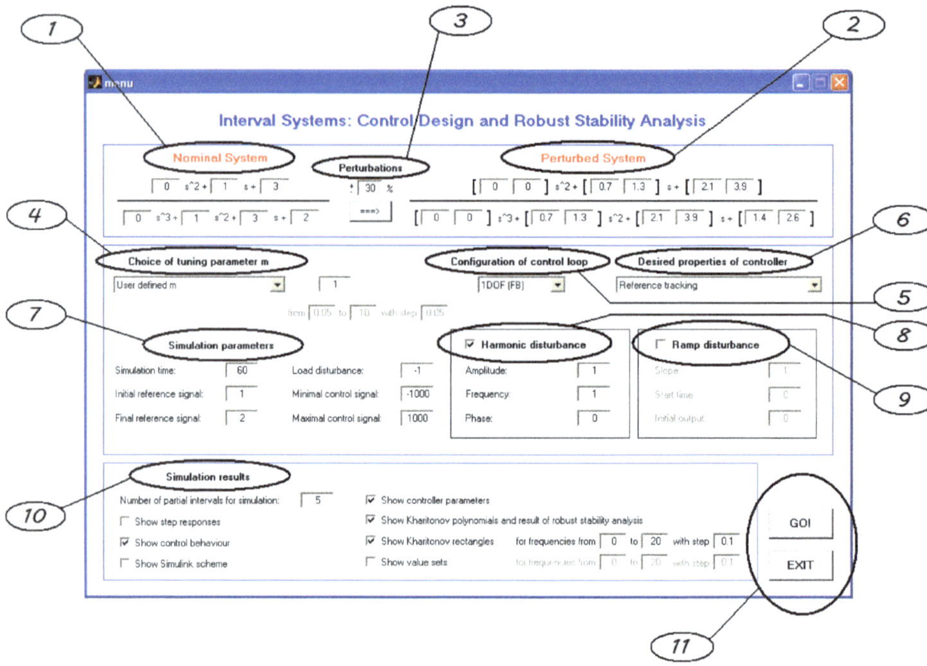

Fig. 1. The window of main menu for the first program

The capability of developed tool is demonstrated on the following example.

2.3 An example of application

The goal of this part is neither to prove the quality of utilized control design method nor to provide the comprehensive survey of all possibilities. It is just to show the basic idea of usage by means of a simple example.

The controlled plant is given as the second order interval system described by uncertain transfer function:

$$G(s,b_i,a_i)=\frac{b_1 s+b_0}{s^2+a_1 s+a_0}; \quad b_1,b_0,a_1,a_0 \in \langle 0.5;1.5\rangle \tag{1}$$

Potentially expressible time constants are assumed to be in seconds. This system with fixed parameters $b_1=b_0=a_1=a_0=1$ is supposed as the nominal one. The simulation conditions were used as follows: All uncertain parameters of the system (1) are divided into 6 partial intervals (sampled into 7 certain values), i.e. the curves corresponding to responses of $7^4=2401$ members of RSS from the family (1) appear in graphs; in control simulations, the reference signal with step change from 1 to 2 is assumed in one third of simulation time and the step load disturbance of the size -1 is injected to the input of the controlled plant during the last third of simulation.

The step responses $h(t)$ of 2402 members of the interval family (2401 systems from RSS + 1 nominal system) are shown in Fig. 2.

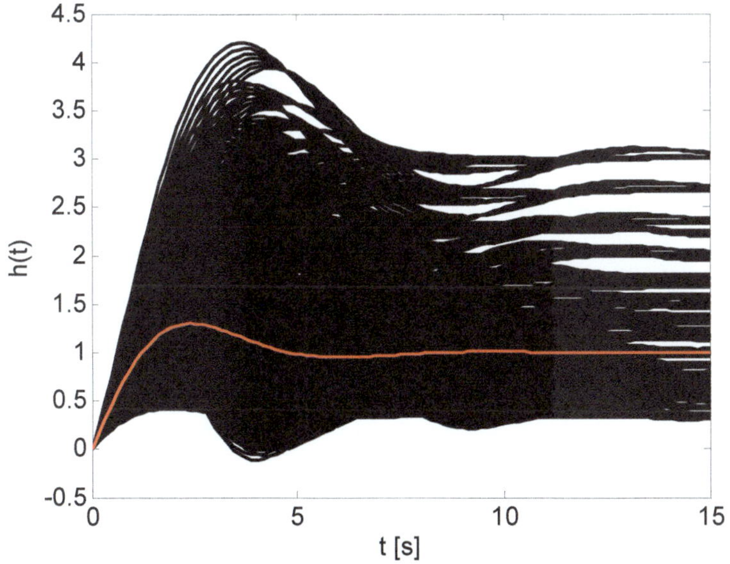

Fig. 2. Step responses of 2402 systems from interval family (1)

The aim is to design the controller which assures the asymptotic tracking of the reference signal and robust stability of the closed control loop, i.e. stability of control system for all members of interval family (1). The selection of $m=1$, 1DOF structure and reference tracking lead to PID controller:

$$C_b(s)=\frac{2s^2+2s+1}{s^2+s} \tag{2}$$

As can be effortlessly verified, the nominal system will be stabilized by this controller in closed loop. The question is if the control circuit is robustly stable. The closed-loop characteristic polynomial has affine linear uncertainty structure:

$$p_{CL}(s,b_i,a_i)=s^4+(a_1+b_1\tilde{q}_2+\tilde{p}_1)s^3+(a_0+a_1\tilde{p}_1+b_1\tilde{q}_1+b_0\tilde{q}_2)s^2+(a_0\tilde{p}_1+b_1\tilde{q}_0+b_0\tilde{q}_1)s+b_0\tilde{q}_0 \tag{3}$$

After substitution:

$$p_{CL}(s,b_i,a_i)=s^4+(a_1+2b_1+1)s^3+(a_0+a_1+2b_1+2b_0)s^2+(a_0+b_1+2b_0)s+b_0 \tag{4}$$

First, the robust stability of (4) is verified via the overbounding method. It means that more complicated uncertainty structure (affine linear in this case) can be "overbounded" by the interval one and this new family is sequentially tested. Unfortunately, this method brings certain degree of conservatism into the analysis due to ignoring the mutual dependencies among coefficients in the original family. As a result, robust stability is investigated only with sufficient (i.e. stronger) and not necessary and sufficient condition. Apart from other things, the overbounding interval polynomial and four related Kharitonov polynomials can be seen in Fig. 3 which represents final result of robust stability analysis from the program.

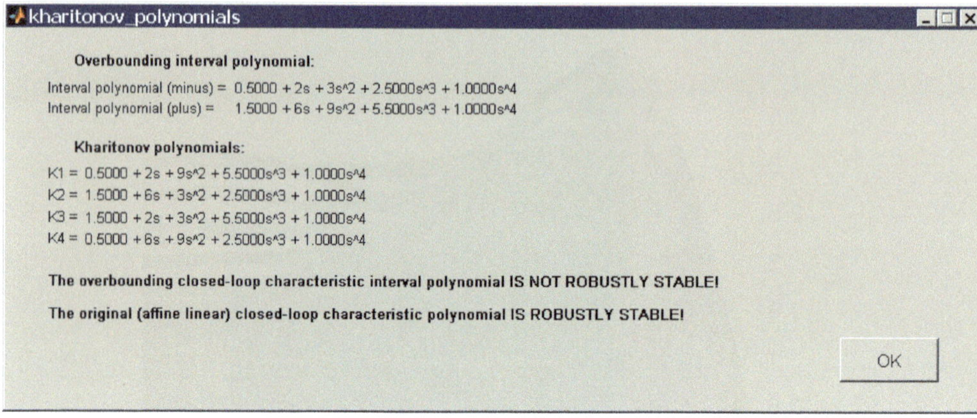

Fig. 3. Results of robust stability investigation from the program

Only two of four Kharitonov polynomials are stable which means that the overbounding polynomial is not robustly stable. Moreover, the Kharitonov rectangles from Fig. 4 (depicted actually only for illustration under frequencies from 0 to 3.5 with step 0.04) with detailed view in Fig. 5 also distinctly indicate robust instability of the overbounding polynomial because they cover the origin of the complex plane. This quite general principle is known as the zero exclusion condition (Barmish, 1994). However, generally it does not point to any conclusion about robust stability of original structure (4) because the mutual dependence among polynomial coefficients has been ignored.

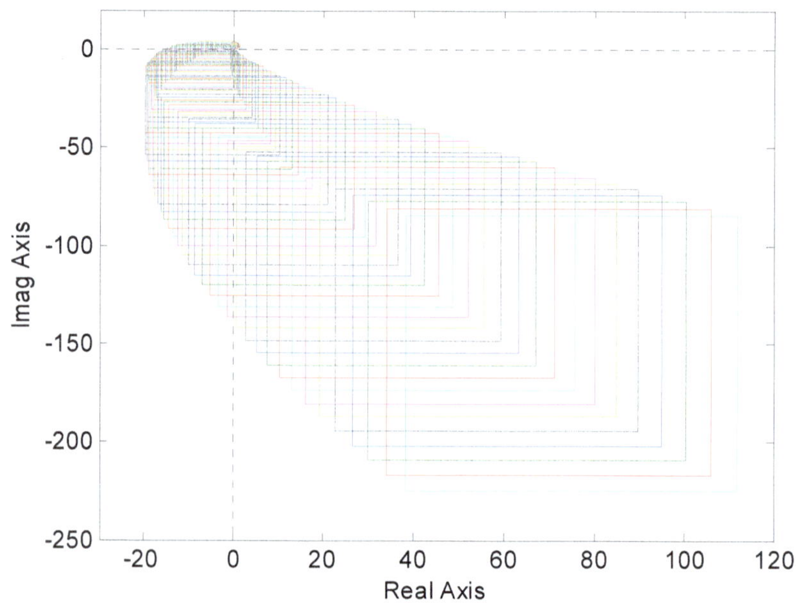

Fig. 4. The Kharitonov rectangles of overbounding interval polynomial - full view

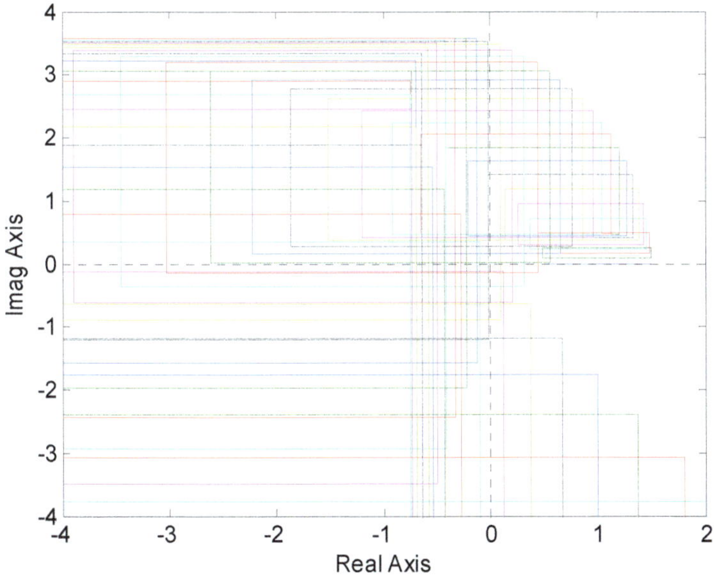

Fig. 5. The Kharitonov rectangles of overbounding interval polynomial - detailed view

Nevertheless, the "true" value sets of the original polytope of polynomials (4) in Fig. 6 with closed look to zero point in Fig. 7 reveal the closed-loop system is robustly stable in fact, because the complex plane origin is excluded from the value sets.

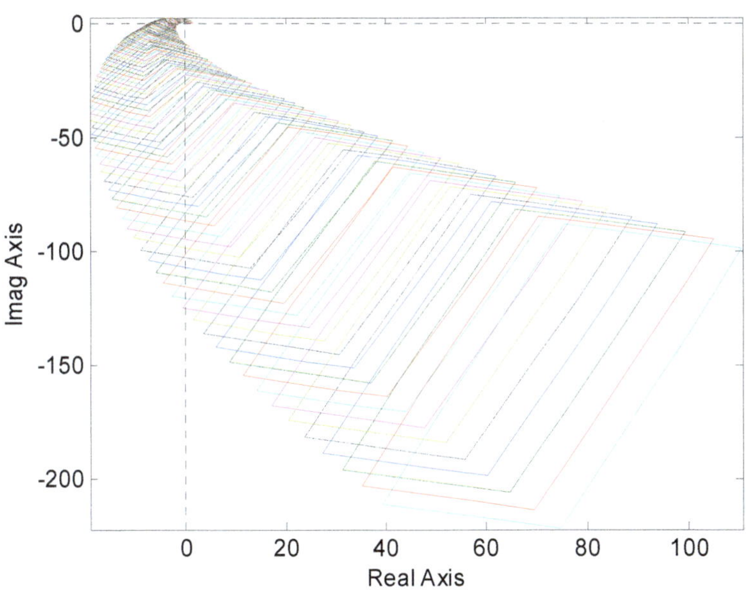

Fig. 6. The value sets of polytope of polynomials (4) – full view

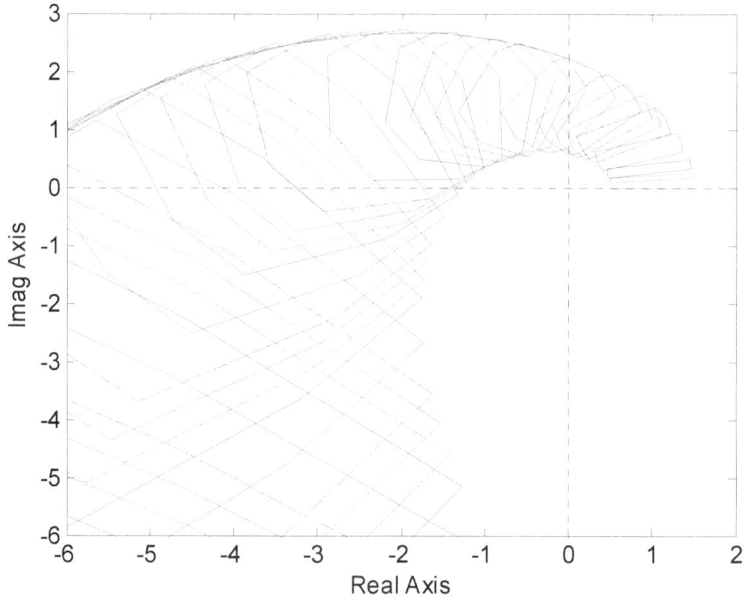

Fig. 7. The value sets of polytope of polynomials (4) - detailed view

Besides, this fact is confirmed by RSS control behaviour itself gained as an output from 1DOF control structure constructed in the Simulink environment. The simple simulation scheme is shown in Fig. 8 while the control responses $y(t)$ can be seen in Fig. 9.

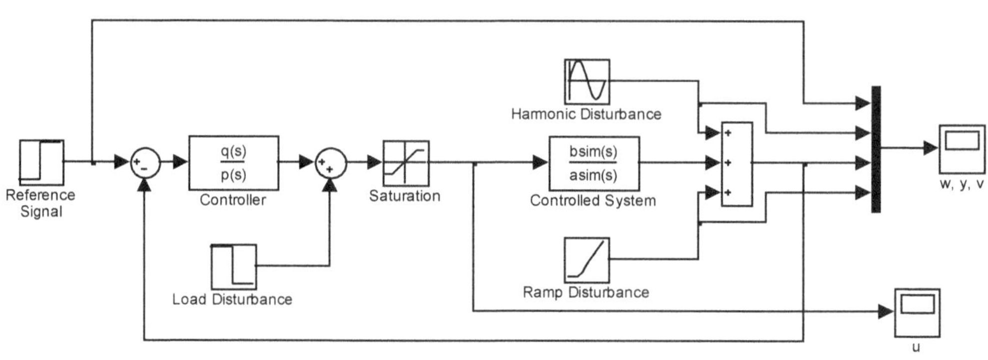

Fig. 8. Simulink scheme of 1DOF control system

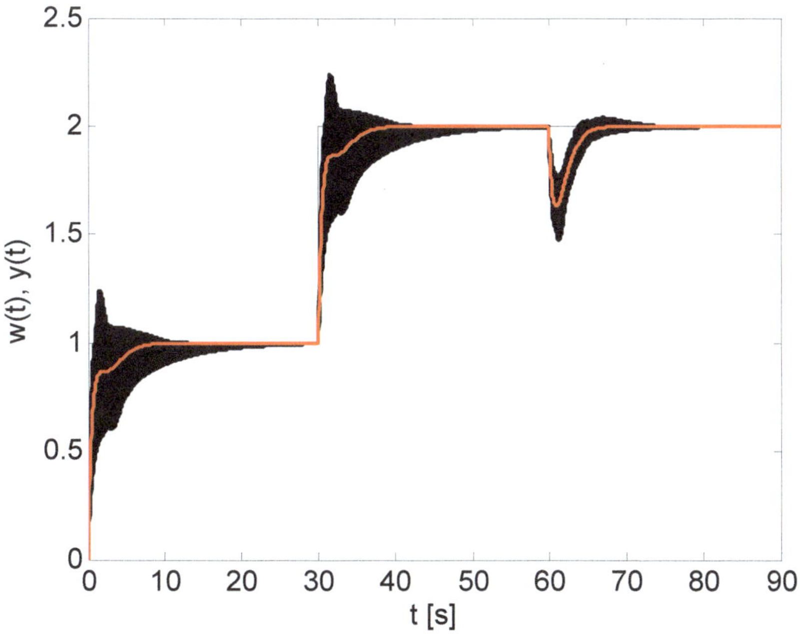

Fig. 9. Control of RSS of interval family (1) and nominal system by controller (2)

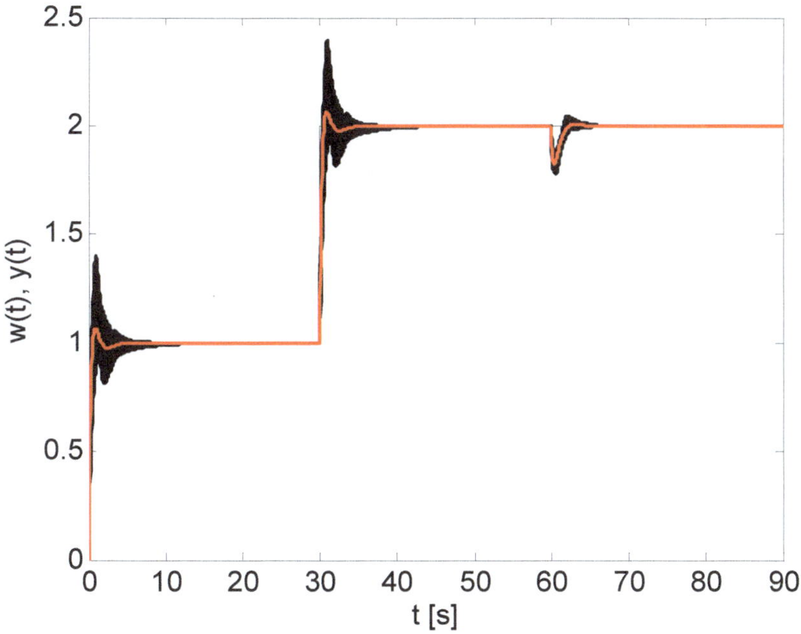

Fig. 10. Control of RSS of interval family (1) and nominal system by controller (7) - 1DOF

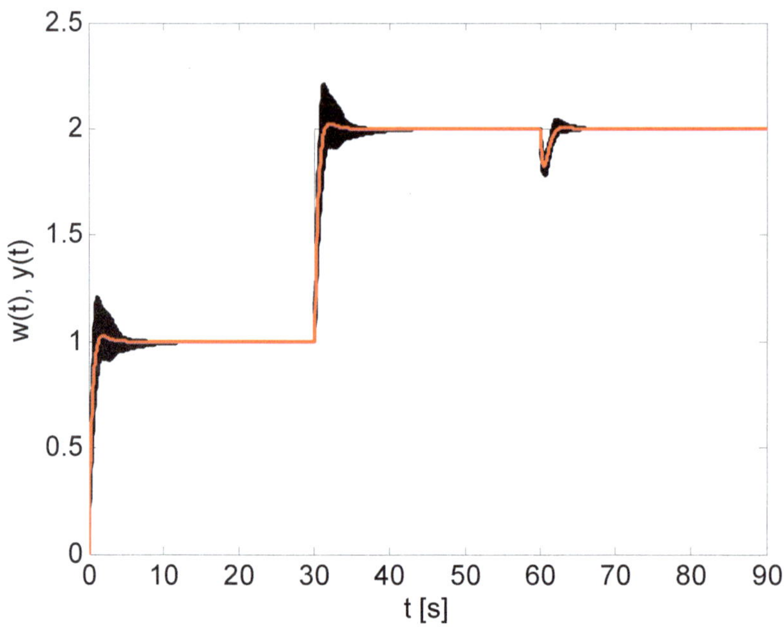

Fig. 11. Control of RSS of interval family (1) and nominal system by controller (7), (8) - 2DOF

The shorter settling time can be obtained by further tuning of controllers by parameter m. Assuming the 1DOF configuration of control system, the selection $m = 1.5$ gives the feedback controller:

$$C_b(s) = \frac{4.0625s^2 + 7.5s + 5.0625}{s^2 + 0.9375s} \tag{5}$$

and the computation for 2DOF structure adds the feedforward part:

$$C_f(s) = \frac{2.25s^2 + 6.75s + 5.0625}{s^2 + 0.9375s} \tag{6}$$

Now, only the final simulations of control behaviour are shown without deeper insight both for 1DOF and 2DOF configurations - see Figs. 10 and 11, respectively. However, the costs for "faster" regulation are more aggressive control signals.

3. Control of time-delay systems using modified Smith predictors

3.1 Theoretical background

The time-delay has been intensively investigated phenomenon during the last decades, because it is very common in many process control applications and its presence in a control loop always brings serious complications. The relatively effective tool for compensation of time-delay term represents the classical Smith predictor which has been known to automation community since 1959 (Smith). However, this control structure has also its disadvantages and limitations.

Some drawbacks of the Smith predictor have been eliminated by improving the idea and creating many modifications of this connection (Watanabe & Ito, 1981; Åström et al., 1994; Mataušek & Micić, 1996; Majhi & Atherton, 1998; Kaya & Atherton, 1999; Hamamci et al., 2001).
All three implemented techniques (Majhi & Atherton, 1998; Kaya & Atherton, 1999; Hamamci et al., 2001) have improved the classical Smith predictor loop using more sophisticated and complicated structure with additional controllers. Naturally, all the methods also use mathematical model of really controlled plant including time-delay term in the inner loop. Moreover, this model is assumed during design of controllers as a nominal system. In practice, however, the really controlled can differ from the ideal assumptions. The example of modified Smith predictor structure is shown in Fig. 12.

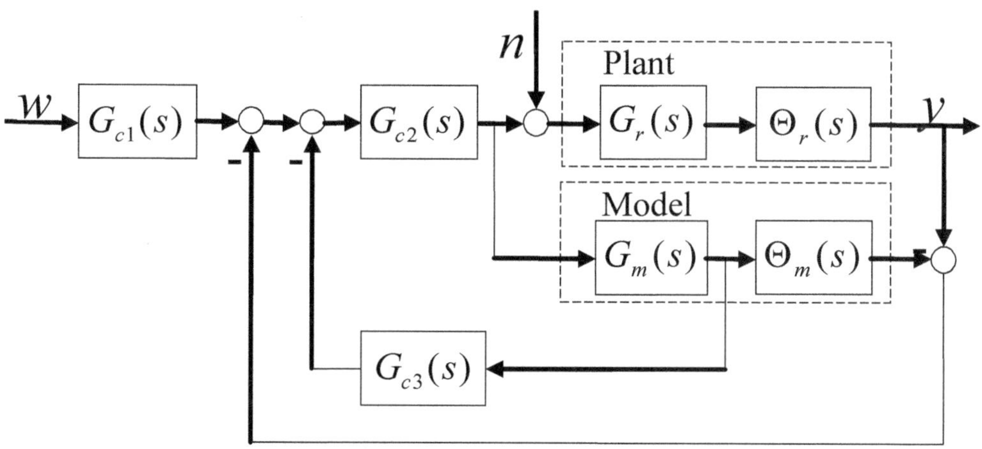

Fig. 12. Example of modified Smith predictor structure (Hamamci et al., 2001)

The controller synthesis itself is based on various approaches and techniques according to the applied modification. For example the standard forms for obtaining the optimal closed-loop transfer function parameters in the meaning of integral squared time error (ISTE) criterion, Nyquist stability criterion, a simple algebraic approach to control system design, coefficient diagram, modification of Kessler standard form, or Lipatov stability analysis have been utilized - see e.g. (Mataušek & Micić, 1996; Majhi & Atherton, 1998; Manabe, 1998; Kaya & Atherton, 1999; Hamamci et al., 2001; Hamamci & Ucar, 2002). The final relations for controller design have been usually pre-derived for first and second order time-delay plants.

3.2 Description of the program

Analogically to the Subsection 2.2, this part briefly presents a simple user-friendly Matlab program (Matušů & Prokop, 2010, 2011b, 2011c), in this case for control of time-delay systems using three various modifications of Smith predictor. The software implementation includes the modification for unstable and integrating processes (Majhi & Atherton, 1998), PI-PD modification for systems with long dead time (Kaya & Atherton, 1999), and modification applying control design by CDM (Hamamci et al., 2001). It can be freely downloaded from the web page (Matušů & Prokop, 2011b). The program is a translated version of the one created under the scope of the Master's Theses (Matušů, 2002). After decompression it can be launched using "go.m" file. The software has been coded in Matlab 6.5.1 - R13SP1 but tested also under several newer versions. The main window of the program GUI (Fig. 13) allows selecting the modification which should be used for a whole control experiment.

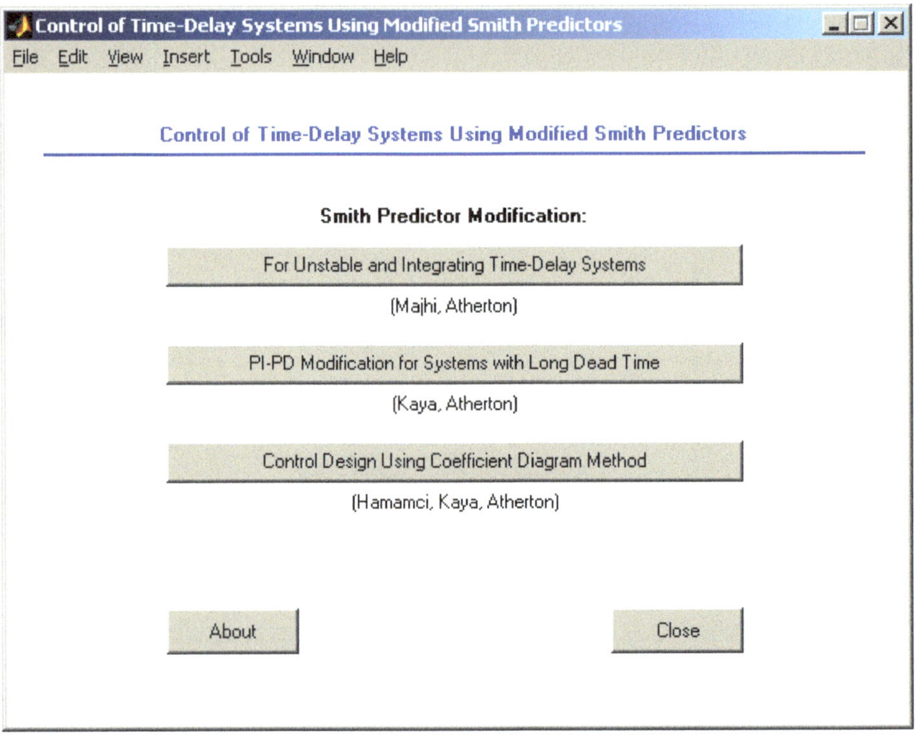

Fig. 13. Initial window of the second program

Subsequently, sort of controlled system (e.g. first order, second order or integrating plant as a special type) can be chosen together with fundamental properties of the experiment (simulation time, reference signal, disturbances) - see Fig. 14.

In the next step, coefficients of the controlled system of specific type and possibly some other additional parameters depending on the used method can be set as illustrated in Fig. 15. However, the program permits not only adjustment of nominal system (considered as a model for control design and in control loops as shown e.g. in Fig. 12), but also of the perturbed system (used as a really controlled plant) with potentially different coefficients.

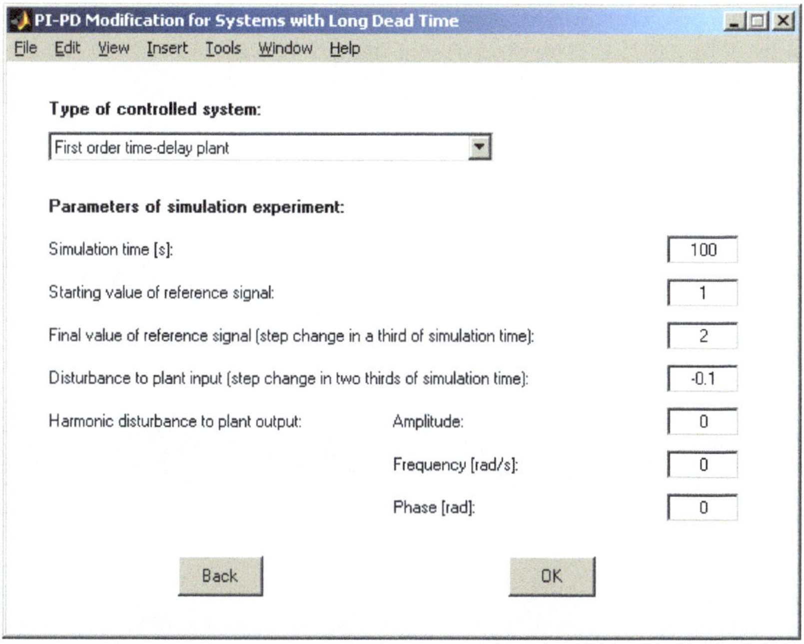

Fig. 14. Basic properties of control experiment

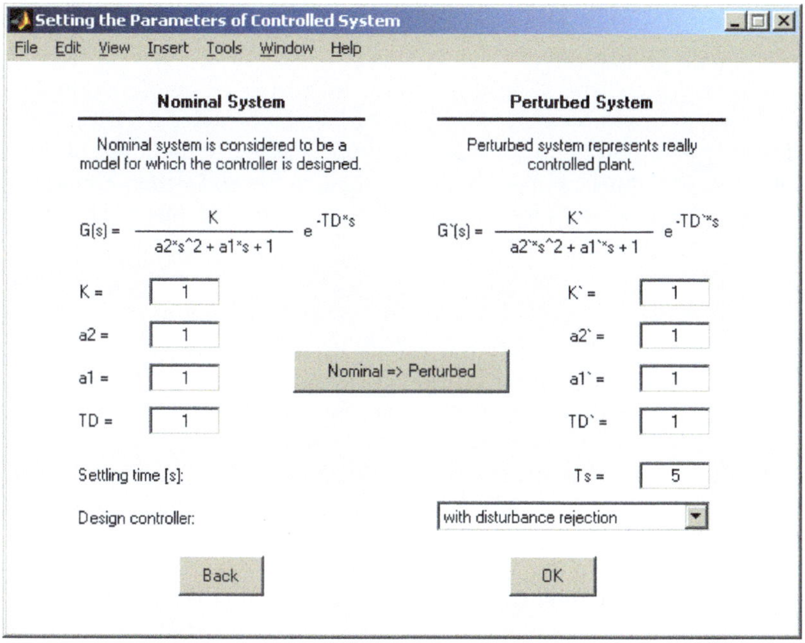

Fig. 15. Definition of parameters for nominal and perturbed system

Finally, the program computes the controllers and opens the Simulink scheme where control behaviour with the preset values can be simulated. An example is shown in Fig. 16.

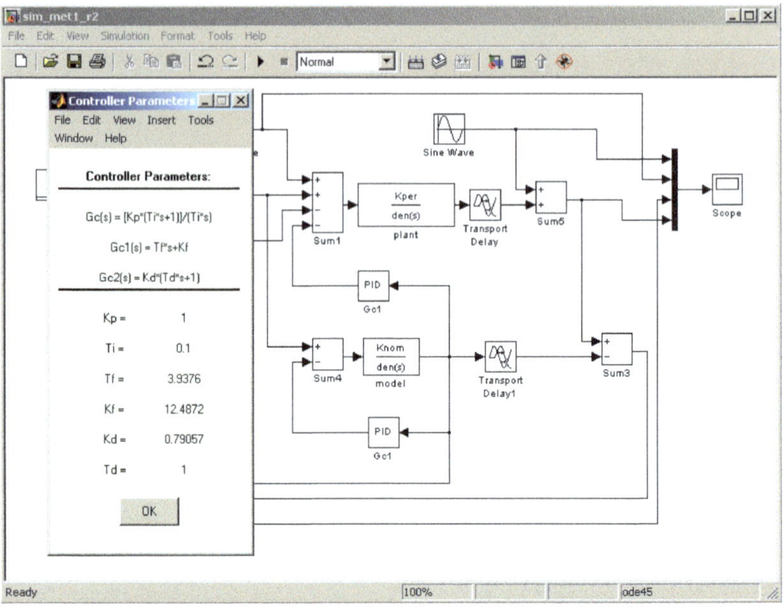

Fig. 16. Display of final controllers and simulation environment

3.3 An example of application

In order to demonstrate the capability of the program, a simple example has been performed. There were assumed the step change of reference signal from 1 to 2 in a third of a simulation time and then the disturbance $n = -0.3$ injected to the input of the controlled plant during the last third of the simulation time. A second order time-delay transfer function with complex poles was considered as a controlled plant. The same transfer function was assumed as a nominal system as well:

$$G(s) = \frac{1}{s^2 + 0,2s + 1} e^{-15s} \tag{7}$$

As in the previous case, the respective time constants are supposed to be in seconds. The modified Smith predictor design by CDM was selected. The consideration of the version with disturbance rejection capability leads to the trio of controllers:

$$\begin{aligned} G_{c1}(s) &= 1 \\ G_{c2}(s) &= \frac{1}{0.05579s^2 + 0.3322s} \\ G_{c3}(s) &= 0.9341s^2 + 1.2929s + 1 \end{aligned} \tag{8}$$

with prescribed settling time $T_s = 3.5\,[s]$.

The control result obtained from the program is visualized in Fig. 17.

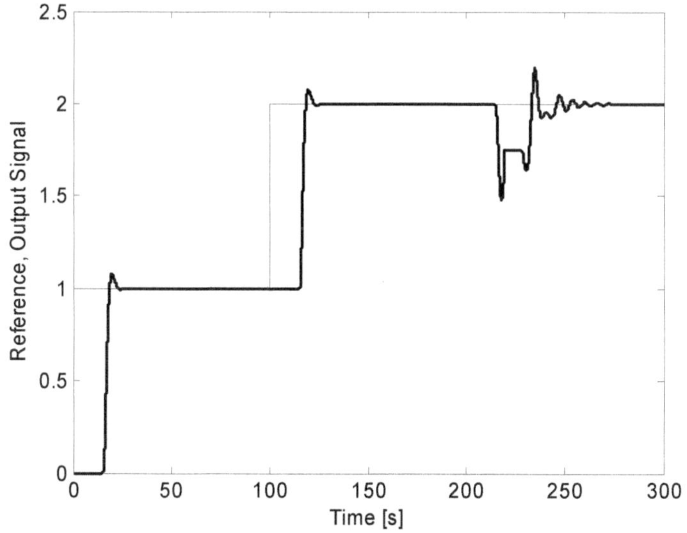

Fig. 17. Control results for system (7)

4. Conclusion

This chapter has presented two freely downloadable Matlab programs which allow user-friendly work with selected control design methods via GUI. The first package is focused on algebraic design of continuous-time controllers under assumption of interval plants while the second one deals with control of time-delay systems using three various modifications of Smith predictor. The described software products can be used both for research and educational purposes.

5. Acknowledgements

The work was supported by the Ministry of Education, Youth and Sports of the Czech Republic under the Research Plan No. MSM 7088352102 and by the European Regional Development Fund under the project CEBIA-Tech No. CZ.1.05/2.1.00/03.0089.

6. References

Åström, K. J.; Hang, C. C. & Lim, B. C. (1994). A new Smith predictor for controlling a process with an integrator and long dead-time. *IEEE Transactions on Automatic Control*, Vol. 39, No. 2, pp. 343-345.

Barmish, B. R. (1994). *New Tools for Robustness of Linear Systems*. Macmillan, ISBN 0-02-306055-7, New York, USA.

Bhattacharyya, S. P.; Chapellat, H. & Keel, L. H. (1995). *Robust Control: The Parametric Approach*, Prentice Hall, ISBN 978-0137815760, Englewood Cliffs, New Jersey, USA.

Hamamci, S. E. & Ucar, A. (2002). A robust model-based control for uncertain systems. *Transactions of the Institute of Measurement and Control*, Vol. 24, No. 5, pp. 431-445.

Hamamci, S. E.; Kaya, I. & Atherton, D. P. (2001). Smith predictor design by CDM. *Proceedings of the European Control Conference 2001*, Porto, Portugal.

Kaya, I. & Atherton, D. P. (1999). A new PI-PD Smith predictor for control of processes with long dead time. *Proceedings of the 14th IFAC World Congress*, Beijing, China.

Kučera, V. (1993). Diophantine Equations in Control - A Survey. *Automatica*, Vol. 29, No. 6, pp. 1361-1375, ISSN 0005-1098.

Majhi, S. & Atherton, D. P. (1998). A new Smith predictor and controller for unstable and integrating processes with time delay. *Proceedings of the 37th IEEE Conference on Decision & Control*, Tampa, Florida, USA.

Manabe, S. (1998). Coefficient diagram method. *Proceedings of the 14th IFAC Symposium on Automatic Control in Aerospace*, Seoul, Korea.

Mataušek, M. R. & Micić, A. D. (1996). A modified Smith predictor for controlling a process with an integrator and long dead-time. *IEEE Transactions on Automatic Control*, Vol. 41, No. 8, pp. 1199-1203.

Matušů, R. & Prokop, R. (2008). Single-parameter tuning of PI controllers: from theory to practice, *Proceedings of the 17th IFAC World Congress*, Seoul, Korea.

Matušů, R. & Prokop, R. (2010). A Matlab Program for Control of Time-Delay Systems Using Modified Smith Predictors. *Annals of DAAAM for 2010 & Proceedings of the 21st International DAAAM Symposium*, pp. 483-484, Zadar, Croatia.

Matušů, R. & Prokop, R. (2011a). Interval Systems - Algebraic Design of Control and Robust Stability Analysis, Available from: http://zamestnanci.fai.utb.cz/~matusu/interval.zip, Accessed: 2011-08-10.

Matušů, R. & Prokop, R. (2011b). Control of Time-Delay Systems Using Modified Smith Predictors, Available from: http://zamestnanci.fai.utb.cz/~matusu/delay.zip, Accessed: 2011-08-10.

Matušů, R. & Prokop, R. (2011c). Implementation of Modified Smith Predictors into a Matlab Program, *Proceedings of the 25th European Conference on Modelling and Simulation*, Krakow, Poland.

Matušů, R. & Prokop, R. (2011d). Single-parameter tuning of PI controllers: Theory and application. *Journal of The Franklin Institute*, Vol. 348, No. 8, pp. 2059-2071.

Matušů, R. (2002). *Control of time-delay systems*. (*Řízení systémů s dopravním zpožděním*). Master's Theses, Faculty of Technology, Tomas Bata University in Zlín. (In Czech).

Matušů, R. (2010). A Software Tool for Algebraic Design of Interval Systems Control. *International Journal of Computational Science and Engineering*, Vol. 5, No. 3/4, pp. 262-268. Inderscience Publishers, ISSN 1742-7185 (Print), ISSN 1742-7193 (Online).

PolyX: The Polynomial Toolbox. (2011). Available from: http://www.polyx.com/, Accessed: 2011-08-10.

Prokop, R. & Corriou, J. P. (1997). Design and analysis of simple robust controllers. *International Journal of Control*, Vol. 66, No. 6, pp. 905-921.

Smith, O. J. M. (1959). A controller to overcome dead time. *ISA Journal*, Vol. 6, No. 2, pp. 28-33.

Vidyasagar, M. (1985). *Control system synthesis: a factorization approach*, MIT Press, Cambridge, MA, USA.

Watanabe, K. & Ito, M. (1981). A process-model control for linear systems with delay. *IEEE Transactions on Automatic Control*, Vol. 26, No. 6, pp. 1261–1269.

Comparison of Two Approaches to Count Derivations for Continuous-Time Adaptive Control

Karel Perutka
Tomas Bata University in Zlin, Faculty of Applied Informatics
Czech Republic

1. Introduction

The control of continuous-time systems can be realized by adaptive controllers. Self-tuning controllers are adaptive controllers which call on-line identification and controllers parameters tuning in one step of computation. Supervision enlarges the area of usage of controllers. It is necessary to count derivations of action and output signals during control, which is usually realized by filters. Settings of filters are directly connected with the model of the system. Another approach allows us to use the regression polynomials instead of filters, because the general form of derivations is known before the control. Without filters, this approach keeps the signal unchanged, but the choice of inappropriate length of time interval for polynomial regression increases the amplitude of noise. The chapter shows two examples of control and suggests the appropriate length of time interval for polynomial regression.

Many processes can be viewed in the point of control as continuous-time systems. The implementation of pseudo-continuous model on continuous-time system is called as hybrid system (De Santis et al., 2009). Mostly these systems are nonlinear and specific method of control is needed (Gregorčič and Lightbody, 2010). This chapter uses the method adaptive control, because adaptive control is often used and gives adequate results (Pasik-Duncan, 2001). At adaptive control, the usage of the appropriate identification method is very important. This paper uses recursive instrumental variable method, but there are several other good methods and papers dealing with identification and parameters tuning (Flores and Pastor, 2005, Tzes and Le, 1996, Coello, 2000). The controlled process in this chapter has multi-inputs multi-outputs and the decentralized controller was used. It is common approach in practice (Martínez-Rosas et al., 2006). Decentralized control can be realized by PID controllers. These controllers are very popular due to their advantages, such as simplicity (Vrančić et al., 2010).

The ideas and results obtained in control can be useful in many different areas, for example in robotics or in production systems. Nice paper about spatial ontology for human-robot interaction was written by Belouaer at al. (Belouaer at al., 2010). A special framework to generate configurations in production systems was written by Kanso et al. (Kanso et al., 2010).

2. Theoretical background

2.1 Self-tuning control

Self-tuning controllers (STC) are based on on-line identification and on tuning the controller parameters with respect to identified changes in controlled systems. The self-tuning controllers can be further divided to the STC with explicit identification and the STC with implicit identification, the STC with implicit identification directly identifies the controller parameters. On the other hand, the STC with explicit identification computes the controller parameters using the parameters of the system model (Bobal et al., 2005).

2.2 On-line identification

When self-tuning controller was used, the scheme of input and output signal modification depicted in figure 1 is applied, because the continuous-time system parameters a_i and b_j are estimated using recursive instrumental variable method. The action (input) signal $u(t)$ is continuously approximated by Lagrange regression polynomial on an interval of given length during entire control. The structure of Lagrange regression polynomial (1) together with its derivation (2), (3) is generally known before the start of identification, only the numerical values of their parameters are needed and counted. It is the alternative way to obtain values of derivations needed for identification. After the polynomial approximation, the approximating polynomial derivation $u^{(i)}{}_L(t)$ is counted. It is sampled in purpose to count the values of subsystem parameters using recursive identification algorithm.

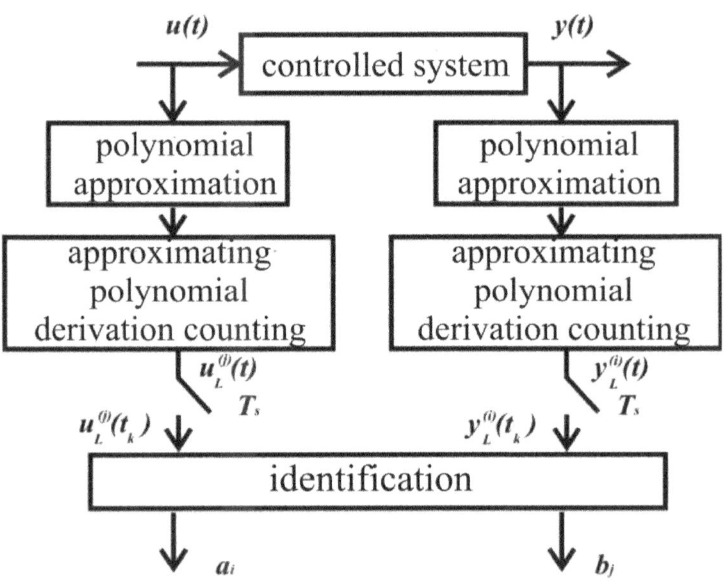

Fig. 1. Scheme of I/O signals modification for STC.

Lagrange polynomial of second order was used in the paper in the form

$$P_2(x) = \frac{(x-b)(x-c)}{(a-b)(a-c)} f(a) + \frac{(x-a)(x-c)}{(b-a)(b-c)} f(b) + \frac{(x-a)(x-b)}{(c-a)(c-b)} f(c) \quad (1)$$

The first derivation is

$$f'(x) \cong P'_2(x) = \frac{2x-(b+c)}{(a-b)(a-c)} f(a) + \frac{2x-(a+c)}{(b-a)(b-c)} f(b) + \frac{2x-(a+b)}{(c-a)(c-b)} f(c) \tag{2}$$

and second derivation is

$$f''(x) \cong P''_2(x) = \frac{2f(a)}{(a-b)(a-c)} + \frac{2f(b)}{(b-a)(b-c)} + \frac{2f(c)}{(c-a)(c-b)} \tag{3}$$

2.3 Recursive instrumental variable

Instrumental variable method is a modification of the least squares method. The least squares method uses the quadratic criterion and the existence of one global minimum. The instrumental variable method does not allow us to obtain the properties of noise, but it has inferior presumptions than the least square method. It is possible to formulate it recursively (Zhu & Backx, 1993).

$$\hat{\Theta}^T(k) = \left(\hat{a}_0, \hat{a}_1, \ldots, \hat{a}_{\deg(a)}, \hat{b}_0, \hat{b}_1, \ldots, \hat{b}_{\deg(b)}, d\right) \tag{4}$$

$$\boldsymbol{\phi}^T(k) = \left[-y(t_k), \ldots, -y_L^{(n-1)}(t_k), u(t_k), \ldots, u_L^{(m-1)}(t_k), 1\right] \tag{5}$$

$$\mathbf{L}(k) = \frac{\mathbf{C}(k-1)\mathbf{z}(k)}{1+\boldsymbol{\phi}^T(k)\mathbf{C}(k-1)\mathbf{z}(k-1)} \tag{6}$$

$$\mathbf{C}(k) = \mathbf{C}(k-1) - \frac{\mathbf{C}(k-1)\mathbf{z}(k)\boldsymbol{\phi}^T(k)\mathbf{C}(k-1)}{1+\boldsymbol{\phi}^T(k)\mathbf{C}(k-1)\mathbf{z}(k)} \tag{7}$$

$$\mathbf{z}(k) = \left[u(t_k), u(t_{k-1}), \ldots, u(t_{k-n-m})\right] \tag{8}$$

$$\hat{e}(k) = \mathbf{y}(k) - \boldsymbol{\phi}^T(k)\hat{\Theta}(k-1) \tag{9}$$

$$\hat{\Theta}(k) = \hat{\Theta}(k-1) + \mathbf{L}(k)\hat{e}(k) \tag{10}$$

2.4 Suboptimal linear quadratic controller

The used suboptimal method was introduced by Dostal (Dostal, 1997). Let us minimize quadratic functional

$$J = \int_0^\infty \left\{\mu e^2(t) + \varphi \tilde{u}^2(t)\right\} dt \tag{11}$$

where $\mu \geq 0, \varphi > 0$ are penalty constants. Stable polynomials g and n are counted as results of spectral factorizations

$$(as)^{*}\varphi as+b^{*}\mu b=g^{*}g,n^{*}n=a^{*}a. \tag{12}$$

Solving the following diophantic equation

$$asp + bq = gn \tag{13}$$

gives the parameters of controller. If the system transfer function has the form

$$G(s)=\frac{b_0}{s^2+a_1s+a_0} \tag{14}$$

The controller is

$$FQ=\frac{q_2s^2+q_1s+q_0}{s\left(p_2s^2+p_1s+p_0\right)} \tag{15}$$

and polynomials g and n are

$$g(s)=g_3s^3+g_2s^2+g_1s+g_0 \tag{16}$$

$$n(s)=s^2+n_1s+n_0 \tag{17}$$

Their coefficients obtained by spectral factorization are in the form

$$g_0=\sqrt{\mu b_0^2} \tag{18}$$

$$g_1=\sqrt{2g_2g_0+\varphi a_0^2} \tag{19}$$

$$g_2=\sqrt{2g_3g_1+\varphi\left(a_1^2-2a_0\right)} \tag{20}$$

$$g_3=\sqrt{\varphi} \tag{21}$$

$$n_0=\sqrt{a_0^2} \tag{22}$$

$$n_1=\sqrt{2n_0-a_1^2-2a_0} \tag{23}$$

2.5 Supervisor

The used supervisor is based on the supervisor introduced by Perutka (Perutka, 2007) and it is used in this paper for the first time.

Supervisor is used for decentralized or decoupled control of multi-input multi-output systems, number of inputs and outputs are the same and denoted as n. Such system is controlled by n sub-controllers. Let us suppose the existence of bits field with n x n dimension. The initial values of the field form the identity matrix. Each row the field corresponds to one subsystem of controlled system.

Step 1. Go through the bits field row by row. The row which gives the highest number after conversion also gives the number of the subsystem in which goes one step of the on-line identification.

Step 2. When the subsystem is set, the last bit in the row of identified subsystem is set to 1 and in remaining rows the last bit is set to 0.

Step 3. One bit left rotation of all rows in bits field.

Step 4. Go through the bits field row by row. The row which gives the lowest number after conversion also gives the number of the subsystem in which goes one step of the on-line identification.

Step 5. Do Step 2.

Step 6. Do Step 3.

Repeat Step 1 to 6 $n/2$-times at even n and $n/2$-times without Step 4 to 6 at last calling at odd n after the change of set-point. After this tuning, run the self-tuning control without supervisor until the new change of the set-point when the supervisor is called.

3. Experimental part

In figures 2-7, there are obtained results of control of two inputs two outputs systems by two controllers. Counting step was 0.2 s. In these figures, the meaning of the symbols is following: w_1 - set-point of first subsystem, u_1 - action signal of first subsystem, y_1 - output signal of first subsystem, w_2 - set-point of second subsystem, u_2 - action signal of second subsystem, y_2 - output signal of second subsystem, $p1_1$, $p0_1$, $q2_1$, $q1_1$, $q0_1$ - parameters of first sub-controller, $b0_1$, $a1_1$, $a0_1$ - parameters of the model of the first controlled subsystem.

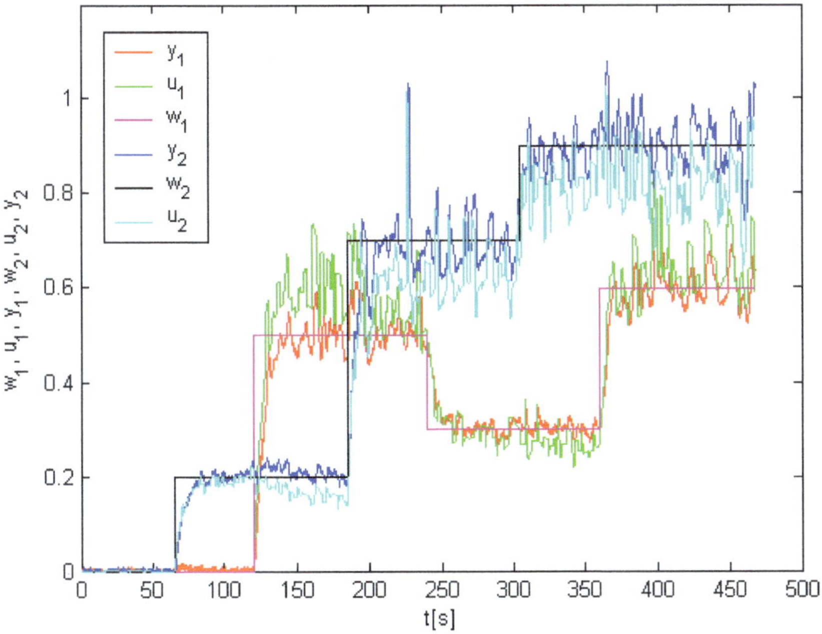

Fig. 2. History of control - too small interval for approximation by Lagrange polynomial.

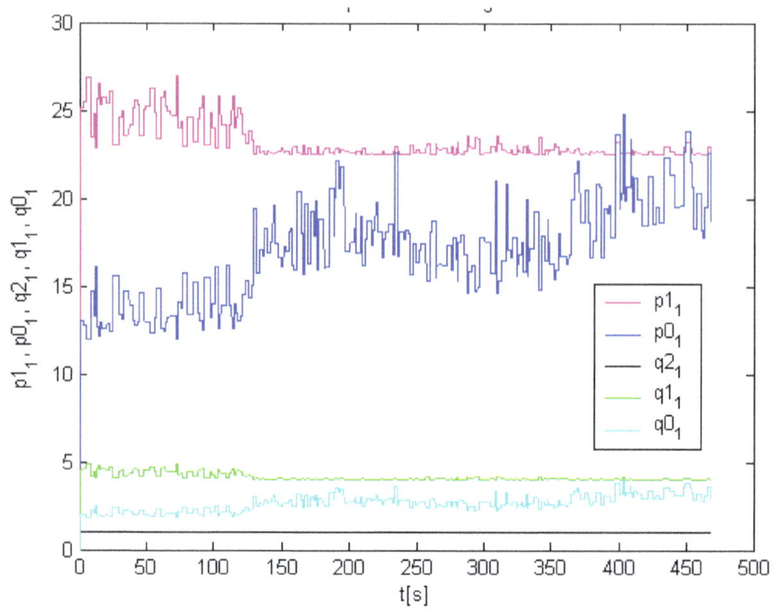

Fig. 3. History of controller parameters for 1st subsystem – too small interval for approximation by Lagrange polynomial.

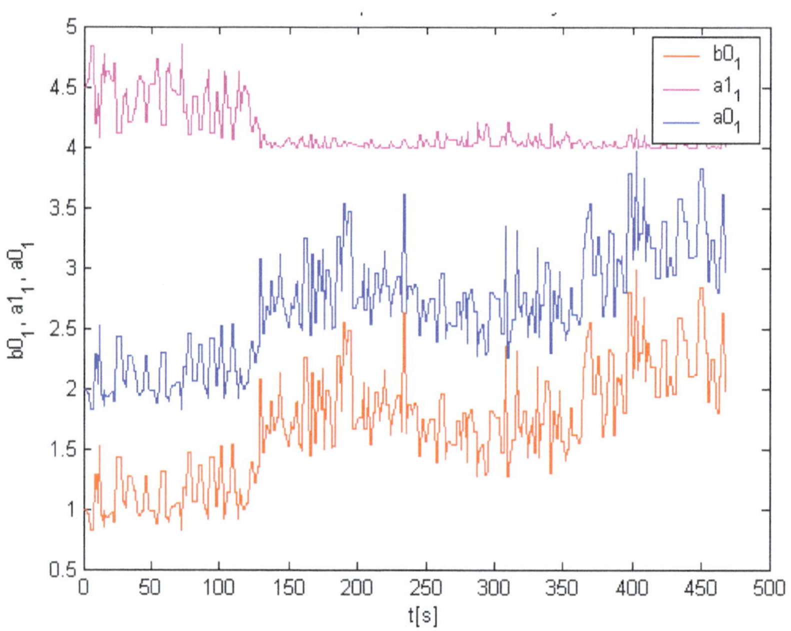

Fig. 4. History of subsystem model parameters for 1st subsystem - too small interval for approximation by Lagrange polynomial.

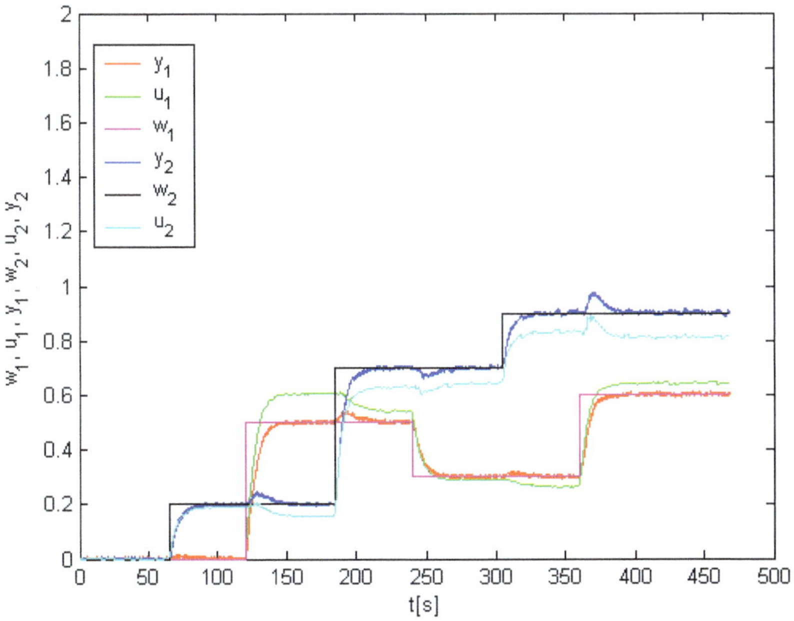

Fig. 5. History of control – adequate interval for approximation by Lagrange polynomial.

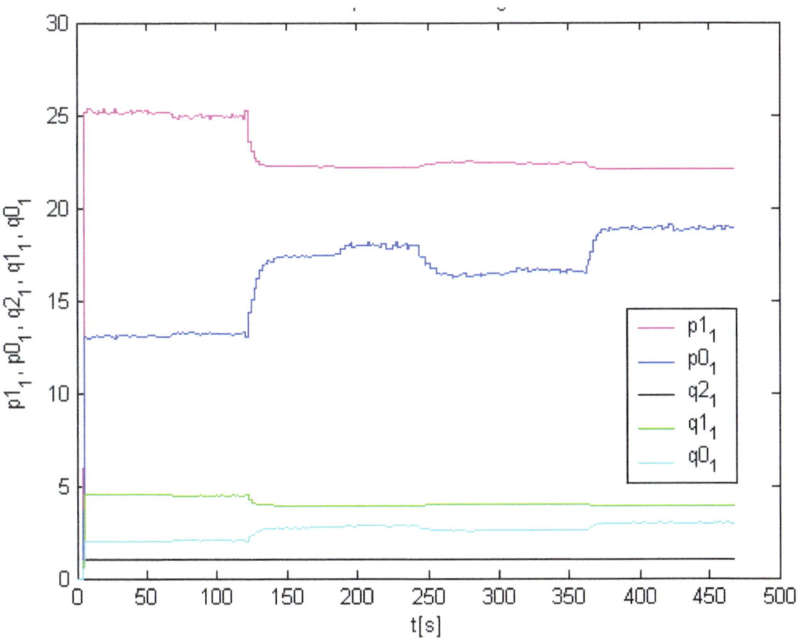

Fig. 6. History of controller parameters for 1st subsystem – adequate interval for approximation by Lagrange polynomial.

Figures 2 - 4 provide the results obtained for time interval of approximation 0.41 s.
Figures 5 - 7 provide the results obtained for time interval of approximation 4.1 s.
From the illustratively shown results, it is clear that it is important to correctly choose the appropriate length of time interval which is used for regression by Lagrange polynomial. By repeating several experiments with different controlled systems it was verified that it is appropriate to use 20 counting steps in the presence of noise.

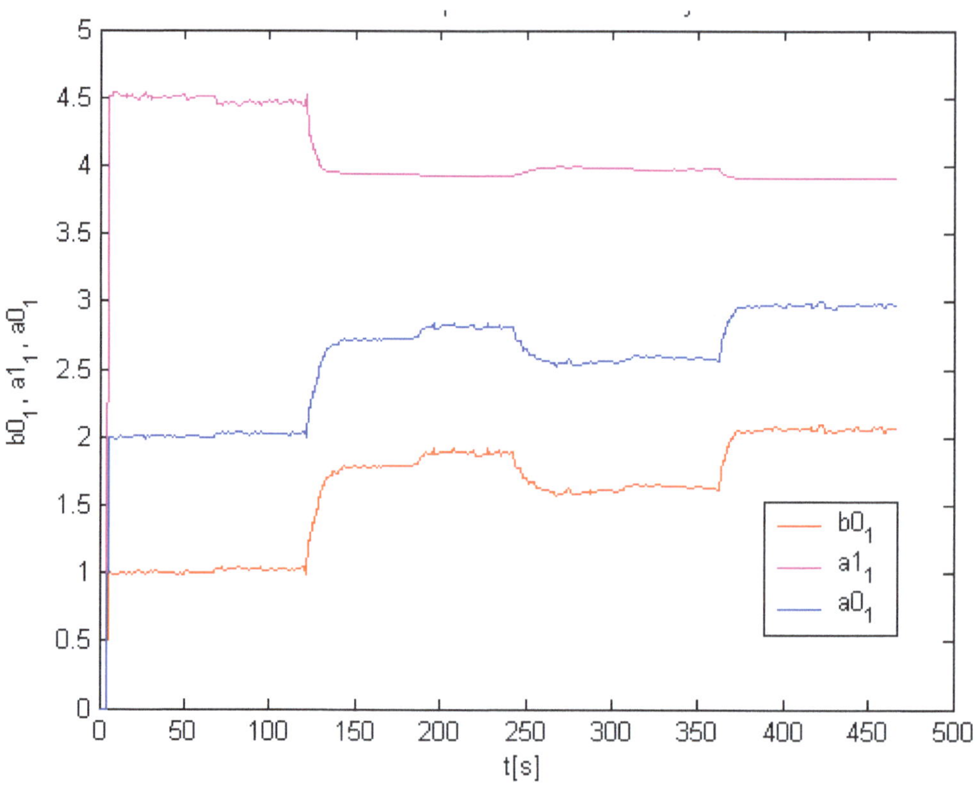

Fig. 7. History of subsystem model parameters for 1st subsystem - adequate interval for approximation by Lagrange polynomial.

4. Conclusions

The chapter presented simulation results of self-tuning control with polynomial regression used for derivations counting. It was shown that inappropriate selection of regression interval makes more noise. The recommended length of the interval was given. Future work will focus on the exact mathematical derivation of the time appropriate interval with the combination of the dynamical filter of noise.

5. Acknowledgements

The author would like to mention MSM7088352101 grant, from which the work was supported.

6. References

Belouaer, L., Bouzid, M., Mouaddib, A.-I., 2010. A Spatial Ontology for Human-Robot Interaction. In *ICINCO 2010, 7th International Conference on Informatics in Control, Automation and Robotics.* SciTePress, Volume 1, pp. 154-159, ISBN 978-989-8425-00-3.

Bobal, V., Böhm, J., Fessl, J., Machacek, J., 2005. *Digital Self-tuning Controllers.* Springer-Verlag London Limited, ISBN 978-1-85233-980-7.

De Santis, E., Di Benedetto, M.D., Pola, G., 2009. A structural approach to detectability for a class of hybrid systems. *Automatica,* 45, pp. 1202-1206.

Dostal, P., 1997. An approach to control of processes of chemical technology. Inaugural dissertation. TU Brno, Brno.

Coello, C.A.C., 2000. Use of a self-adaptive penalty approach for engineering optimization problems. *Computers in Industry,* 42, pp. 113-127.

Flores, J.J., Pastor, N., 2005. Time-Invariant Dynamic Systems identification based on the qualitative features of the response. *Engineering Applications of Artificial Intelligence,* 18, pp. 719-729.

Gregorčič, G., Lightbody, G., 2010. Nonlinear model-based control of nonlinear processes. *Computers and Chemical Engineering,* 34, pp. 1268-1281.

Kanso, M., Berruet, P., Philippe, J.-L., 2010. A Framework Based on a High Conception Level to Generate Configurations in Production Systems. In *ICINCO 2010, 7th International Conference on Informatics in Control, Automation and Robotics.* SciTePress, Volume 1, pp. 244-248, ISBN 978-989-8425-00-3.

Martínez-Rosas, J.C., Arteaga, M.A., Castillo-Sánchez, A.M., 2006. Decentralized control of cooperative robots without velocity-force measurements. *Automatica,* 42, pp. 329-336.

Pasik-Duncan, B., 2001. On stochastic adaptive control of continuous-time systems. *Nonlinear Analysis,* 47, pp. 4807-4818.

Perutka, K., 2007. *Decentralized Adaptive Control.* Thesis. Zlin, Czech Republic: UTB Press, 2007.

Tzes, A., Le, K., 1996. Application of Frequency Domain Adaptive Infinitive Impulse Response Filtering for Identification of Flexible Structure Dynamics. *Mechanical Systems and Signal Processing,* 10, pp. 65-91.

Vrančić, D., Strmčnik, S., Kocijan, J., de Moura Oliveira, P.B., 2010. Improving disturbance rejection of PID controllers by means of the magnitude optimum method. *ISA Transactions,* 49, pp. 47-56.

Zhu, Y., Backx, T., 1993. *Identification of Multivariable Industrial Processes for Simulation, Diagnosis and Control*. Springer-Verlag Ltd., London, United Kingdom, ISBN 3-540-19835-0.

Part 3

Multimedia

10

Video Compression from the Hardware Perspective

Grzegorz Pastuszak
Warsaw University of Technology
Poland

1. Introduction

Many advanced multimedia applications require image compression technology with ever higher compression ratios and better visual quality. The need for the real-time high-efficiency video compression usually involves the use of hardware accelerators. In general, the development of architectures mapped into integrated circuits allows simultaneous processing of various data. On the other hand, the hardware framework suffers from limitations on the algorithm flexibility due to timing dependencies coming from the designed dataflow. Thus, the development of efficient video codecs in integrated circuits should take into account the algorithm details of the video codec. The following sections address various aspects of the video-compression design at the hardware architecture level. Section 2 analyzes the video coding dataflow and the design efficiency regarding timing and resources. To illustrate challenges in the hardware design, Section 3 reviews architectures of main modules of the H.264/AVC hardware encoder. The implementation results are given in Section 4.

2. High-performance coding

The real-time performance means that the encoder (decoder) must process all input (produce all output) video frames/fields/macroblocks in a limited amount of time. The section analyzes the codec structures in terms of timing properties and resource consumptions.

2.1 Dataflow

Video systems for the compression of greyscale visual information operate on the three-dimensional signal. An additional dimension is added to index colour and auxiliary components. Colour components refer to one of some colour spaces such as RGB, YUV, and YCbCr.

The dataflow in the encoder of visual data is depicted in Fig. 1. A video encoder consists of four main functional parts related to temporal modelling, spatial modelling, quantization, and binary coding. Frame (or field) in a video sequence can be processed in two basic modes. The first is called INTRA and exploits only spatial modelling, as for images. The second is called INTER and uses both modelling parts.

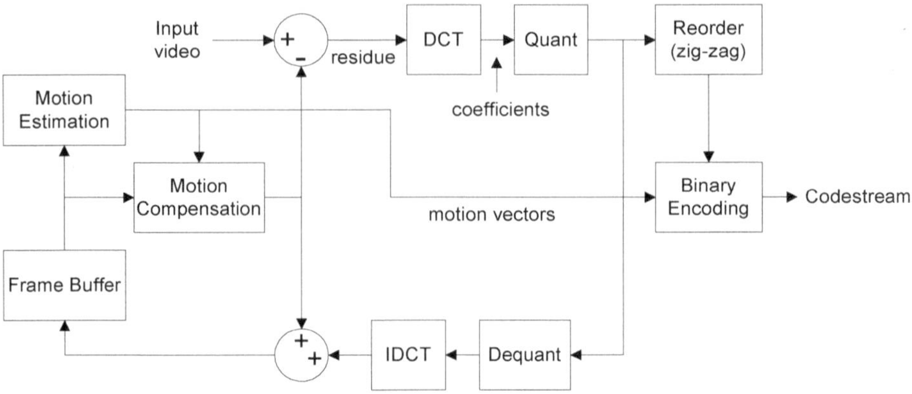

Fig. 1 Block diagram of the video encoder

The temporal model attempts to reduce temporal redundancy by exploiting similarities between neighbouring frames, usually by constructing a prediction of the current frame. The prediction is formed from one or more frames preceding or following the current one. When a selected reference frame is a previously encoded frame, the current one is referred to as a P-frame (see Fig. 2). When both a previously encoded frame and a future frame are chosen as reference frames, the current one is referred to as a B-frame. For a selected frame(s), the motion estimation (ME) module compares allowable pixel blocks (e.g., macroblocks) in the current frame with its surrounding area in the previous frame(s) and attempts to find the best match. The matching area (the prediction) is subtracted from the current macroblock in the motion compensation module. The difference between positions in the current and referred frames identifies motion vectors (MVs). If the motion estimation and compensation process is efficient, the remaining residual data should contain only a small amount of information. The temporal model outputs a residual frame and a set of parameters, typically the set of motion vectors.

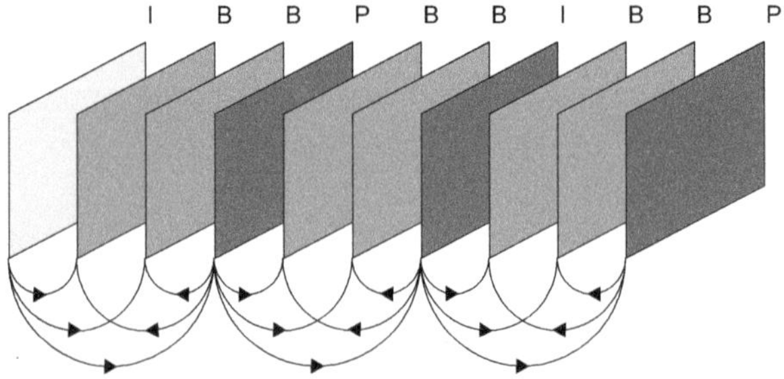

Fig. 2. I/P/B frames in a video sequence.

The spatial model exploits correlations between neighbouring samples within one frame to reduce spatial redundancy. This can be achieved by applying transform and/or prediction. The transform converts the samples into another domain in which they are represented by spatial frequency coefficients. Typically, the transforms operate on a two-dimensional block

of pixels rather than on a one-dimensional signal. Their ability to concentrate the signal energy enables few coefficients to recreate a recognizable copy of the original block of pixels. Apart from transform techniques, the spatial redundancy can be reduced using the prediction from neighbouring pixels within the same frame (interpolation and extrapolation).
For a typical block of pixels, most of the coefficients produced by the transform are close to zero. The quantization reduces the precision of each coefficient so that the near-zero coefficients are set to zero and only a few significant non-zero coefficients are left. Note that the quantization removes less important information.
The I- and P-frames must be stored in the buffer to be used as references when the INTER frames are encoded. The content of frames buffered in the encoder should be identical to the content of frames buffered in the decoder. Therefore, instead of simply copying frames into the buffer at the encoder side, they undergo some operations as in the decoder. In particular, to create a reconstructed frame, the quantized coefficients are rescaled, inverse transformed, and added to the motion-compensated reference block. These operations make up the feedback loop in the encoder. When the INTER frame is encoded, the motion estimator uses frames stored in the buffer to determine the best matching area for motion compensation.
The last step in the video coding process is binary coding that produces the output codestream. Inputs to the binary coder include transform coefficients for the residual data, motion vectors, frame pointers, block sizes, and other control information. The variety of these parameters, correlations between them, and their statistics affect the algorithm of binary coding, especially its complexity. The algorithm can adopt one or more coding methods. Finally, the type of binary coding depends on the application.

2.2 Timing

The section will analyze the number of clock cycles the codec can allocate to pixel-domain coding units. Moreover, the codec structure will be related the processing latency.

Pixel resolution	Time resolution	Throughput [MB/sec]	Max clock cycles per MB
576x720	25	40600	2461
480x640	30	36000	2777
720x1280	25	90000	1111
720x1280	30	108000	925
1080x1920	25	204000	490
1080x1920	30	244000	408

Table 1. Summary of timing requirements for different video formats

In order to satisfy real time requirements, the encoder throughput should be high enough. In practice, the required throughput depends on the video resolution related to time and pixel domains. They are measured in frames per second (fps) and pixel area, respectively. Additionally, subsampling of chroma components can affect the performance. As the video compression processes pixels in 16x16 pixel macroblocks, it is convenient to use the number of macroblocks per second to specify the throughput. Having a specified architecture, the performance depends on the clock frequency. In particular, the throughput is proportional to the frequency. Table 1 shows average macroblock throughputs required for different resolutions and the average number of clock cycles allocated to each macroblock at 100 MHz. In practice, the hardware encoder performance should have a computation margin to compensate for wait states caused by initializations (e.g. probability models, rate control), the fullness of the output stream buffer, etc.

Apart from clocking the video codec core, it is important to provide the sufficient bandwidth to the external memory used to buffer original and reference frames. Particularly, each macroblock involves read access to one 16x16 original pixel block and some (N+5)x(M+5) reference pixel blocks. Note that N and M are the horizontal and vertical sizes of the reference area, respectively. The increase by five is the overlap which results from the subpixel interpolation. It is possible that the codec accesses to some smaller reference areas when a macroblock is partitioned and the partitions have different motion vectors and/or reference frames. Using more reference frames proportionally increases the number of read accesses to the reference area. As each reconstructed macroblock must be stored in this area, one 16x16 write access is performed for a macroblock. At the encoder side, input pixels should be stored in the external memory prior to reading original macroblock pixels, whereas the reconstructed frames are read and formed into output pixel stream at the decoder side. Thus, both sides need similar bandwidth to provide a pixel interface. If the bandwidth is not wide enough, the codec can encounter wait states decreasing its performance. In order to optimize communication with the external memory, one must employ efficient access scheduling between multiple write and read ports.

The video codec latency comes mainly from buffering input and output streams. In the encoder, the input pixel stream must be first stored in the memory line by line. If the number of pixel lines is sufficient to form 16x16 macroblocks, read access can start. In the case of emerging H.265 video standard, the traditional processing based on macroblocks is generalized to larger-size coding units (16x16, 32x32, and 64x64). As a consequence, the required number of buffered pixel lines increases accordingly. If the latency is not crucial parameter, the input buffer can keep more frames, i.e., the delay between writing and reading of the same pixels can be significant. In contrary to pixel streams, the amount of data in the code streams varies in time. Apart, from the bit-rate instability, transmission conditions change. When the bandwidth of the transmission channel between the encoder and decoder is limited, the buffer fullness also varies in terms of the amount of both code-stream and corresponding-pixel data. As the decoder buffer can underflow, the delay between decoding and displaying should be set to avoid situations when there are no decoded pixels to display in the output buffer.

Efficient hardware video codecs exploits the macroblock-level pipeline. The pipeline stages are distinguished with reference to mutual dependencies of processing blocks. In practice, the encoder embeds at least three stages associated with the motion estimation, internal loop (intra prediction, transforms, quantization, and reconstruction), and entropy coding in parallel with the deblocking filter. In the decoder, it is enough to exploit two macroblock-level stages since the motion estimation is not present.

The internal loop in the encoder involves some computation cycles for each macroblock when the Intra mode is analyzed. Particularly, the prediction for Intra 4x4 and 8x8 blocks is computed with reference to reconstructed pixels of blocks adjacent to the current one to the top and left side. Therefore, the processing of a block of the same size in the in the loop can start when the reconstruction for the top and left neighbours is finished. Owing to the number of blocks within the macroblock, the total number of clock cycles sacrificed to the

Intra 4x4 mode is equal to 16xN in the straightforward approach. N denotes the number of clock cycles between starting the prediction and finishing the reconstruction. Computations for other Intra and Inter (chroma/luma) modes can be interlaced with those for the Intra 4x4 blocks to reduce the number of clock cycles. This schedule does not have to decrease the total throughput as there are usually significant time gaps within all N-clock periods. Moreover, it is possible to schedule the processing so that some pairs of Intra 4x4 blocks can be computed immediately one by one without waiting for the reconstruction, i.e., reconstructions do not affect each other (Roszkowski & Pastuszak, 2010).

2.3 Resources

The section will review practical limitations on the amount of resources in available technologies and relate them to the complexity of video codecs. In general, it is possible to design the dataflow with very-high throughputs. In practice, the design should minimize the resource consumption due to the cost of silicon area and power consumption. When Application Specific Integrated Circuits (ASIC) are taken into account, encoder architectures (with the Inter prediction) reported in scientific literature consumes above 500K gates (see Table 2). For the Intra encoders the resource consumption can be significantly reduced below 100K gates. Note that the gate unit is equivalent to the basic two-input NOR/NAND gate. Additionally, designs embed some on-chip memories used as buffers with relatively quick data access. On contrary to the ASIC technology, the Field Programmable Gate Array (FPGA) devices embed other logic units which group the functionality of several gates. However, a simple mapping between the number of gates and logic units is difficult as it depends on the design, synthesis tools, and specific technologies. Due to the amount of logic resources, only the designs limited to the Intra mode can be easily mapped to FPGA technologies. The decoders are much simpler as they do not embed mode selection algorithms (Roszkowski et al., 2010).

Design	technology	Gate count	On-chip Memory [bit]	Max clock frequency [MHz]	Clock cycle per macroblock	features
Y. W. Huang, et. All (2005)	TSMC 0.25 μm	85K	14336	54	1300	Baseline SDTV, Intra
Lin Y.-K., et. All (2009)	TSMC 0.13 μm	94.7K	14720	140	560	Baseline 1080p, Intra
Lin Y.-L. S et. All (2010)	TSMC 0.13 μm	1697K	87040	158	632	High 1080p, Inter SR: 64x64
Liu Z. et. All (2009)	0.18 μm	1140K	887193	200	672	Baseline 1080p, Inter SR: 196x128
Chen Y.-H. et. All (2009)	TSMC 0.13 μm	452K	138854	54	~1330	Baseline D1, Inter SR: 64x32

Table 2. Comparison of different architectures

Important modules coupled with the hardware video codec with the support for Inter pictures are the external memories. They are used to store reference pictures and buffer original pictures (in the encoder). For high resolutions, a wide data width should provide a sufficient bandwidth. In practice, it can be achieved using several DDR(1/2/3) memories, where the address/control bus is common, and the data bus distributed between memory chips (to increase data width). The memories and associated connections occupy the board area. Furthermore, the coupling with the external memories requires the memory controller with the scheduler to support some different ports. In practice, the controller embeds some on-chip memories to provide burst data access. Although, these resources are not taken into account when comparing different designs, their area cost can be significant.

3. Architecture design

The multimedia compression employs the sequence of processing steps, and each of them must apply separate approaches to optimize performance and resource consumption. Firstly, each processing block operates on different type of data at input/output ports. Secondly, the type of an operation involves specific timing dependencies and requires specific amount of resources. Thirdly, the block-level pipeline should be balanced in terms of throughput to utilize maximally all hardware resources. In the area of integrated circuit design for video technologies, most efforts concentrate on the development of standardized codecs from MPEG and H.26x series. The latest standard H.264/AVC allows the best compression ratio at the cost of computationally-intensive algorithms. Following subsections describe main processing blocks in the developed H.264/AVC video codec. This review allows the identification some challenges when facing the vide compression in the hardware framework.

3.1 Motion estimation

Block diagram of the developed ME system is presented in Fig. 3. The system is composed of the motion vector generator, compensator, the bank of 64 memories (fine search area and original data), the coarse-level full-search (FS) estimator, the interpolator, and the external memory controller. The architecture employs two-level hierarchical ME procedure. Thus, at the first stage, the coarse FS module performs FS on the whole search area (SA) subsampled with 16:1 ratio. To reduce the noise influence on initial MV accuracy, each pixel of the coarse SA is obtained by averaging of 16 pixels of the reference frame (Jakubowski 2008). The search range of the coarse FS is [-64, 63] pixels at most in both horizontal and vertical direction. When the coarse FS is completed, the interpolator fetches fine 40x40 reference samples from the external memory and generates quarter-pel ones within [-8, 7] range in both directions around the initial MV obtained from the coarse FS. The interpolator accepts eight column-oriented samples in a clock cycle. Therefore, processing of one colour component takes at least 200 clock cycles. Since every eight samples at the input corresponds to 128 ones at the output, memory write ports work at the doubled clock frequency.

Samples generated by the interpolator are loaded into the Fine Search Area space in SRAM. Thus, any search point inside the fine SA can be checked instantly with quarter-pel accuracy using the same hardware as for integer-pel MVs. For the sake of limited resources, ordinary SAD is used for evaluation of sub-pel MVs instead of sum of absolute transformed differences which requires the Hadamard transform. When interpolated fine SA is available in the Fine Search Area SRAM, the MV generator can perform adaptive ME according to the Multi-Path-Search algorithm described in (Jakubowski 2008). The MV generator sends MVs to the memories to obtain predictions. Based on these predictions the compensator

calculates residua and SAD values. The MV generator can determine the next step of the adaptation algorithm with reference to SAD values.
The compensator architecture is based on the pipeline design. It operates on 8×8 partitions and employs a SAD tree with four pipeline stages to generate SADs for all partition modes. Original and reference data are transferred from the local memories with double clock rate in the alternating way. Thus, in a single cycle of master clock, 64 samples of original and reference 8×8 blocks are fed to the SAD tree. Hence, to obtain SAD for the whole 16×16 MB, four clock cycles are necessary. Since during SAD calculation the next MV can be processed, every four clock cycles a new MV can be sent to the compensator. With such a setup, it is particularly beneficial to send MVs in long series, since it reduces the average time of single MV processing and increases the hardware utilization (avoiding wait states). Apart from the inter prediction, the compensator computes residua for intra predictions, which are first written to memories using 16x16-sample port.

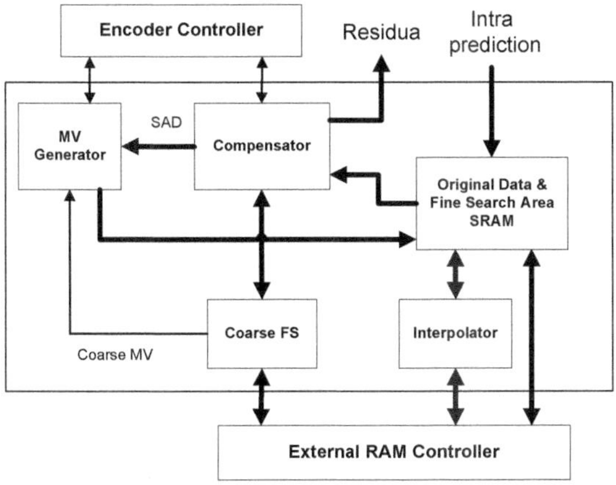

Fig. 3. Block diagram of motion estimator.

3.2 Intra prediction

High Profile of H.264/AVC standard defines three different kinds of INTRA prediction modes to be used for luma samples, and separate modes for chroma samples. Modes to be used for the luma sample prediction are: 4x4, 8x8, and 16x16, and are named after block sizes they operate on. The most commonly-used prediction modes are 4x4 ones. There are nine 4x4 modes, and eight of them are directional extrapolations of reconstructed samples from two neighbouring blocks (see Fig. 4). The ninth DC mode assigns the average of all reconstructed samples neighbouring with the current 4x4 block to predicted values. The 8x8 prediction modes are simple extensions of the 4x4 ones to blocks of the larger size. Therefore, there are also nine 8x8 modes, labelled identically as 4x4 ones. Except for the block size, the only difference comes from the prefiltering process. In particular, reference samples neighbouring with currently processed block undergo filtering before they are used for the prediction. Two of the directional modes: horizontal and vertical are the simplest since the prediction is equal to the copy of samples located to the left and above of the processed block, respectively. The remaining modes require some more complicated calculations according to the equations defined in H.264/AVC standard. Particularly, predictors are determined using two simple equations where the result is the weighted average from two or three reference samples.

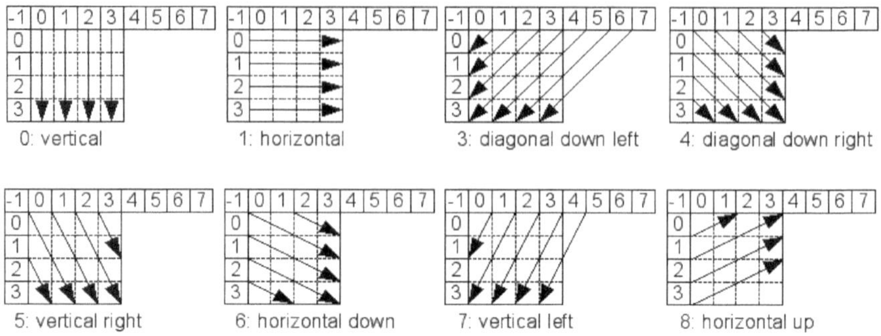

Fig. 4. Intra prediction modes for 4x4 blocks

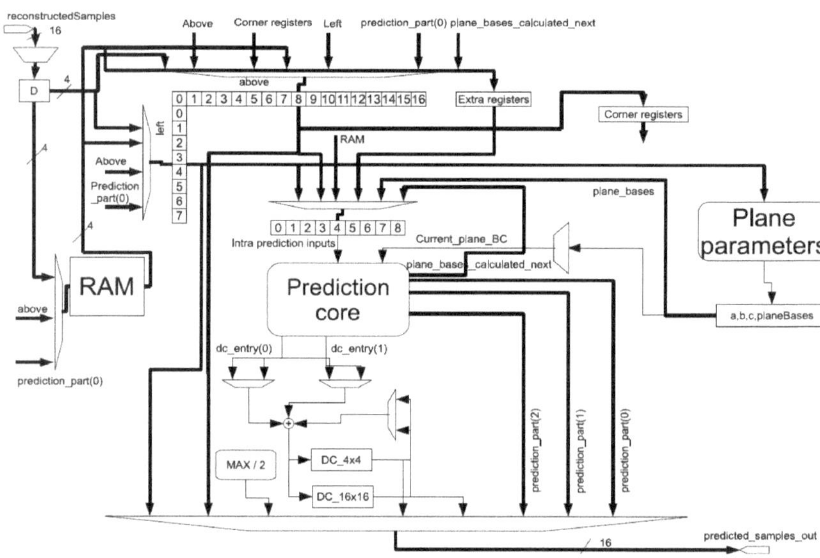

Fig. 5. Intra prediction block diagram

There are four 16x16 prediction modes defined by the H.264/AVC standard. Three of them: horizontal, vertical, and DC are simple extensions of corresponding 4x4 modes to 16x16 blocks. The fourth is the plane mode, the most computationally intensive one.

There are four chroma prediction modes defined for 4:2:0 and 4:2:2 sub-samplings. In fact, they are 16x16 luma prediction modes adapted to chroma block sizes. For 4:2:0 and 4:2:2, the prediction block size is 8x8 and 8x16 samples, respectively. In the case of 4:4:4 sampling scheme, there is no sub-sampling, and chroma predictions are obtained as the luma ones.

The Intra prediction architecture is described in details in (Roszkowski & Pastuszak, 2010). The architecture incorporates two important sub-modules that can be distinguished in the INTRA prediction module. These are: the neighbouring-sample buffer and the INTRA prediction arithmetic core. The first sub-module is responsible for tracking which 4x4 block is to be processed next and the selection of neighbouring samples as the reference. The second sub-module calculates all prediction modes for the 4x4 block selected by the first one. Fig. 5 presents the neighbouring sample buffer sub-module. The most important part is the on-chip dual-port RAM module. It keeps reconstructed samples neighbouring with the currently processed macroblock and reconstructed samples inside the macroblock, which are needed to calculate the prediction for next 4x4 blocks. The raster order of macroblocks involves keeping the whole frame line in the RAM to provide adjacent samples from the top-neighbouring macroblock. Since both 4x4 and 8x8 predictions are computed in the interleaved manner, reconstructed samples for the two modes must be stored, which increases the memory space. Each memory cell keeps four adjacent samples.

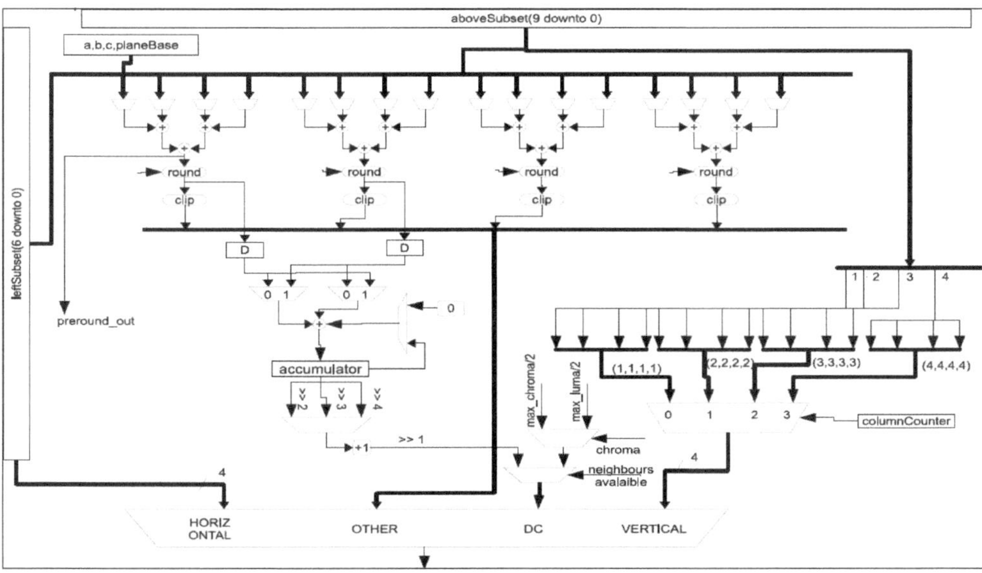

Fig. 6. Intra prediction arithmetic core

Plane prediction mode parameters are calculated in a separate sub-module in parallel with the calculation of 16x16 or chroma vertical and horizontal prediction modes. This allows a significant complexity reduction of the calculations of plane mode parameters as the multiplications can be replaced by the series of shift, addition, and accumulation operations. The input values to the prediction core are kept in nine intermediate registers. The rest of the module consists of the two levels of adders and multiplexers (see Fig. 6). The first and second levels of adders are responsible for the computation of the prediction values using three and two reference samples, respectively. As the result of the calculation, 15 different prediction values are obtained, out of which only up to 10 are valid for a 4x4 block. Those 10 are selected by the output multiplexer (MUX). The DC mode requires the reconfiguration of the adder structure, which is accomplished by multiplexers coloured dark grey in Fig. 7. The new configuration, together with the extra adder, allows the calculation of the prediction of the whole 4x4 block in one clock cycle. The prediction for 8x8 and 16x16 blocks, done by the accumulation, takes 2 and 4 clock cycles, respectively. The remaining multiplexers are used to reconfigure the core for the plane prediction.

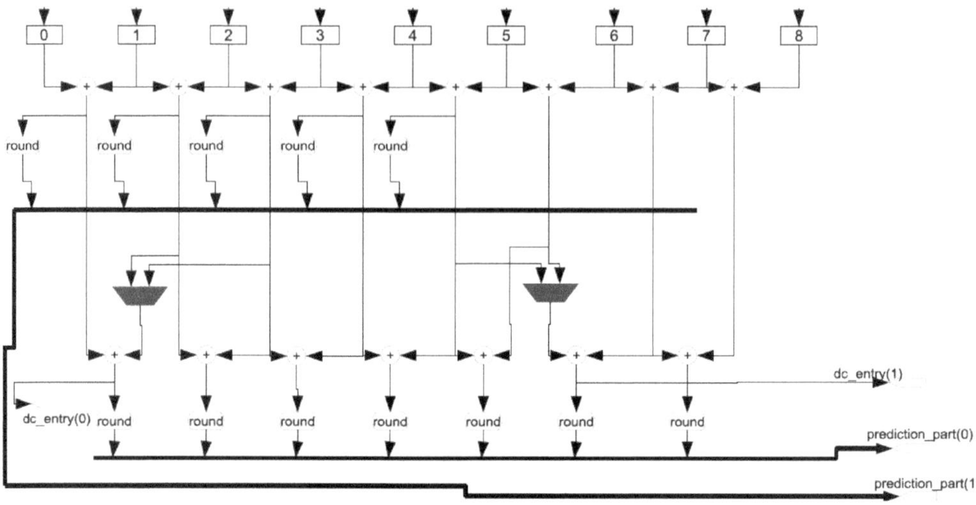

Fig. 7. Intra prediction modes for 4x4 blocks

3.3 Transforms

The primary transform applied in H.264/AVC is an exact-match integer 4×4 spatial block transform, which approximates DCT. The forward and inverse 4x4 transforms are shown in the Equation 1 and 2, respectively.

$$T_{FORWARD\_4x4} = \begin{bmatrix} 1 & 1 & 1 & 1 \\ 2 & 1 & -1 & -2 \\ 1 & -1 & -1 & 1 \\ 1 & -2 & 2 & -1 \end{bmatrix} \quad (1)$$

$$T_{INVERSE\_4x4} = \begin{bmatrix} 1 & 1 & 1 & \frac{1}{2} \\ 1 & \frac{1}{2} & -1 & -1 \\ 1 & -\frac{1}{2} & -1 & 1 \\ 1 & -1 & 1 & -\frac{1}{2} \end{bmatrix} \quad (2)$$

A secondary transform (Hadamard) performed on DC coefficients of the primary transform (for chroma DC coefficients and also luma in the 16x16 mode) allows for even more compression in smooth regions. Both transforms are similar, i.e., the secondary uses only 1 and -1 values in the matrix. For High Profile, the encoder can adaptively select between a 4×4 and 8×8 transform size for luma. The forward and inverse 8x8 transforms are shown in the Equation 3 and 4, respectively. As can be seen, the inverse matrix is a transposed version of the forward one.

$$T_{FORWARD\_8x8} = \begin{bmatrix} 1 & 1 & 1 & 1 & 1 & 1 & 1 & 1 \\ \frac{3}{2} & \frac{5}{4} & \frac{3}{4} & \frac{3}{8} & -\frac{3}{8} & -\frac{3}{4} & -\frac{5}{4} & -\frac{3}{2} \\ 1 & \frac{1}{2} & -\frac{1}{2} & -1 & -1 & -\frac{1}{2} & \frac{1}{2} & 1 \\ \frac{5}{4} & -\frac{3}{8} & -\frac{3}{2} & -\frac{3}{4} & \frac{3}{4} & \frac{3}{2} & \frac{3}{8} & -\frac{5}{4} \\ 1 & -1 & -1 & 1 & 1 & -1 & -1 & 1 \\ \frac{3}{4} & -\frac{3}{2} & \frac{3}{8} & \frac{5}{4} & -\frac{5}{4} & -\frac{3}{8} & \frac{3}{2} & -\frac{3}{4} \\ \frac{1}{2} & -1 & 1 & -\frac{1}{2} & -\frac{1}{2} & 1 & -1 & \frac{1}{2} \\ \frac{3}{8} & -\frac{3}{4} & \frac{5}{4} & -\frac{3}{2} & \frac{3}{2} & -\frac{5}{4} & \frac{3}{4} & -\frac{3}{8} \end{bmatrix} \quad (3)$$

$$T_{INVERSE\_8x8} = \begin{bmatrix} 1 & \frac{3}{2} & 1 & \frac{5}{4} & 1 & \frac{3}{4} & \frac{1}{2} & \frac{3}{8} \\ 1 & \frac{5}{4} & \frac{1}{2} & -\frac{3}{8} & -1 & -\frac{3}{2} & -1 & -\frac{3}{4} \\ 1 & \frac{3}{4} & -\frac{1}{2} & -\frac{3}{2} & -1 & \frac{3}{8} & 1 & \frac{5}{4} \\ 1 & \frac{3}{8} & -1 & -\frac{3}{4} & 1 & \frac{5}{4} & -\frac{1}{2} & -\frac{3}{2} \\ 1 & -\frac{3}{8} & -1 & \frac{3}{4} & 1 & -\frac{5}{4} & -\frac{1}{2} & \frac{3}{2} \\ 1 & -\frac{3}{4} & -\frac{1}{2} & \frac{3}{2} & -1 & -\frac{3}{8} & 1 & -\frac{5}{4} \\ 1 & -\frac{5}{4} & \frac{1}{2} & \frac{3}{8} & -1 & \frac{3}{2} & -1 & \frac{3}{4} \\ 1 & -\frac{3}{2} & 1 & -\frac{5}{4} & 1 & -\frac{3}{4} & \frac{1}{2} & -\frac{3}{8} \end{bmatrix} \quad (4)$$

To simplify computations, all the transforms should be decomposed into two or three stages using butterfly structures. Actually, the standard defines the inverse transforms in this form. Thus, rounding operations must be performed in the decomposed form to keep the specification consistency. For a block, appropriate matrix is applied on each row and then on each column to obtain the 2D transform.

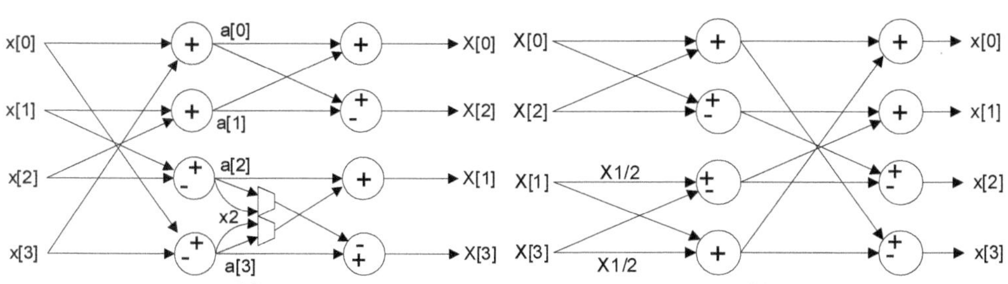

Fig. 8. Diagram of the forward and inverse transforms for 4x4 blocks

The best way to implement a transform is to use its decomposed form. Such a decomposed form is depicted in Fig. 8 for the 4x4 blocks and in Fig. 9 for 8x8 blocks. The forward 4x4 transform in Fig. 8.a supports both the approximate DCT and the Hadamard transform. Particularly, additional multiplexers enable a small reconfiguration of the connectivity. The transforms for 8x8 blocks are more complex. They consist of four processing stages, whereas two processing stages are used for 4x4 blocks.

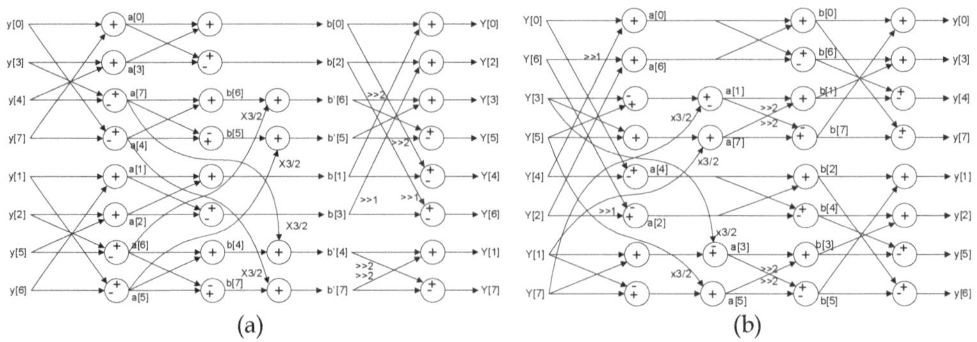

Fig. 9. Diagram of the forward and inverse transform for 8x8 blocks

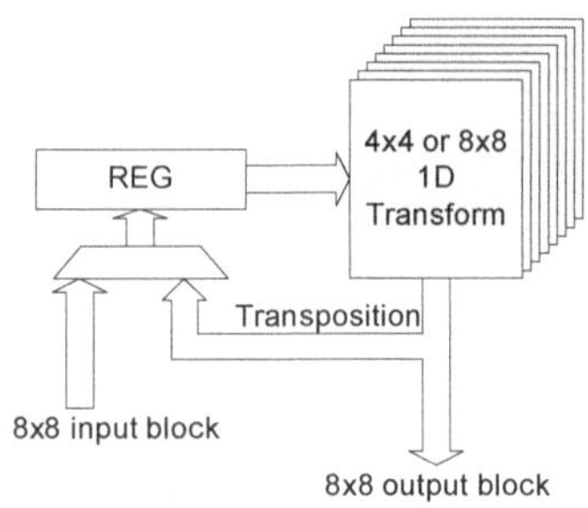

Fig. 10. Block diagram of the transform module

When multiple transforms are to be supported, the encoder can simply embed dedicated modules, each of which supports one transform type. To keep dataflow regularity in the forward or inverse transform, two modules for the four-element transform (4x4 blocks) and one for the eight-element transform (8x8 blocks) should be employed. The selection between two transform types is done by multiplexers placed at the output stage. Such a design is inefficient in terms of hardware resources since only one branch is used at one time. Thus, the efficient solution should utilize the same resources with the overhead as little as possible. The transform architecture with higher throughput can be easily designed by employing eight parallel eight-point transform logic units, as shown in Fig. 10. The result is computed in two clock cycles, 1D transform is performed in one cycle. More details about sharing resources between the two transform types can be found in (Pastuszak, 2008.a).

3.4 Quantization

The forward and inverse transform matrices are not orthogonal. To achieve this feature in the whole processing, quantization step sizes are modified. As a consequence, the step size depends on the position in the coefficient block. Actually, the quantization and dequantization are accomplished by the multiplication and shifting operations. Equations 5 and 6 show formulas for the quantization and the dequantization, respectively.

$$X_q(i,j) = sign\{X(i,j)\}(|X(i,j)|A(Qp\%6,i,j) + \tfrac{1}{3}2^{11+L+Qp/6}) >> (11 + L + Qp/6) \quad (5)$$

$$X_r(i,j) = (X_q(i,j)B(Qp\%6,i,j) + 2^{L-Qp/6}) << (Qp/6 - L) \quad (6)$$

In these equations, L is equal to either 4, 5, or 6 and depends on the transform size. Functions A and B include values of multiplicands for each location in the block. The values depend on the quantization parameter (Qp) and the transform size. Note that the position inputs identify the coefficient location in the rectangular structure (4x4 or 8x8). Block diagrams in Fig. 11 show dataflow of the quantizer and the dequantizer. In contrary to the dequantization, the quantizer embeds the addition of a fraction dependent on the coefficient sign. Since the units map one input into one output, it is easy to parallelize them to increase the throughput.

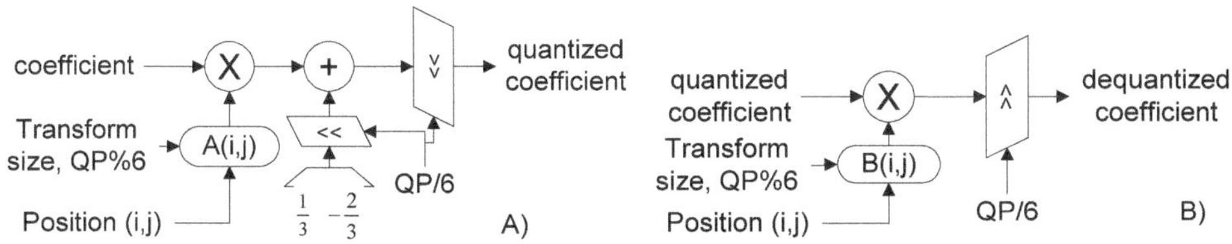

Fig. 11. Block diagram of the quantization (A) and dequantization (B)

3.5 Reconstruction

Following dequantization, reconstructed residuals are added to the prediction (intra or inter) to obtain reconstructed samples. As prediction samples in the encoder are computed in the motion estimation and intra prediction units, their bypassing to the reconstruction stage involves significant storage resources and write conflicts. An alternative is to refer to original residuals (registered at the transform input) and original samples in two successive clock cycles. In the first cycle, this approach allows the computation of the reconstruction error equal to the difference between original and reconstructed residuals. The reconstruction error subtracted from original samples gives reconstructed samples in the second cycle. In the high-throughput datapath, many subtractors can be utilized to perform the parallel reconstruction (see Fig. 12). To avoid underflow/overflow, the result is limited to the pixel sample range in the following pipeline stage.

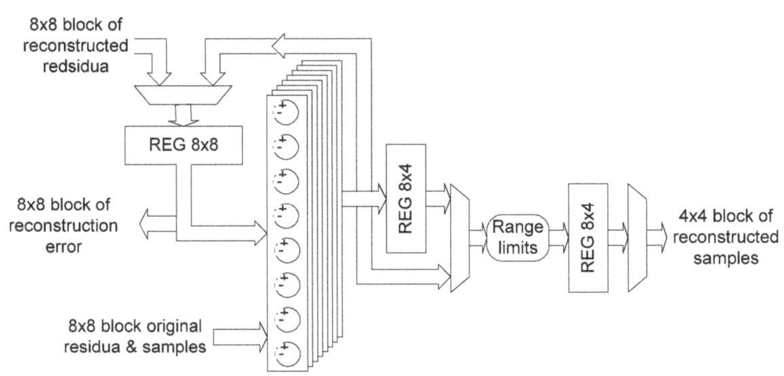

Fig. 12. Block diagram of the reconstruction unit

3.6 Mode selection

The simplest way to select the coding mode is to compute Sum of Absolute Differences (SAD) for each tested prediction and to select the case with the minimal SAD. This approach does not provide the optimal mode selection. A more advanced approach refers to actual code-stream rates and distortions. However, this involves much more computations and storage resources. The cost measure for a given mode is based on the cost functions, according to the following equations:

$$J_1(R,D) = D + R * \lambda \tag{7}$$

$$J_2(R,D) = D * \lambda^{-1} + R \tag{8}$$

Note that λ is the Lagrangian multiplier whose value is adjusted to the desired compression ratio. The J_1 and J_2 cost measures are expressed in distortion and rate domain, respectively. In the developed architecture the second measure has been selected as the multiplication is performed only once after obtaining the distortion. The distortion is computed based on the reconstruction error (see previous section). Particularly, the error for each sample should be squared, and the results can be summed within 4x4 partitions. Such Sum of Squared Errors (SSE) can be multiplied by the Lagrangian multiplier. As the developed architecture operates on 8x8 blocks, the distortions for four 4x4 subpartitions are summed, and only one multiplication circuit is enough for the assumed throughput of 32 samples/coefficients per clock cycle. The cost for larger partitions can be obtained by summing costs for smaller ones. The second factor in the cost function is the rate measured in bit units. To estimate actual rates, the analysis of quantized coefficients following the binarization schemas is indispensible. There are two entropy coding modes with different schemas. When the CABAC is used, coefficients are coded using Exp-Golomb schema before arithmetic coding. Although the estimation of rates based on single coefficient values is easy, the probability adaptation can affect the estimation accuracy. On the other hand, the CAVLC binarization is based on the concatenation of successive codewords. Thus, the total rate is the sum of codeword rates. Since the CAVLC adapts binarization schemas while coding coefficients within 4x4 blocks, the estimation of coefficient rates involves the signal chain between 16 subcircuits corresponding to each coefficient. To shorten critical paths, the subcircuits should be placed in successive pipeline stages.

The block diagram of the mode selection module is depicted in Fig. 12. The parallelism employed in the developed encoder enables the repetition of quantization and transformation for different coding options to select the best one. In particular, it is assumed that the pipeline can process 8x8 blocks at the average throughput of 32 samples/coefficients per clock cycle. Hence, the module is able to check four 8x8, two 16x8, and two 8x16 partitions in successive eight clock cycles. The 16x16 partition (not partitioned macroblock) is analyzed in the separate path that simply aggregate costs of four successively analyzed 8x8 partitions. Addition of side cost (e.g., motion vectors, intra directions, macroblock/submacroblock types) allows a more reliable cost comparison. Actually, motion vectors and intra directions are coded using the prediction from the top and left neighbours. The dedicated memory (context) keeping picture line data allows the reference to the top neighbours, excluding cases when the reference partition belongs to the same macroblock.

As the mode selection for a macroblock and its partitions takes some time, it is necessary to buffer quantized coefficients for some different modes. When the macroblock mode is selected, quantized coefficients comprising a 4x4 block are accessed concurrently at the entropy coder side, so that they are read in form 16x8-bit memory buffer. This parallel access results from the fact that such an order is at the write port. The analysis path uses pointers to identify addresses of 8x8 blocks stored in the buffer. Additionally, a vector of three-bit registers (kept in the write stage) identifies how many references to an 8x8 partition at a given address are valid. 8x8 partitions are written at four address identified by one pointer, and each address corresponding to a 4x4 block is distinguished by two bits based on its location. If intra and inter blocks are written, the corresponding register is set to one and four, respectively. If a reference is no longer valid, the pointers select which register should be decremented (discarded pointer). Four references match the case when an 8x8 partition contributes to the macroblock mode selection for four portioning types. Actually, each partition can have a different motion vector and reference picture selected based on the cost minimization. While the final macroblock mode is not selected, the best partition mode for both transform sizes and some quantization parameters should be looked for. This requires additional storage resources to save pointers, costs, motion vectors, and reference pictures (partition cost buffer and 16x16 cost buffer with pointers). Also intra modes should have storage space assigned. Taking into account the throughput, it can be seen that the analysis of partitions larger than 8x8 requires the pipeline registers carrying coding mode parameters and costs. This correspond to the first part (partition cost) of the mode selection block diagram in Fig. 13.

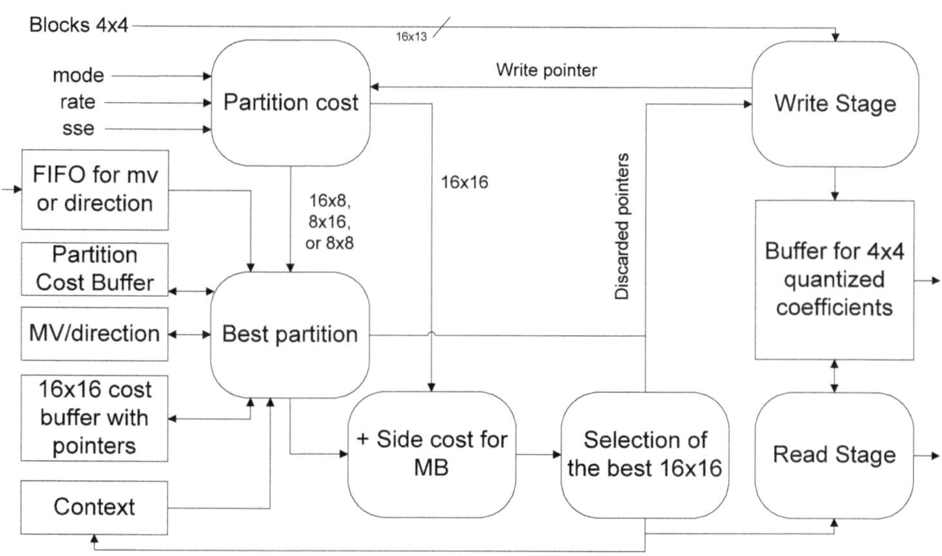

Fig. 13. Block diagram of the mode selection module

3.7 Entropy coding

In H.264/AVC two modes are employed for binary coding: Context Adaptive Binary Arithmetic Coding (CABAC) and Context Adaptive Variable Length Coding (CAVLC). The first mode provides higher compression efficiency at the cost of computational complexity.

The following subsections review the processing blocks for the two modes. More details can be found in the reference (Pastuszak, 2008.b).

3.7.1 Variable length coding

Since residual coefficients comprise the largest part of the codestream, exploiting correlations between them considerably improves the compression efficiency. Five types of syntax elements are processed in the CAVLC mode:

- For luma blocks, the total of non-zero coefficients and trailing ones (series of high-frequency coefficients equal to one) are coded as one element by the use of four look-up tables (three VLC tables and one 6-bit fixed table, each having 64 entries). The tables are selected adaptively based on the number of non-zero coefficients in the neighbouring 4x4 blocks. Besides, there are three additional tables for chroma blocks.
- Sign coding does not require context modelling, since one bit per non-zero coefficient is enough to convey this information.
- The code for each coefficient level is made up of a prefix and a suffix. The length of the latter one is initialized to either 0 or 1 and incremented every time when consecutive levels exceed predefined thresholds. This adaptation is due to the observation that statistically values of coefficients increase while passing from high to low frequencies.
- The total of zero-valued coefficients (total_zeros) preceding the last non-zero coefficient in the coding order refers to some VLC tables. One table is selected based on the number of non-zero coefficients coded earlier.
- The number of zeros preceding each non-zero coefficient (run_before) is encoded in reverse order. The adaptation is performed by the selection of codes dependent on the number of zero-valued coefficients left to be coded in this order.

The developed architecture of the H.264/AVC binary coder embeds the binarization unit as a part sufficient to support the CAVLC mode and perform the binarization in the CABAC mode. The binarization unit embeds four pipeline stages, as depicted in Fig. 14. Most of registers incorporated to the architecture are shared in both coding modes. Input data are submitted through dedicated ports, each of which matches one type of syntax element.

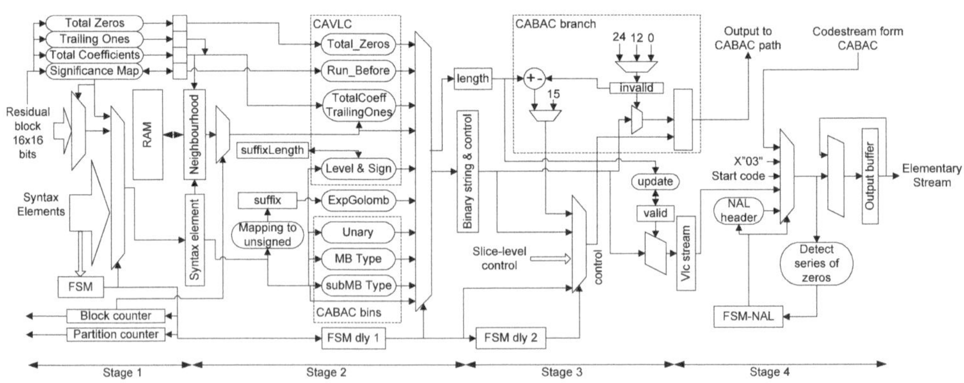

Fig. 14. Block diagram of the double-mode binarization unit

The binary coder processes syntax elements in the order defined in the standard. The order depends on selected options and previous data. Therefore, the architecture incorporates a

Finite State Machine (FSM) to determine the type and the order of the processed data. Transitions of the FSM depend on the values of syntax elements available on parallel input ports. The first stage selects one input port and loads corresponding data to the syntax-element buffer on the basis of the state of the FSM, counters, and a significance map. The FSM determines the type of the syntax element, whereas the counters point one subunit of a given macroblock such as a partition and a block (4x4). One FSM used in two modes simplify the design as states and transitions are almost the same. The main difference in transitions comes from the order of syntax elements within a 4x4 block. In the CAVLC mode, each block is scanned two times (i.e., non-zero coefficient levels precede runs of zero coefficients) whereas in the CABAC mode, just one scan is enough.

The first stage analyzes each 4x4 block to compute the number of non-zero (Total Coefficients) and zero-valued (Total Zeros) coefficients, the number of trailing ones, and the significance map. The significance map consists of 16 bits, where each bit is set active when the corresponding coefficient is non-zero. This allows the selection of coefficients to be processed. When a coefficient is selected, the corresponding significance indicator is set inactive. In the CAVLC mode, the first stage performs also the reference to a total of non-zero coefficients and trailing ones for the upper- and the left-neighbouring blocks. The referred numbers are used to compute the average (nC) forwarded to the second stage.

Raster scanning of macroblocks involves the use of an on-chip memory to convey references between rows. The memory incorporated into the architecture has the bit width equal to 48. This value matches the accumulated length of reference registers on one macroblock edge. The number of entries determines the maximal frame width in macroblocks and is set to 128 allowing HDTV resolutions.

Although context-formation rules for the CABAC differ from those for the CAVLC, it is possible to share storage elements in both modes. Thus, in the CABAC mode, the architecture keeps motion vectors differences instead of non-zero coefficients in the neighbourhood registers and the on-chip memory. However, the storage space is doubled since four six-bit MVD can be used for the smallest 4x4 partition. Nevertheless, sharing enables the efficient reduction of hardware resources. Additionally, the control subcircuit is common to both modes.

The second stage maps syntax elements onto their binary representation using the set of primitives (subcircuits) implemented as a combinational logic. Apart from a binary string, the primitives produce the corresponding length. For a given syntax element, a one-cycle delayed FSM selects the outcome of one primitive. The primitives support Unary, Exp-Golomb, macroblock, and submacroblock binarizations. The second stage includes dedicated subcircuits for adaptively-coded syntax elements in the CAVLC mode (i.e., 4x4 residual blocks).

The third stage forwards all code strings produced in the second stage to one of two paths. The first path, which supports the CABAC mode, assembles a binarized representation of a syntax element along with control data into 16-bit words and submits them to the context formatter. Each syntax element allocates bits in a specific way. The control information data includes the number of valid bits, indicators of the last syntax element in a series (e.g., coefficients), and the information about the neighbouring subunits within the current macroblock (e.g., coded block flag). When a binary string is long, it is divided into parts conveyed in successive output words to the CABAC path. A

relevant part is selected using the barrel shifter driven by the register which identifies the number of released bits (invalid). In practice, some particular values are allowed, such as 0, 12, and multiplications of 7.
The second path, which supports VLC binarization schemas, concatenates code strings to form a codestream. The concatenation is performed in the VLC buffer and code strings are appended in successive clock cycles using a barrel shifter. Particularly, the shifter is driven by the number of valid bits kept in a separate register. It is increased by the length of a code string and decreased by the number of bits (eight-byte units) forwarded to the next stage.
The last fourth stage combines codestreams produced by the binarization and CABAC paths and encapsulates them into Network Abstraction Layer units. Note that data are accepted only from one path at a time depending on the selected mode and the processing state. The encapsulation amounts to adding one-byte header and the start code byte sequence at the beginning of each slice and sequence/picture headers. Additionally, an emulation prevention three byte (0x03) has to be inserted into the codestream when there is a forbidden byte sequence encountered. To facilitate the insertion process, previous pipeline stages (including CABAC path) are halted for one clock cycle. A dedicated subcircuit is responsible for the detection of the forbidden byte sequence. The subcircuit searches for 22 zero-valued bits starting from byte-aligned positions. All the processes are controlled by a dedicated FSM.

3.7.2 Arithmetic coding

The CABAC keeps up to 1024 probability models to increase the coding efficiency. Each type of syntax elements corresponds to a set of probability models pointed by different context labels. Each model is a Finite State Machine (FSM) that consists of the value of the more probable symbol (MPS) and the probability of the less probable symbol (LPS). The two variables are initialized based on the quantization parameter Qp with reference to the initialization set and the frame type. The FSMs are updated according to pre-defined adaptation rules. Context labels are computed as a sum of an offset ordered to a syntax element and an increment. Some increments are generated by referring to two adjacent macroblocks (16x16) or blocks (8x8 or 4x4) located on the left and the top of the current one. For other kinds of context labels, increments are formed on the basis of the previous bin value and the position in the binary string.
The main process in the CABAC is the recursive subdivision of a probability interval. In order to subdivide a probability interval length (range) into two subranges, probability estimates are determined on the basis of the probability model. The length of the first subinterval (LPS) is equal to the probability estimate, whereas that of the second one (MPS) is obtained by subtraction of the estimate from the current interval length. Depending on LPS/MPS coding, one of these subintervals is selected as a new interval length and renormalized to have the non-zero bit in the MSB position. While coding LPS, the subtraction outcome is added to the interval base (low). Successive renormalization shifts for the interval length trigger analogous modifications of the interval base. Bits released from MSB positions of the interval base drive the codestream formation process.
As some contexts are generated with reference to two adjacent macroblocks located to the left and on the top of the current one, the information relevant to form future contexts is

stored in registers and a double-port RAM memory, respectively. Access to the memory is performed on the macroblock basis. In the memory, 29 bits are required.

The architecture of the context formatter embeds one processing stage with an additional output stage as shown in Fig 15.a. Input data are produced by the binarization block and stored in the FIFO buffer. Loading of these data into registers is controlled by the FSM. Transactions of the FSM are driven by the counter (COUNT) and values of bits in the binarized representation. For each binarized syntax element, the counter determines the position of the bit for which context is generated. In fact, the position indicates the number of bits that have already been processed. On the basis of the state of the FSM, the context offset corresponding to a given syntax element is generated. Several offset-increment pairs are generated and stored in a small buffer. The adjustment of the context generation ratio is achieved by reading two pairs from the buffer. Having processed a syntax element, the input registers (CUR REG) are reloaded by the data for the following syntax element. If the information in the left-neighbouring registers (LEFT REG) is no longer referenced, the registers are successively rewritten by states of relevant registers for the current macroblock. This information is also stored in the context memory when all data for the current macroblock are released.

The block diagram of the CABAC initialisation unit is depicted in Fig. 15.b. The unit sets states of the CABAC probability model prior to submitting context-symbol pairs form the context formatter. To perform this task, one pair consisting of an index and a binary value of MPS is generated in each clock cycle. Although the initialisation procedure stops the main coding routine of the CABAC, associated time intervals have a small impact on the throughput. The initialisation unit applies three pipeline stages. The first stage generates the address to the 4Kx16-bit ROM memory used to keep initialisation parameters for four sets of parameters (one for INTRA and three for INTER) for High Profile (460 contexts). The second stage computes the internal variable denoted as *preState* on the basis of the quantization parameter Qp and parameters read from the memory. The computation is accomplished with the use of the multiplication and addition units. Apart from this, the subtraction of an offset value from the address taken from the previous stage provides the context label. The third stage maps the *preState* variable onto a MPS value and an index.

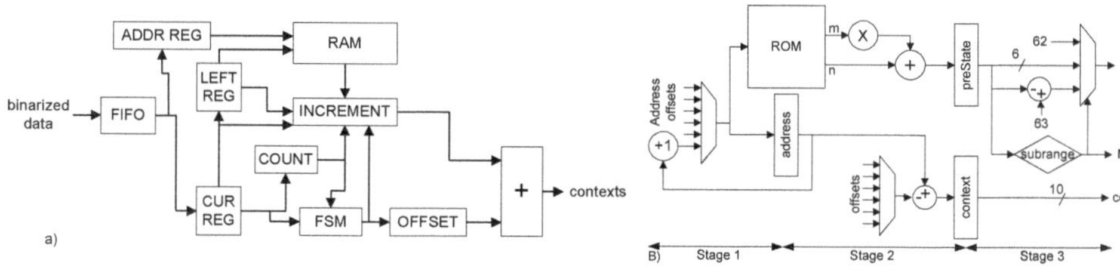

Fig. 15. Architectures of the context formatter (a) and the initialization unit (b)

The architecture of the arithmetic coder core with the enhanced bypass mode applies 9 pipeline stages (see Fig 16). This allows the minimization of critical paths and the adaptation to timing constraints resulting from reading the probability state memory. The first delay stage for input data is introduced to adjust input data to those read data from the probability state memory (addressed by the context label). As a consequence, the second stage receives simultaneously values of contexts, symbols, indices, and the most probable symbols. On the basis of these variables, the circuit calculates a new indices and new values of the most probable symbols and stores them into the memory. Moreover, there are control signals to indicate either LPS or MPS coding. The memory operates at the doubled frequency to overcome problem of the simultaneous access to two entries corresponding to contexts ordered to symbols submitted in the same clock cycle of the main clock. If any same context labels are submitted in consecutive clock cycles, the first stage takes actual indices and MPS values from the following stage to keep the data consistency.

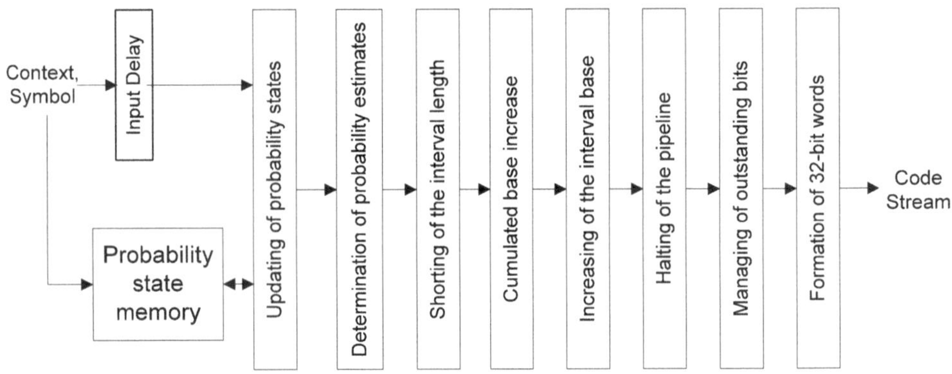

Fig. 16. Arithmetic Coding Pipeline

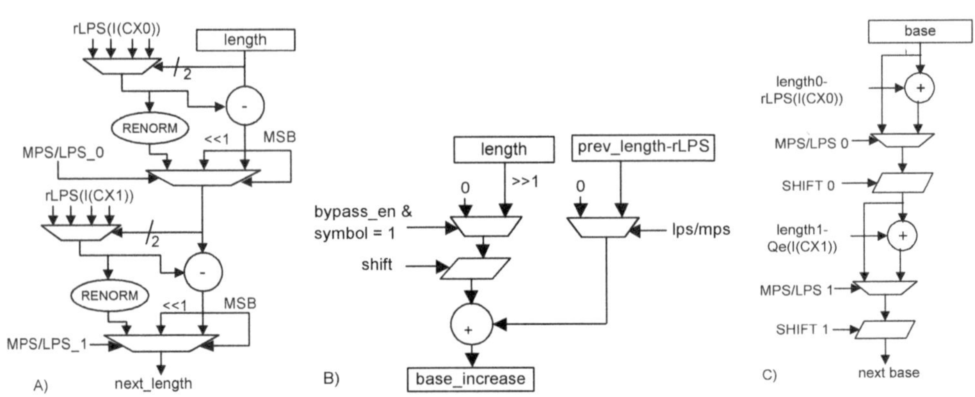

Fig. 17. Arithmetic Coding Stages 4th-6th

The LPS/MPS signal, along with the old index value, is forwarded to the third stage, which calculates probability estimates rLPS using four LUTs. The next stage reduces the interval length as shown in Fig. 17.a. The fifth stage computes the cumulated variables corresponding to the regular and bypass-mode symbols as shown in Fig. 17.b. They are used to increase the base register at the sixth stage (see Fig. 17.c). Bits released from this register are formed into codestream at the eight stage. Here, the outstanding bit counter collects series of ones and looks for a zero-valued bit or a carry to produce a part of the codestream.

It may occur that the number of bits to release is greater than the buffer size in the following stage. Such an event implicates the insertion of wait states, which stop all preceding pipeline stages, and the context-formation unit. A hold signal is driven directly by a register to optimize the clock rate. This involves a one-clock-cycle delay, which in turn imposes the use of an additional seventh stage to prevent loses of data between stopped and unstopped registers. The final tenth stage collects codestream into 32-bit words and releases them outside of the CABAC block.

3.8 Deblocking

The deblocking filter is applied to minimize artefacts on block/macroblock boundaries along both horizontal and vertical edges. The filtering is a two-phase non-linear operation that affects samples adjacent to boundaries and sometimes also their direct neighbours. Both phases are similar. In the first phase, the horizontal filter operates on vertical edges, whereas the vertical filter operates on horizontal edges in the second phase. The deblocking-filter data path is shown in Fig. 18. The module accepts one sample per clock cycles and the same throughput is at the output. Samples are carried by the pipeline registers. When a block edge samples are in q0 and p0 registers, the filter is activated (writing samples form the filter logic to registers). Since macroblocks are coded in the raster order, it sis necessary to incorporate a dedicated memory to buffer four picture lines (line of MB) for the filtering horizontal edges between macroblocks. One macroblock memory (MB1) is used to transpose the horizontally-filtered samples before the vertical filter. Another one (MB2) keeps left neighbouring samples form the previous macroblock.

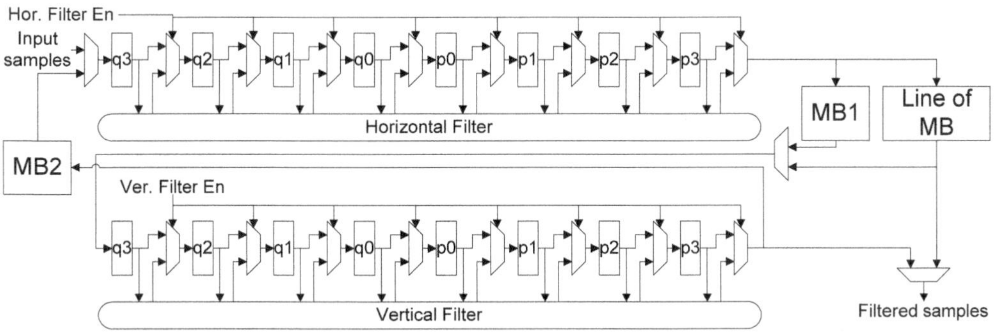

Fig. 18. Dataflow in the deblocking filter

There are four filter strengths, and the selection depends on two variables written into the codestream, the quantization parameters (alpha and beta), and edge type (macroblock or block). Horizontal and vertical filter logic embeds all the functionality that modifies samples based on the filter strength value. In particular, the non-linear filter logic analyzes input samples according to predefined formulas and compares the result with thresholds determined by the filter strength. If the threshold is exceeded, the filter is activated.

4. Implementation results

There are many complete video coding solutions developed by the scientific teams and commercial companies. The performance and resource cost is summarized in Section 2.3 for some of H.264/AVC encoders. This Section provides the implementation results of the developed architecture for key modules and compares them with other works.
Table 3 summarizes the resource consumption for modules described in Section 3. Note that the full encoder architecture needs more resources for the control and additional buffering between some modules. Moreover, the real hardware implementation requires some communication interfaces, i.e., the external memory controller, the codestream port, and the configuration port. Maximal clock rates obtained for the architecture are equal 100 MHz and 250 MHz for Aria and TSMC technologies, respectively.
Compared to other designs (see Section 2.3.), the developed architecture needs more on-chip memoires. The higher memory consumption results from the buffers incorporated to support the mode selection based on the rate-distortion analysis. This feature makes the architecture suitable for FPGA devices equipped with a significant amount of on-chip memories. Compared other designs, the logic consumption is relatively low when taking into account the encoder capability. Particularly, it can support High Profile options and HDTV at 200 MHz. Moreover, the advanced mode selection based on the rate-distortion criteria allows a better compression ratio for a given bit rate.

Module	Aria II [ALUT]	TSMC 0.13 μm [gate]	Memory [Kbit]
INTER PRED.	18756	140413	1300
INTRA PRED.	4599	23197	64
DCT	5279	37178	0
IDCT	5468	65869	0
QUANT	32xDSP+5236	78038	0
DEQUANT	32xDSP+2221	35421	0
RECONSTR.	3175	23420	0
DEBLOCKING	2395	17910	26
MODE SEL.	1xDSP+10646	39333	148
ENTROPY	66xDSP+6682	33206	105
ENCODER	72419	673256	2250

Table 3. Resource consumption for main modules of the hardware video encoder

5. Conclusion

The complexity of the state-of the-art video compression is high. The real-time performance requires the use of most advanced IC technologies to support high-definition resolutions. Additionally, the need for buffering at different processing steps requires on-chip and external memories. The improvement in the compression efficiency requires more resources to tests many prediction modes and perform the rate-distortion analysis. The architecture described in the chapter is still developed. Particularly, it includes more advanced methods for the mode selections (alternative distortion measures, different quantization parameters, adaptive quantization), multi-view coding, and the robust rate control.

6. Acknowledgment

The work presented was developed within the research project LIDER/05/8/L-2/10/NCBiR/2011 founded by the Notional Centre for Research and Development, Warsaw, Poland.

7. References

Chen Y.-H., Chen T.-C., Tsai C.-Y., Tsai S.-F., & Chen L.-G.; Algorithm and Architecture Design of Power-Oriented H.264/AVC Baseline Profile Encoder for Portable Devices, *IEEE Transactions on Circuits and Systems for Video Technology*, vol.19, no.8, pp.1118-1128, ISSN 1051-8215, Aug. 2009

Jakubowski, M. & Pastuszak, G.; (2008). Data Reuse in Two-Level Hierarchical Motion Estimation for High Resolution Video Coding, *Proceedings of SIGMAP 2010 International Conference on Signal Processing and Multimedia Applications*, pp. 159-162, Athens, Greece, July 26-28, 2010

Lin Y.-K., Ku C.-W., Li D.-W., & Chang, T.-S.; (2009). A 140-MHz 94 K Gates HD1080p 30-Frames/s Intra-Only Profile H.264 Encoder. *IEEE Transactions on Circuits and Systems for Video Technology*, Vol.19, No.3, (March 2009), pp. 432-436, ISSN 1051-8215

Lin Y.-L. S, Kao C.-Y., Kuo H.-C., & Chen J.-W., VLSI Design for Video Coding: H.264/AVC Encoding from Standard Specification to Chip, *Springer*, 2010, ISBN 978-1-4419-0958-9

Liu Z.; Song Y., Shao M., Li S., Li L., Ishiwata S., Nakagawa M., Goto S. & Ikenaga, T.; HDTV1080p H.264/AVC Encoder Chip Design and Performance Analysis, *IEEE Journal of Solid-State Circuits*, vol.44, no.2, pp.594-608, Feb. 2009

Pastuszak, G.; (2008). Transforms and Quantization in the High-Throughput H.264/AVC Encoder Based on Advanced Mode Selection, *Proceedings of ISVLSI 2008 IEEE Annual Symposium on VLSI*, pp. 14-17, Montpellier, France, April 7-9, 2008

Pastuszak, G. (2008). A High Performance Architecture of the Double-Mode Binary Coder for H.264.AVC. *IEEE Transactions on Circuits and Systems for Video Technology*, Vol.18, No.7, (July 2008), pp. 949-960, ISSN 1051-8215

Roszkowski, M.; Abramowski, A.; Wieczorek, M. & Pastuszak, G. (2010). Architecture design of the hardware H.264/AVC video decoder. *Journal of Electronics and Telecommunications*, Vol.55, No.3, (3/2010), pp. 291-300, ISSN 0867-6747

Roszkowski, M. & Pastuszak G.; (2010). Intra Prediction Hardware Module for High-Profile H.264/AVC Encoder, *Signal Processing - Algorithms, Architectures, Arrangements, and Applications* (SPA 2010), Poznań, Poland, 23-25 September 2010.

Y. W. Huang, B. Y. Hsieh, T. C. Chen, and L. G. Chen, "Analysis, fast algorithm, and VLSI architecture design for H.264/AVC intra frame coder," *IEEE Trans. Circuits Syst. Video Technol.*, vol. 15, no. 3, pp. 378–401, Mar. 2005.

11

DRM & Security Enabling Mechanisms Leveraging User Centric Multimedia Convergence

Anastasios Fragopoulos, John Gialelis and Dimitrios Serpanos
Industrial Systems Institute (I.S.I.)
Greece

1. Introduction

There have been considerable efforts to have audiovisual multimedia systems and applications converge, in home environments with homes as spaces of convergence, and for nomadic users with advanced mobile devices as points of convergence. These trends are important but also have limitations that have to be addressed and overcome, i.e. home-centric systems fail to account for increased mobility and desire to provide continuous service across spatial boundaries outside the home; device-centric convergence, e.g. in 3G phones, supports nomadic use but provides a very limited user experience as no single device and interface will fit many different applications well; furthermore, in both cases there are a lot of different security aspects that arise and have to be taken in care. The trend in our era, is to move beyond home and device-centric convergence toward truly user-centric convergence of multimedia, where the user is acting as the point at which services (multimedia applications) and the means for interacting with them (devices and interfaces) converge.

One of the biggest challenges that we are facing in the deployment of architectures in such environments is related to, on the one hand, the security and protection of digital contents that interchanged between users in such pervasive ubiquitous computing environments and on the other hand with the provision to the users with security mechanisms that allow them to perform secure transactions (e.g. authentication, privacy protection, secure data transfer, etc.) in those environments. Moreover protection of Intellectual Property (IP) is a necessity in modern multimedia architectures. The concerns of content creators for loss of revenues constitute a strong obstacle to the wide deployment of architectures that involve distribution of IP protected digital content. Today the end-users are equipped with different types of small devices that allow them to be the digital contents creators, thus creating digital content that wish to share with third parties. In most cases, the end-users would like to have mechanisms which would give them the possibility to protect the content which have created and possibly to set their own usage rights over it, thus specifying towards third users how their digital content shall be used.

Digital Rights Management (DRM) mechanisms constitute of various technologies that have been developed and deployed by content providers, creators, distributors, in order to protect their digital media from illegally, unauthorized and without the appropriate rights usage of their products, while let them to use possibly unsafe media like Internet for delivering their products with less hesitation and anxiety about non-legitimate usage of their content. Moreover, the increasing capabilities of embedded systems combined with their decreasing cost have enabled their adoption in a wide range of personalized entertainment services, applications and services, leading thus to a user-converged networked multimedia environments. Furthermore, as dynamicity in networks, embedded security and interoperability in DRM systems become the critical aspects in such networked ecosystems, new emerging frameworks for secure, user-converged digital content delivery are required, leading to specific treatment for the design and deployment of DRM systems that take in care the resource-demanding nature of security in embedded systems, (Fragopoulos & Serpanos, 2005), (Fragopoulos et al., 2009). Thus, it is a requirement to provide integrated DRM mechanisms in such services and applications that target delivery of IP protected content to a large base of clients. Considering the technical problems that result from add-on security solutions to independently developed network services, the design and deployment of architectures with security and DRM as inherent requirements will lead to secure solutions that will increase the trust placed by content providers on the system and thus, it will lead to wider availability of services to a larger population.

In this book chapter, we describe extensive research that has been done in the areas of DRM management and embedded security, towards the design and deployment of an integrated architecture that exposes security functionalities and DRM mechanisms, focusing on user-centric and nomadic environments. In that context we have taken specific care how we could adapt our architecture in order to cope with mobility management issues, while providing interoperability towards other similar architectures.

The architecture that we describe, (a) provides the capability to the user to act as content creator who set its own usage rules to his content, thus protecting digital content from unauthorized usage from non-legitimate users; (b) operates over heterogeneous network technologies; (c) provides to the end-user friendliness; (d) provides adequate security mechanisms, under possible attacks; and, (e) is adapted to mobility frameworks, providing secure access to DRM protected multimedia digital contents, (enabling DRM in session migration – high mobility environments).

Towards implementing licensing for DRM-protected multimedia contents, we have focused mainly to the usage of newly proposed MPEG-21 standard, (International Standards Organization [ISO], 2005), (Burnett et al., 2006), which has been recently proposed for primary use in the area of multimedia world, allowing the seamless, interoperable, transparent and universal delivery of multimedia digital contents to the end-users in a dynamic environment. More specific, our research work involved usage of two parts of the standard, (a) Part 4, MPEG Intellectual Property Management and Protection (IPMP), which provides mechanisms for protection of digital item, since security problems may arise from the fact that, the digital item's description, i.e. its structure, contents, attributes and metadata, is a clear XML document and it is easily visible to anyone and vulnerable to non-authorized usage; and, (b) Part 5, MPEG Rights Expression Language (REL), which provides a simple XML-based data model which allows to the content creator to meta-describe the license that describes the usage rules over a specific digital content. The use of MPEG-21, leads to a DRM scheme that is adaptive to the end-user needs, i.e. different users must have different usage rights over the same digital content, while also characterized by interoperability.

2. Security & Digital Rights Management (DRM) in user-converged multimedia environments

There have been considerable efforts to have audiovisual systems and applications converge, especially in home environments where homes can be considered as spaces of convergence, and for nomadic users with advanced mobile devices as points of convergence. State-of-the-art research initiatives have as primary scope to progress beyond home and device-centric convergence toward truly user-centric convergence of multimedia, having as long term vision "The User as the Multimedia Central", i.e. the user as the point at which services (multimedia applications) and the means for interacting with them (devices and interfaces) converge, (see following figure for the depiction of this vision).

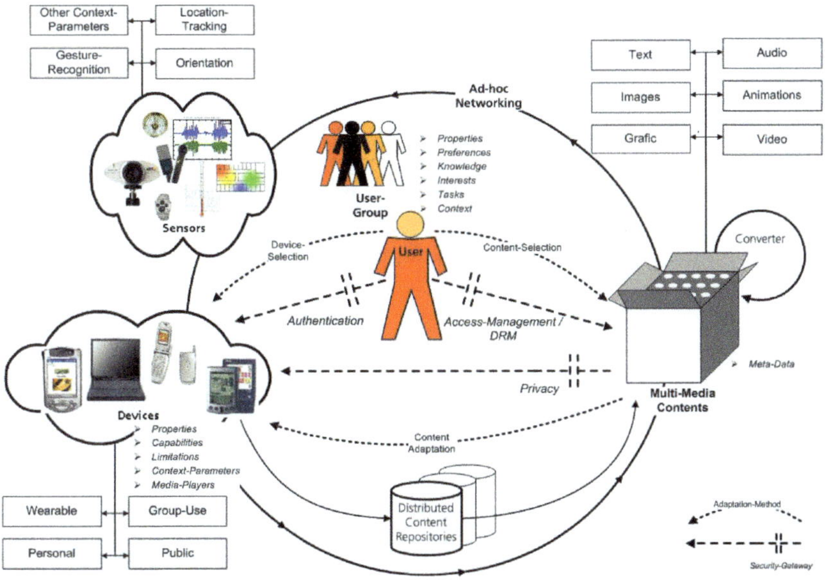

Fig. 1. User-converged multimedia services, (Intermedia, 2008)

As it can be seen to the previous figure, the user has access to different kinds of devices for interaction. From all accessible devices in a specific user context, the user selects his choice of devices himself and other devices are selected by the system automatically. The devices are used for multi-modal interaction and multimedia output. Based on the characteristics of the user(s) and the devices, the system selects the most appropriate content to be delivered. The content is available on distributed repositories including servers in the environment, networked repositories or small repositories hosted on small personal devices. When the best fitting content was found, it is delivered to the best fitting device that was selected beforehand. Potentially, the content needs to be transformed (e.g. to another media-type, another quality, or just in screen size) to fit the preferences of the user, the network bandwidth and the capabilities of the output device. In order to ensure private use of content and secure information transfer, the user needs to authorize to get access to different devices, the content selection process must consider DRM issues and the content presentation on the target device has to prevent displaying any private content on public displays.

One of the biggest challenges that we have to face in the deployment of architectures in such environments is related on the one hand, with the security and protection of digital contents that interchanged between users of such architectures and on the other hand with the provision to the users with security mechanisms that allow them to perform secure transactions (e.g. authentication, privacy protection, secure data transfer, etc.) in those environments. Intellectual Property (IP) protection is a mandatory request in modern multimedia architectures. The concerns of content providers for loss of revenues constitute a strong obstacle to the wide deployment of services that involve distribution of IP protected content. But the traditional content providers are not the only ones, if we consider the fact that today the end-users are equipped with different types of small devices that allow them to be the digital contents creators. In most cases, the end-users would like to have mechanisms which would give them the possibility to protect their artefacts and possibly to set their own usage rights over it, thus specifying toward third users how their digital content shall be used. Thus, it is a requirement to integrate DRM solutions in services and applications that target delivery of IP protected content that resides into the home network of the user. Considering the technical problems (system weaknesses) that result from add-on security solutions to independently developed network services, the approach of our effort to develop architectures with security and DRM as inherent requirements will lead to secure solutions that will increase the trust placed by content providers on the system and thus, it will lead to wider availability of services to a larger population.

2.1 The necessity of DRM

The rapid evolution of technologies for communication between computing systems in comparison with, (a) the high availability of broadband services at high data rates, (b) the great improvements in technology of hardware components of computers, and (c) the usage of Internet as a communication media for various entities, facilitate the convenient distribution of digital multimedia contents between users, most of the times illegally and without the necessary licensing.

The proliferation of the pre-mentioned technological achievements allow to the content creators, distributors, etc. to convey their profit instantly, while the end users to have the requested digital content in their hands, as soon as they requested and paid for it. But, after the contents creators and distributors send their goods to the end users they lose control over it. These combined with the fact that the end users have various digital technologies in their hands, which allow them to copy, and possibly redistribute digital contents, leads to significant loss of revenue for the digital contents creators and other interesting parts. So, in order to avoid such occurrences, the companies have developed and deployed various mechanisms which are used as countermeasures against unauthorized usage of digital content by non-legitimate users. Such mechanisms are defined under the umbrella of "Digital Rights Management", (DRM). *Digital,* since it refers to mechanisms for protection of digital contents; *Rights,* since the mechanisms allow the usage of contents under specific rights, which are set by the creators and distributors; *Management,* since the parties that deploy DRM over specific content have the ability to handle and manage its usage logically, even after the digital content has been distributed.

In general, Digital Rights Management (DRM) mechanisms constitute of various technologies that have been developed and deployed by content providers, creators, distributors, in order to protect their digital media from illegally, unauthorized and without the appropriate rights usage of their products, while let them to use possibly unsafe media like Internet for delivering their products with less hesitation and anxiety about non-legitimate usage of their content. In a typical DRM system, digital content is accessed by the end-users according to access conditions and terms that are specified into a license that accompanies the content. In general, licenses are provided by the license creators, who create them according to directives specified by the content owners/providers/creators. Various methods and a numerous DRM systems have been proposed, and developed for protection of Intellectual Property (IP), either from the academic community or by digital content industry. What characterizes most of them is the lack of interoperability, i.e. different content providers use different non-standardized non-interoperable DRM systems which may create great problems in contents usage by legitimate users.

But, what exactly means interoperability in the context of DRM? Gasser and Palfrey (Gasser and Palfrey, 2007) define interoperability looking issues regarding the different DRM stakeholders, i.e. users, content creators/owners and vendors, stating that, "In the context of DRM the term interoperability encompasses consistent functioning of the overall system including security and access, such that the system is able "mutually to use" information in the form of usage rules, content and technical measures "in all the ways in which they are intended to function". This would apply even when content from different interoperable services is used and when such content is used on different interoperable devices. For the consumer, interoperability means he can choose different devices and use them with different services. For the content producer or content aggregator interoperability means he is not locked in to one distribution channel that forms a gatekeeper to the marketplace. For the device and ICT developer, interoperability means that his products can be used with different content services - and that a gatekeeper does not form around a specific DRM technology."

A more formal definition of DRM interoperability has been given by Heileman et al, (Heileman et al, 2005), who mention into their survey paper that, "It seems that everyone has a notion of what interoperability means, which generally revolves around the idea of "things_ working together. A slightly more formal definition related to technology is: "*The ability of one technology to interact with another technology in order to implement some useful functionality. It is possible to make nearly any two DRM technologies work together in a manner that satisfies this definition. Specifically, by building translation services, it is often possible to make one DRM regime work with a different DRM regime. At the current stage of development of DRM markets, this approach to interoperability makes sense. However, in order to facilitate the continued development of DRM markets, more detailed notions of interoperability of DRM technologies must be developed. In this sense, the real issue is not interoperability per se, but rather the level of interoperability that allows better DRM solutions to be created...*"

2.2 Communication & networking technologies

Wireless technologies represent a rapidly emerging area of growth and importance for providing either ubiquitous access to a backbone wired network or formulate autonomous ad hoc wireless networks. The Access Point (AP)-infrastructured wireless networks architecture is based on at least one AP providing a server function. All kind of communication between all wireless nodes should pass through this AP. This AP might be connected to a wired backbone network as well. On the other hand, mobile Ad hoc Networks (MANETs) are autonomous networks consisting of routing nodes (or some routing nodes with other nodes that do not route) that are free to move about. They may be connected to a larger network e.g. the Internet, or operate as an isolated intra-network. Wireless networks can be categorized according to the extent of their coverage area into: Wireless Local Area Networks (WLANs), Wireless Wide Area Networks (WWANs), and Wireless Personal Area Networks (WPANs).

Recently, industry has made significant progress in resolving some constraints to the widespread adoption of wireless technologies. Some of the constraints have included bandwidth and high infrastructure as well as service cost. Wireless technologies can support and provide cost-effective solutions. Wireless is being adopted for many new applications: to connect computers, to allow remote monitoring and data acquisition, to provide access control and security, and to provide a solution for environments where wires may not be the best solution.

As networks become more and more complicated and applications more and more demanding, a very common network topology for state-of-the-art multimedia is a hybrid wired/wireless architecture. Hence, the need for interoperability of heterogeneous networks with hybrid structure is in doubtfully a major requirement, when integrating communication scenarios for indoor and outdoor applications.

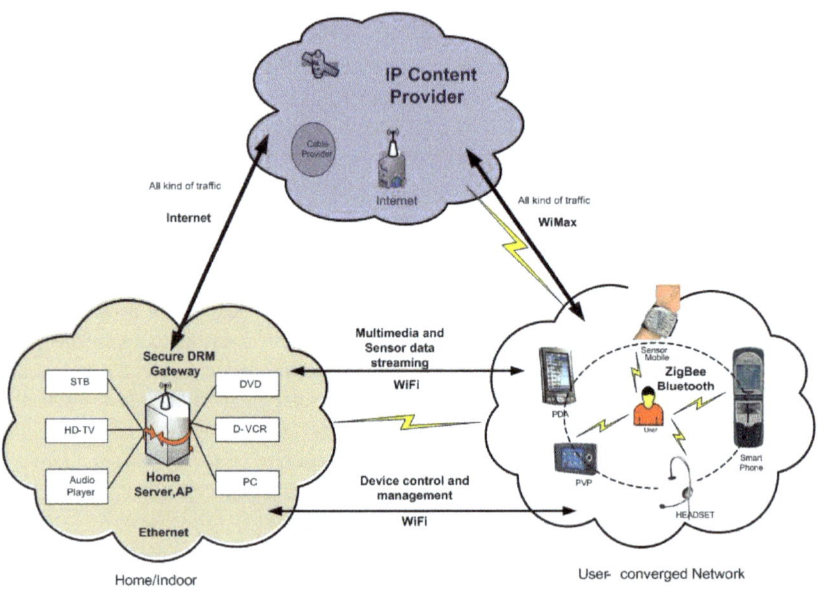

Fig. 2. Depiction of an architecture that consists of the Home Network, the user-converged network and the content provider.

On the other hand, when dealing with a hybrid wired / wireless network, questions arise regarding Quality-of-Server (QoS) and power awareness issues especially concerning the wireless part of the hybrid network. Integration of QoS and power awareness in wireless networks is nowadays a growing research area as high throughput, timeliness and power efficiency is demanded by several home and other applications. Thus, trade-offs especially between QoS parameters and power consumption must be considered to a network with mobile nodes. Thus, the objective is to define and depict an integrated architecture of a DRM system in a user - centric framework involving indoor and outdoor heterogeneous networking conditions for multimedia and sensor applications.

When dealing with multimedia and sensor applications for both indoor and outdoor networking environments, the user-centric approach should be taken into consideration, (see Figure 2 for a depiction of a user-converged architecture). Dynamic networking, multimedia sharing, content adaptation, personalized interfaces and data security are only some of the main challenges when dealing with the user-centric approach. This approach involves aspects of both integration and convergence. The term integration refers to the fact that different networking techniques should be incorporated in order transparent network communication to be achieved. On the other hand, convergence in such systems corresponds to seamless content sharing among multiple devices connected to each other. Moreover, as long as both indoor and outdoor networking is concerned, the user-centric approach can be split out into the home-centric and device-centric convergence. Multimedia and sensor convergence allows the end-user to adopt various communication interfaces as they become available, transparently, without any interruption.

According to the given parameters, all data are adapted to the available device characteristics.

The framework network architecture, in terms of networking infra-structure, consists of two physical media, one wired and one over the air. The heterogeneous topology corresponds to the wireless WiFi, Bluetooth, ZigBee or WiMax standards and to the wired Ethernet and Internet based protocols.

Technically speaking, in an indoor/home environment, the user, with the aid of a PDA, can communicate with various multimedia and sensor devices as well as the home server through the WiFi (IEEE 802.11b and g) technology, supporting multimedia and sensor content applications. WPAN standards like Bluetooth (IEEE 802.15.1) or ZigBee (IEEE 802.15.4) are more appropriate for connecting to a PDA several wireless devices such as, oximeter sensor, for heath care monitoring or head sets for entertainment, within the body area of the final user. On the other hand, in an outdoor environment, the user has the ability to communicate with the home server through a WiMax (IEEE 802.16) access point and a Home Gateway internet-based link. In that point of view, both ad hoc and infra structured topologies are engaged in regard to wireless communications. As holds for all multimedia and sensor applications or possible communication scenarios, the user perceptive quality defines tight QoS requirements that the heterogeneous network should support.

2.3 General requirements

Considering the previous mentioned aspects of security, DRM and communication technologies, we can make some initial assumptions regarding the design and deployment of such an architecture which will expose security & DRM mechanisms to its end users. To be more specific, our architecture has to fulfil the following basic requirements,

1. Protect digital content from unauthorized usage from non-legitimate users, which are not authorized to have access to it;
2. Operate over heterogeneous wired and wireless network technologies;
3. Provide to the end-user friendliness, i.e. the part of the DRM system that handles access, management and rights of digital content should be very friendly and easy to be accessed by the end-users;
4. Provide adequate security mechanisms, under possible attacks;

5. Define the extent to which our system shall be interoperable with other systems. More specifically, we stress into this requirement since the interoperability is a major requirement of the modern DRM architectures in the context of the multimedia world;
6. Provide extensibility and renewability, i.e. we would like our system to be easily adaptive and flexible under changing requirements, new ideas, new business models, etc;
7. Provide trustworthiness, which means that the DRM system must reassure that it will behave according to the terms and rights that have been agreed;
8. Protect the privacy of the participating entities in value chain, i.e. a misuse of a DRM system could lead to user's privacy breaches;

2.4 The DRM process model

As we have mentioned, one of the key issues in user-converged multimedia architectures is related to the Intellectual Property of contents being distributed. As a matter of fact, multimedia distribution across several kinds of devices and networks, which can communicate among them, facilitates the convenient distribution of digital artifacts like audio, video, etc., among users, most of the times illegally and without the necessary licensing. Digital Rights Management (DRM) mechanisms consist of various technologies that have been developed and deployed by content providers, creators, distributors, in order to protect their digital media from usages that may be illegal, unauthorized and without the appropriate rights of their products. At the same time, DRM facilitates the use of possibly unsafe media like the Internet for delivery of these products with less hesitation and anxiety about non-legitimate usage of their content. This problem is even more stressed in Personal Area Networks, where content distribution among users and their usage generally takes place in an autonomous and uncontrollable form, due to the likely absence of a public connection.

One of the bases of our architecture is the identification of the stakeholders in the DRM chain, in order to define and understand more deeply the security requirements of the proposed DRM architecture. We can identify three major partners in the chain, i.e. the content creator, content (license) provider and the user, see following figure, (Fig.3),

Fig. 3. The three key stakeholders in a typical DRM system.

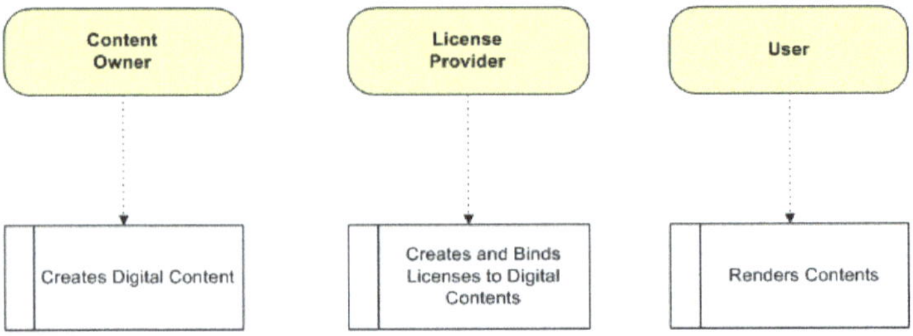

Fig. 4. A DRM architecture stakeholders and their main roles and functionalities.

The role of the content creator is to generate digital content and to gain profit by providing it to intermediate resellers like, for example, content distributors, who will finally resell it to the end-users. The user access the digital content according to the directives (access rules) that have been set by the content owners and implemented in accordance with the digital content by the license providers. Schematically the role of each part can be seen at the following figure, (Fig.4).
Furthermore, trying to identify in more depth some of the functionalities of each part, we can identify, (a) the license creator has to generate a license according to the directives of the content creator, while after the creation of the license it has to bind - encapsulate it to the digital content, thus creating a packaged content which delivered to the end user; (b) the end user must be equipped with a DRM client that will be comprised of a part that process the packaged content and a part that is responsible for the rendering of the content. Schematically, this process model can be seen at the following figure, (Fig.5),

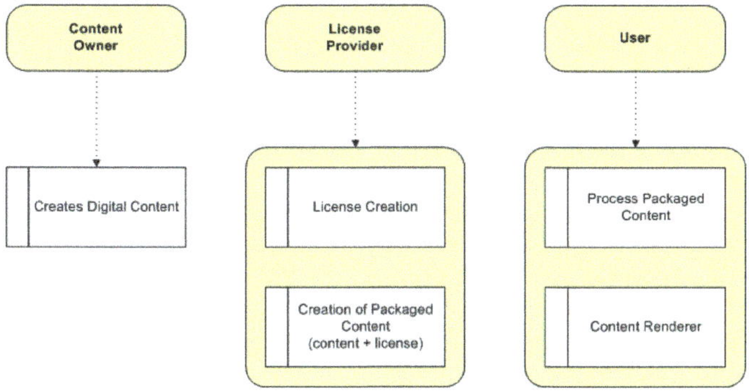

Fig. 5. DRM model that incorporates the packaged content and the breakdown of the content rendition component.

In the context of user's environment, the content owner could be the user by himself, who creates his own digital contents and would like to protect them and possibly share them in a controlled manner with other users, in his vicinity. Summarizing the previous we can identify a security requirements concerning contents owner view, as follows,

a. The digital content shall be accessed by the end user through a device, which shall be able to enforce the digital rights that have been set into the license, which has been provided by the license provider and accompanies the content. Analyzing this a bit furthermore, we can identify the following,

i. License generation for digital content shall be done only by authenticated license creators, i.e. authorization of license creators

ii. The content has to be protected from possible disclosure while communicated towards third parties, i.e. confidentiality of data.

b. The digital content shall be accessible only by a multimedia rendered that shall be capable to "decode" the license that accompanies the content, which comes from the license creator and defines the terms of contents' usage.

c. There must be set mechanisms that assure the integrity of communicated data between the license creator and the end-user, while there must not be any delays in the provision of the data. More specifically,

i. In the client's side renderer, there must be a part that is assumed to be trusted, which shall be responsible for storing the digital content, the license and any other sensitive data. Moreover, that part shall order the rendering of the content and shall also "monitoring" the contents consumption under the terms that have been specified by the license.
ii. Availability to the end user to have access to the digital content and the license, whenever he/she wants
iii. Correct reception of digital content/license by the end user, (integrity mechanisms)
d. The architecture, especially in the client's side, must be equipped with mechanisms tolerant to possible attacks, constraining the impact of the attacks,
i. Prevent spreading of an attack in the whole architecture, i.e. identify possible mechanisms that isolate attacks to a specific part of the architecture, not allowing it to affect the rest parts of the architecture,
ii. The architecture must be easily adaptable to new security requirements that may come to the near future
e. Protection of user' privacy and personal data, i.e.
i. The license creator stores only the private and personal data that the end user permits,
ii. No third party may have access to any personal data of the end user, without agreement between end user and the third party.

Regarding the end user, the following requirements must be satisfied,

i. Authentication between the end user and license creator, both sides, i.e. the license creator has to be authenticated towards the end user, while the end user has also to be authenticated to the license creator,
ii. Assurance that the license creator has set to the license only the terms that have been agreed between the two parties, (end user and license creator),
iii. The data for the authentication of the end user are only available to the end user,

2.5 DRM system – internal components

A typical DRM system involves three main parts, a. the content provider, who simply offers contents under specific rights; b. the DRM platform, which comprises the main part of the system, providing DRM functionalities i.e. secure storage, security primitives, contents provision interface, licensing translation - phrasing, etc; and c. the client's system, which refers to the end user's device that consumes the digital content.

Going one step further, decomposing the DRM platform, we can identify its main subsystems and primary functionalities as it can be seen in figure 6.

The following parts are identified: a. Content provision module (CPM) is the interface between the content provider and the DRM platform. The connection between them can be achieved through a trusted channel e.g. SSL protocol, which shall be established between content provider and content provision module; b. Secure storage, is a secure and trusted environment into which the content shall be stored after it's transferred to the DRM platform. In general, the content is stored in an encrypted form. It may be possible to store and other sensitive information like cryptographic keys, licensing information, etc; c. Licensing Translation Mechanisms, (LTM) which refers to the part that is responsible for the translation of the licenses that accompany the digital contents. The license defines the certain terms and conditions under which the digital content may be used and it is defined by the content provider. This module may include functions that allow the generation of licenses for a specific content by the instructions that set by the content provider. The licenses are stored at the licenses repository, which can be assumed as a secure environment; d. Content preparation module (PREP), which is the part of the DRM platform that prepares the digital content for distribution to the end users. It performs various functionalities like content's encryption, transcoding of the content depending on user's preferences, enrichment of content with licenses and other metadata. The license is "mixed" in a secure manner with the digital content and the end user receives the digital content as a single secure container; e. Content Distribution Mechanisms, (CDM), which is the part that handles the distribution of digital content through various distribution channels; f. Authorization module, which is responsible for the authentication - authorization of the end user. As soon as the user is identified as a legitimate one the authorization module sends to the user decryption keys.

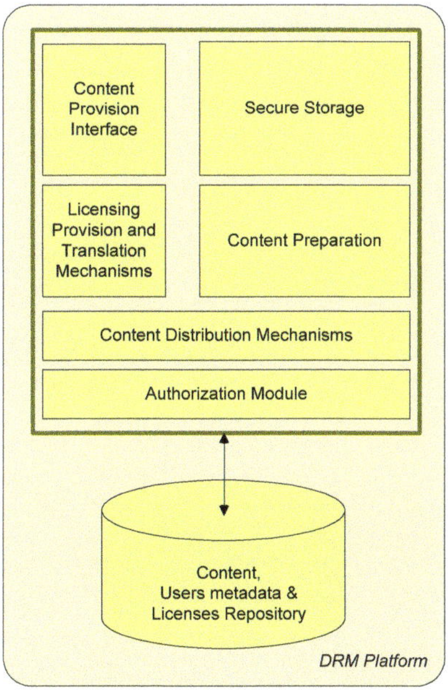

Fig. 6. The main parts and functionalities of the DRM Platform.

Furthermore, (Becker et. al, 2003), it has to be identified the flow of information in typical DRM system, between the content provider, the DRM platform and the end user, (see following figure, for a schematic view).

In step 1, the content provider contacts the CPM establishing a channel for transferring the digital content to the DRM platform; it also sends to the LTM module the license for content's usage or necessary information for its generation. The digital content is stored temporarily to the secure storage space, (step 3) and the license may be store at the licenses repository. In step 4, the PREP module prepares the digital content to be distributed, i.e. applying watermarking, encryption, transcoding - if necessary and licenses embodying. After that the digital container is transferred to the end user's client system through delivery networks, with the help of CDM module, (steps 5, 6). This container is stored in a secure storage space, which resides into client's system. The user interrogates the authorization module of DRM platform, requesting access for the received digital content. If the DRM platform identifies the user as a legitimate one, it sends to him a "grant access" and some security information related with the specific digital content, e.g. keys for content's decryption. Then the content is decrypted into the client's system and it's rendered according to the usage rights that set by the accompanying license. The user's client system must be equipped with mechanisms that allow the enforcement of digital rights that defined by the license; as an extension, the DRM platform may be kept informed about the rights enforcement.

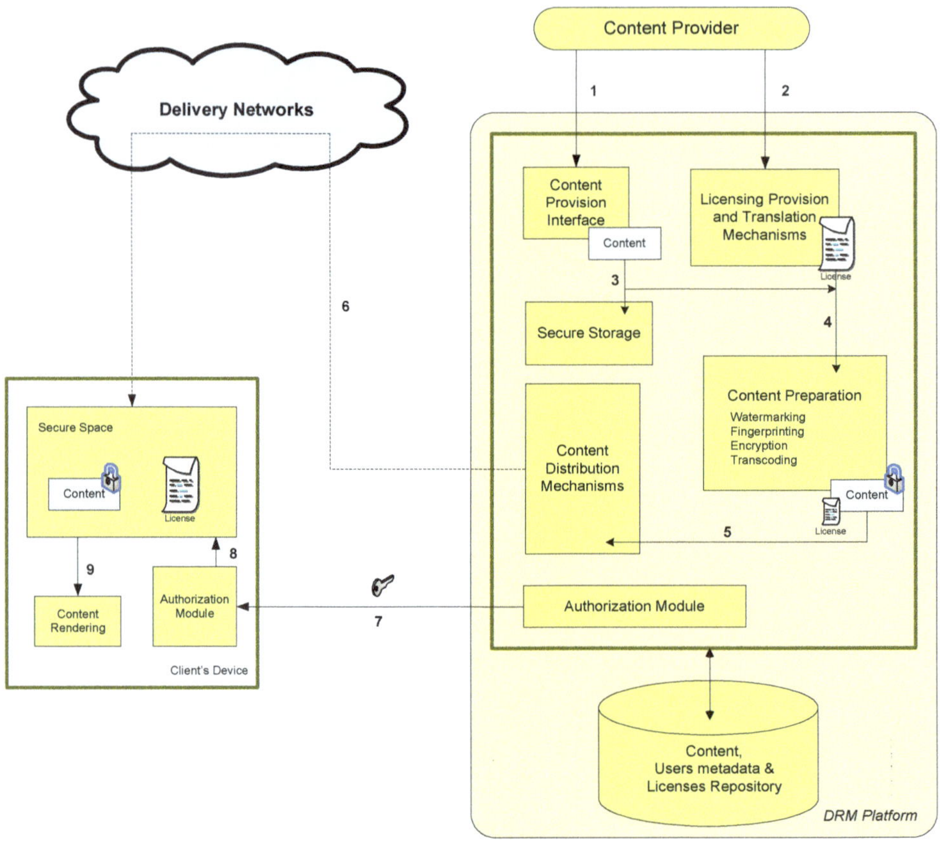

Fig. 7. How information flows in a typical DRM system.

3. MPEG-21 multimedia framework standard

MPEG-21 (International Standards Organization [ISO], 2005), (Burnett et al., 2006) is a standard that defines mechanisms and tools as means of sharing digital rights, permissions and restrictions that are set from content creators over digital contents regarding their contents usage from consumer. It is an XML-based standard that is designed to communicate machine-readable license information in a ubiquitous, unambiguous and secure manner between peer entities.

MPEG-21 has been proposed for primary use in the context of multimedia world, allowing the seamless, transparent and universal delivery of multimedia content to the end-user, thus solving any interoperability issues.

It is based on two essential concepts: the definition of a fundamental unit of distribution and transaction, which is the Digital Item, and the concept of users interacting with them. Digital Items can be considered the kernel of the Multimedia Framework and the users can be considered as the entities that interact with the digital items inside the Multimedia Framework. At its most basic level, MPEG-21 provides a framework in which one user interacts with another one, and the object of that interaction is a Digital Item. Due to that, we could say that the main objective of the MPEG-21 is to define the technology needed to support users to exchange, access, consume, trade or manipulate Digital Items in an efficient and transparent way.

3.1 MPEG REL

In general, a Rights Expression Language is a XML-based machine-readable language that allows the declaration of rights and permissions concerning the usage of digital contents. The MPEG REL, (ISO, 2005), provides flexible and interoperable mechanisms to support transparent and augmented use of digital resources throughout the DRM chain, thus protecting the digital resources and the rights and conditions, which are specified for them. For instance, it provides mechanisms in support of publishing, distributing, and consuming digital content such as electronic books, digital movies, digital music, broadcast content, interactive games, computer software, and other creations in digital form. It also supports specification of access and usage controls for digital content in cases where financial exchange is not a term of use, and supports exchange of sensitive or private digital content and personal information. The MPEG REL may provide guaranteed end-to-end interoperability, consistency, and reliability among different systems and services. In order to achieve this, it offers richness and extensibility in declaring rights, conditions, and obligations; ease and persistence in identifying and associating these with digital content; and flexibility in supporting multiple usage/business models.

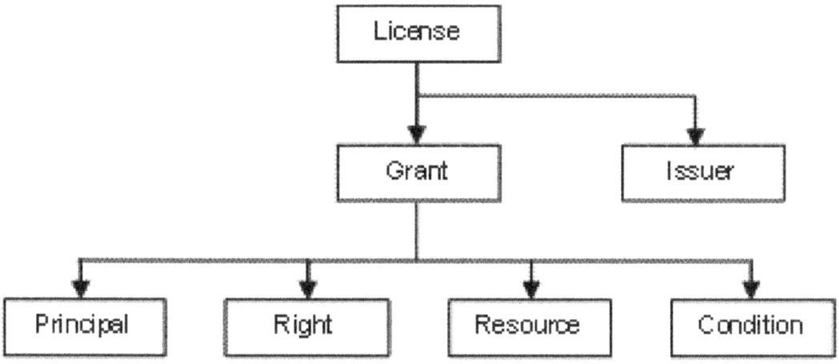

Fig. 8. A schematic view of a simple license that consists of a grant and an issuer.

The MPEG REL adopts a simple data model for many of its key concepts and elements. This data model for a rights expression includes four basic entities. The basic relationship among

these entities is defined by the MPEG REL assertion "*grant*" By itself, a grant is not a complete rights expression that is transferred from one party to another. A full rights expression is called a license and it consists of one or more grants and an issuer, which identifies the entity who has issued the license. Structurally, a grant consists of the following elements, (a) the principal, (b) the right, (c) the resource, and (d) the condition. In the following figure, we can see a schematic view of a license,
The MPEG REL defines the *r:license* element for the representation of a license into a digital item and comprises of two basic parts, (a) the *r:grant* element, that refers to whom the license is granted to and describes the license's main parts, and (b) the *r:issuer* element, which is optional and it contains information that identifies the issuer of the license. Structurally, a grant consists of the following basic elements, (a) the *r:principal* element, which defines to whom the grant is issued; (b) the *r:right* element, which defines an act of the principal to digital item; (c) the *r:resource* element, that defines the object on which the right in the grant applies to; and, (d) the *r:condition* element, that must be met before the right on the resource can be applied.
In the context of REL enhancement, it has been proposed (Delgado et al. 2005) an extension of the REL, in which it is adapted the use of a new element *protectedResource,* (see figure 9 for an abstract view of the elements), which give us the possibility to include into the license some sensitive encrypted information, for example, the key that is used for content's encryption, thus allowing key distribution to the remote end. As we have mentioned, this element comprises information that is protected with some form of (symmetric key and/or public key) encryption and it includes the following elements, (a) the *r: digitalResource* that specifies the digital content; (b) the *xenc:EncryptedData* a in which information about encryption of the resource is provided; (c) the *xenc:EncryptedKey* which contains information about encryption of the key that is has been used to encrypt the resource. Moreover, in the digital resource element, we can specify a hash code of the encrypted digital content, which can be used for the verification of the integrity of the received file to the remote end.

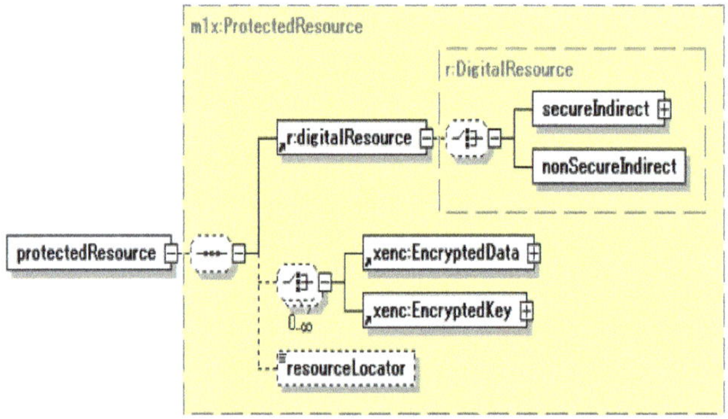

Fig. 9. Structural view of the protected Resource element.

In figure 10 we present an XML-view of a simple license, (comprising the basic elements), which has been extended with *protectedResource* element. Moreover, in figure 16, we present a detailed XML-view of *protectedResource* element, which is used to include into the license, information about the encryption key that we have used to encrypt the digital resource song1.mpg.enc, with the use of 128-bit AES algorithm; furthermore, the encryption key has been encrypted with the use of public key transport algorithm RSAES-PKCS1-v1_5.

```
<r:license>
      <r:grant>
            <r:keyHolder>
                  . . . . .
            </r:keyHolder>
            <mx:play>
            <m1x:protectedResource>
                  . . . . .
            </m1x:protectedResource>
      </r:grant>
      <r:issuer>
            . . . . .
      </r:issuer>
</r:license>
```

Fig. 10. An XML view of a simple MPEG-21 license, where the *protectedResource* element is included.

3.2 MPEG-21 intellectual property management and protection

The security problems may arise from the fact that, the digital item's description, i.e. its structure, contents, attributes and metadata, is a clear XML document and it is easily visible to anyone and vulnerable to non-authorized usage. Due to that fact, the MPEG-21 includes a part named Intellectual Property Management and Protection (IPMP), (ISO, 2006), which provides mechanisms for protection of digital item. More specifically, MPEG-21 IPMP in conjunction with the MPEG-REL (Rights Expression Language) provide a framework that enables all users in the digital contents delivery chain to express their rights and interests in digital items and to have the assurance that those rights and interests will be persistently and reliably managed and protected across a wide range of networks and devices. The core notion in MPEG-21 IPMP is related with the IPMP tools that are used to protect the digital item. Those tools are not pre-described by the standard, but each user, vendor, etc., may define and implement his own set of tools, which perform basic security functions like encryption/decryption algorithms, authentication and data integrity mechanisms, watermarking, fingerprinting. With the use of MPEG-21 IPMP components, we may protect the whole Digital Item or a part of it through the encapsulation of the original DIDL elements that we want to protect, with additional information (IPMP Info) that refer to mechanisms and tools for the protection of the original elements. MPEG-21 IPMP defines a new set of IPMP DIDL elements, which have the same role and semantics as an element defined in DIDL. The structure of an IPMP DIDL element can be seen to the following figure, (Fig.11).

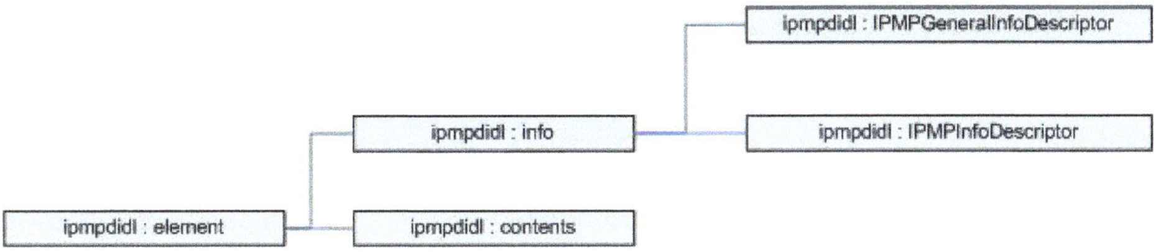

Fig. 11. Structure of an IPMP DIDL element.

The *ipmpdidl:info* element contains information about protection and usage rules of the digital content, which may be categorized to, (i) information about protection of the whole digital item, which is included in the child element *ipmpdidl : IPMPGeneralInfoDescriptor* and (ii) information about protection of a certain part of the digital item content, which may be categorized to, *ipmpdidl : IPMPInfoDescriptor*, see figure 11. Both pre-mentioned child elements have two purposes of existence,

a. to describe the tools that are used for digital items protection, and
b. to provide a set of licenses that accompany the content and define its usage rules.

4. Security-related aspects

In a user-centric environment that is enhanced with Digital Rights Management (DRM) mechanisms, we can identify different types of security requirements which have to be taken in care on the design and architecture implementation. Basically, we can identify tow different directions that may identify the security requirements that we have to take in care, named, (a) application-level security requirements; and, (b) embedded systems security requirements.

4.1 Application-level security aspects

A variety of security aspects can be identified when one tries to design and implement computing systems architecture. Those aspects are more emphasizing in the case of architectures that serve DRM functionalities. Our proposed architecture fulfils the basic security requirements, and in all cases we have tried to use state-of-the-art algorithms and methodologies. The basic security requirements that have to be fulfilled can be summarized as follows, (Schneier, 1996)

a. *Confidentiality*: it must be assured that stored or transmitted data are well protected from possible disclosure. In our approach sensitive information is stored, handled and transmitted in an encrypted manner.
b. *Integrity*: the data that have been transmitted are the original ones and have not changed medially; hashing methodologies are used when it is necessary.
c. *Authentication*: refers to the capability of mutual identification between various parties in a transaction. In our case we have used various types of authentication techniques, like public-key cryptography and digital signatures.
d. *Access Control*: it means that only legitimate users must have access to specific computing resources. This is the basic objective of our DRM architecture.

4.1.1 Authentication mechanisms

Authentication refers to methods and mechanisms which allow to an entity to prove to a remote end its identity, i.e. in a transaction between two end-users over a possibly unsafe communication network, there must be mechanisms that assure that each part can be authenticated by a remote end. In a DRM system, for example, a device or a user must be authenticated to the content provider's network in order to be able to participate in various transactions related with digital content consumption. User's authentication can be achieved depending whether the user: (a) knows something, e.g. a PIN, a password; (b) posses something, e.g. a token, a smart card, etc.; or (c) has something inherent, e.g. a biometric characteristic. Password based authentication mechanisms are quite weak, so we shall not make any further comments. A strong authentication method is based on challenge-response protocols, where an entity (claimant) can prove its identity to a remote end by exhibiting the possession of a secret strongly associated with the claimant without the necessity to reveal it explicitly to the remote end. Challenge-response can be achieved with symmetric key techniques, public-key techniques and zero-knowledge protocols.

4.1.2 Encryption/decryption mechanisms

We define two main categories of cryptographic algorithms that are basically used for encryption/decryption of data; a. the symmetric algorithms and b. the asymmetric (public-key) algorithms. Roughly speaking, symmetric algorithms use the same key for encryption and decryption, while in asymmetric algorithms two different keys are used, a private one for encryption and a public one for decryption. Each participating entity must possess a pair of keys (a private key, *PKprv* which is held secretly from common knowledge and a public key *PKpub* which is publicly known).

Symmetric algorithms have the disadvantage of the use of a single key between entities that take part in encryption/decryption transactions, having as their main problem the efficient and secure key distribution between participating entities. Widely used symmetric algorithms are Triple Data Encryption Standard (DES), RC2, IDEA, Advanced Encryption Standard (AES - Rijndael), (Schneier, 1996), (Menezes, 1996). In a typical DRM system the digital contents are encrypted / decrypted using symmetric cryptographic algorithms. In the context of our work, we have implemented optimized AES algorithm.

On the other hand, the asymmetric algorithms (often called public-key algorithms) lack the problem of key distribution, since each part has a set of keys (private - public) and whenever it encrypts data it uses the public key of the part to whom encrypted data shall be transferred, while the recipient of encrypted data can decrypt those data using his private, thus deriving plaintext message, see following figure for a schematic depiction of pre-mentioned. Those algorithms rely on hard mathematical problems thus requiring high computational and processing capabilities, which make them inappropriate for encrypting/decrypting huge amount of data. Well known public key algorithms are the RSA (Rivest - Shamir - Adhelman) algorithm, the Diffie - Hellmann algorithm, and Elliptic Key Algorithms, (Schneier, 1996), (Hankerson et. al, 2004). In the context of our work, we have implemented the RSA algorithm.

4.1.3 Privacy issues

As we have argued before, whatever security mechanism is employed by a system, by any means it should take care and protect user's privacy. Especially, in the deployment of secure DRM systems, where there is an explicit relationship between end users and contents providers, with the latter ones having as their primary goal the protection of their digital assets, protection of end user's privacy is an issue that requires special treatment. As a recent example of a DRM mechanism that led to breach of user's privacy, we shall make a note in the Sony BMG case (Roush, 2006); SONY BMG in its effort to build a DRM mechanism to protect its commercial digital assets (CD's), embedded in each CD a "specific" program which it's proved to act as a rootkit , allowing hackers and non-legitimate users to gain access to end users computing devices without their permission, leading thus to violation of user's privacy. This was one of the biggest technological blunders in the history of modern computing leading SONY to recall its commercial products and publicly apologize for the scandal.

The Encyclopaedia of Cryptography [Henk et. al., 2005] defines privacy as an entity's ability to control how, when and to what extent personal information about the entity shall be communicated to third parties, while Anderson (Anderson 2001) defines privacy as the secrecy for the benefit of an individual entity, where secrecy refers to generic mechanisms that do not allow unauthorized usage and access of data and resources. There are various ways, mechanisms and techniques that we can employ to protect user's privacy. Adams (Adams 2006) classified the privacy's technologies for online environments into four levels; in level 1, he distinguishes technological and societal techniques, whether it based on technology (computers, devices and software) or humans, respectively; in levels 2 and 3, he classifies techniques based on actions that taken part from the participating entities in a scheme into which privacy's issues may arise, i.e. an entity may provide its personal information towards a third party, or an entity that holds a user's personal data may perform something with those data; finally, in level 4, Adams categorizes privacy techniques according to the threat model under consideration.

4.2 Embedded security aspects

The increasing capabilities of embedded systems combined with their decreasing cost have enabled their adoption in a wide range of applications and services, from financial and personalized entertainment services to automotive and military applications in the field. Importantly, in addition to the typical requirements for responsiveness, reliability, availability, robustness and extensibility, many conventional embedded systems and applications have significant security requirements. However, security is a resource-demanding function that needs special attention in embedded computing. Furthermore, the wide deployment of small devices which are used in critical applications has triggered the development of new, strong attacks that exploit more systemic characteristics, in contrast to traditional attacks that focused on algorithmic characteristics, due to the inability of attackers to experiment with the physical devices used in secure applications. Thus, design of secure embedded systems requires special attention, (Fragopoulos et al., 2009). Secure embedded systems must provide basic security properties, such as data integrity, as well as mechanisms and support for more complex security functions, such as authentication and confidentiality. Furthermore, they have to support the security requirements of applications, which are implemented, in turn, using the security mechanisms offered by the system.

In user-centric environments, the devices that are carried by the users, basically, are comprised of small and highly constrained embedded systems. It is the increasing capabilities of embedded systems, which combined with their decreasing cost, have enabled their adoption in a wide range of applications and services, from financial and personalized entertainment services to automotive and military applications in the field. Importantly, in addition to the typical requirements for responsiveness, reliability, availability, robustness and extensibility, many conventional embedded systems and applications have significant security requirements. However, security is a resource-demanding function that needs special attention in embedded computing. Furthermore, the wide deployment of small devices which are used in critical applications has triggered the development of new, strong attacks that exploit more systemic characteristics, in contrast to traditional attacks that focused on algorithmic characteristics, due to the inability of attackers to experiment with the physical devices used in secure applications. Thus, design of secure embedded systems requires special attention.

4.3 Trusted computing aspects

Trusted Computing[1] gives us the possibility to implement a Digital rights management system which would be very hard to circumvent. For example, in case that a user downloads a video, remote attestation could be used so that the video file could refuse to play except on a specific music player that enforces the creators' rules that specified into the accompanying license; this means that specific media players would be able to play user's music. Sealed storage could be used to prevent the user from opening the file with another player or another computer. The music would be played in protected memory, which would prevent the user from making a non-authorized copy of the file while it is playing; and secure I/O would prevent capturing what is being sent to the rendering module output. Circumventing such a system would require either manipulation of the computer's hardware, capturing the analogue (and possibly degraded) signal using a recording device or a microphone, or breaking the encryption algorithm. Utilizing Trusted Computing aspects could lead to new business models for use of software (services) over Internet. By strengthening the DRM system, one could base a business model on rendering programs for a specific time periods or "pay as you go" models. For instance one could download a music file which you only could play a certain amount of times before it became unusable, or the music file could be used only within a certain time period. Furthermore, Trusted Computing Components can be the core blocks of embedded DRM and security mechanisms in small, portable, limited-resources devices that are used by nomadic users. In the following part of this sub-section, we briefly describe some already proposed DRM architectures which utilize as core blocks Trusted Computing Environments aspects.

(Messerges and Dabbish, 2003) proposed DRM architecture for use in highly constrained environments like mobile phones, setting as a basic demand that the part of the DRM policy has to be implemented with the use of a trusted system. Their architecture comprises of a. the DRM manager, which is responsible for authentication of licenses and digital contents, for enforcement of digital rights and for content decryption; b. Security Agents, are in close co-operation with the security hardware and provide: (a) Memory management (secure storage), (b) implementation of basic cryptographic operations, and (c) Key management. After digital content's decryption is sent to a trusted agent, which is an application that is trusted to access and manipulate decrypted content data. The overall picture is depicted in figure 12.

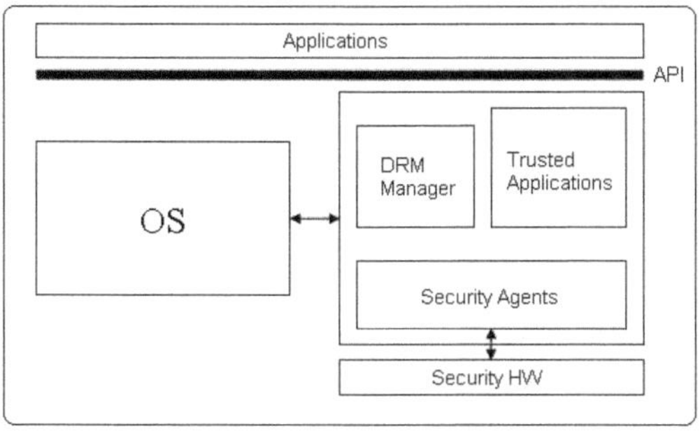

Fig. 12. A DRM architecture, which utilizes Trusted Computing Systems, (Figure has been adapted from (Fragopoulos, 2005))

[1]Trusted Computing Group, http://www.trustedcomputinggroup.org/

Based on the notion of Trusted Computing Systems, Lipton in (Lipton et al, 2002) and Serpanos in (Serpanos and Lipton, 2001) proposed the use of a special hardware agent, named spy, for protection of Intellectual Property in several environments. Their basic idea relies on the fact that even if the content provider side is secure, one cannot trust the client, which, in general, is considered as an un-trusted source. They proved that the existence of a special, tamper-proof hardware component that resides on the client system, as shown in Figure 13 is necessary to ensure protection of content provider Intellectual Property.

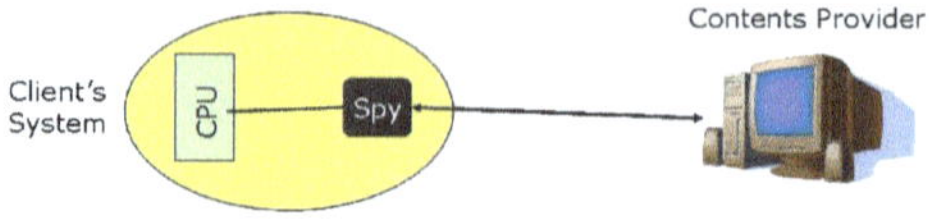

Fig. 13. Spy component as a trusted system for protection of Intellectual Property, (Figure has been adapted from (Fragopoulos 2005)

The spy is a tamper-proof hardware module, acting as a passive I/O system, allowing detection of I/O activity, having limited memory and computational power. It works in master mode in the client system and is monitoring the CPU. In a Video-on-Demand environment, for example, the user downloads the video from the content provider and stores it temporarily in the RAM of client system. There is an application that renders the digital content and it is assumed that it is the only application that runs on the system. If the user tries to make non-legitimate use of content, e.g. copying or transmission of the reproduced (decrypted) data, then some kind of I/O activity must take place at the client system, such as a disc access or a network transmission; such activity will be noticed by the spy, which, in turn, will notify the content provider in order to stop video transmission. Figure 14 shows the operation of this simple protocol, which performs the previous mentioned actions.

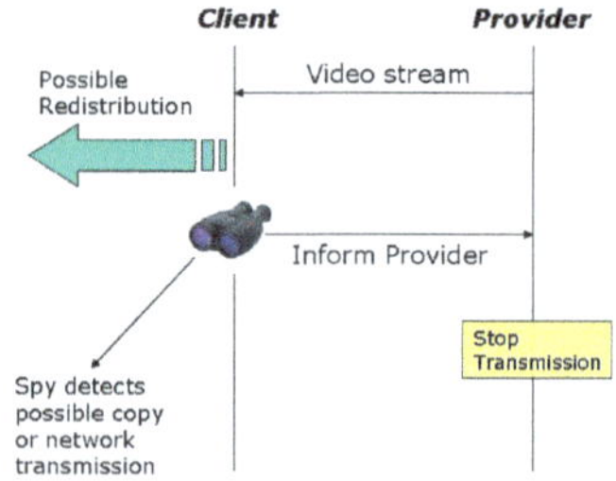

Fig. 14. A simple protocol demonstrating spy operation. (Figure has been adapted from (Fragopoulos, 2005)

4.4 Inter-networked embedded systems security, privacy and dependability aspects

Security, privacy and dependability (SPD) for systems and services that are built from integrated and interoperating heterogeneous services, applications, systems and devices is a very important aspect which must be ensured. Such systems and services must be robust in the sense that an acceptable level of service is available despite the occurrence of transient and permanent perturbations such as hardware faults, design faults, imprecise specifications, accidental operational faults, and deliberate, malicious, attacks.

The main goal is to address the upcoming impact of the Internet of Things to security, privacy, and dependability, from the early stages of design up to final deployment thus, creating new market opportunities by enhancing security, privacy and dependability so as to increase people's confidence in applications, systems, devices and infrastructures that were considered vulnerable or untrustworthy in the past.

Therefore, one target is to enhance security of Embedded Systems (ESs) as stand-alone or networked systems, i.e. at both the node and the network level while special focus should be given to developing technologies for: efficient, reliable, adaptable, and dependable ESs:

- ESs that defend against malicious attacks from intruders, maintain the confidentiality of sensitive data and protect intellectual property.
- Efficient and reliable communications and dependable networks for and utilizing ESs.

Another target is to develop appropriate ES technologies enabling protection of critical public infrastructure, such as transportation/communication/utilities networks and public building/areas. In this respect, special focus will be given on developing ES technologies to:

- Improve mobility of people and goods while preserving privacy;
- Provide support for critical applications, such as protection of infrastructures.

Solutions should contribute to one or more of the following specific objectives:

- Definition of a common conceptual framework to address the requirements for security, privacy and dependability.
- Instantiation of this framework with architectures, components, methods, interfaces and communications, tools and tool chains, to enable the design, development, analysis, validation, and deployment, as well as certification (or qualification).
- Trusted service platforms supporting the governance of the Internet of Things.
- Seamless and secure interactions and cooperation of ESs over heterogeneous communication infrastructures.

Indicative application examples imposing increased SPD requirements

- Telemetric monitoring of vital parameters of patients with chronic diseases is recognized to improve their medical condition and hence their quality of life. As a result of this, there are plenty of products and solutions for personal health monitoring available today that acquire physiological data in real-time. In order for such systems to be widely acceptable and utilized by the medical community and the patients, they must be developed satisfying the security requirements imposed by real-time data communication and protection of sensitive physiological data and measurements, data integrity and confidentiality, and protection of the monitored patient's privacy.
- Wireless Sensor Networks, which can be used as backbone networks in order to convey different types of digital contents that are locally generated by public utility applications. Such environments must be developed satisfying the basic security requirements for real-time, secure data communication, and protection of data and measurements, data integrity and confidentiality.

In such environments, it is imperative to design and deploy efficient and effective network architectures as well as a generic, if possible, communication interface targeted to connect external application networks with different types of "smart" embedded devices satisfying the basic security requirements for real-time, secure data communication, and protection of sensitive data and measurements, data integrity and confidentiality, and protection of the users' privacy. The architectures and the interface must consider the limited resources of the interconnected embedded systems, especially in light of the significant resources required

for implementing security in which, in general, are quite resource-hungry leading thus to significant technical problems.
By utilizing MPEG-21 standard's primitives, which define mechanisms and tools as means of sharing digital rights, permissions and restrictions for digital content from content creator to content consumer, we have shown that protection of transmitted "sensitive" information and enhancement of privacy is accomplished, since there is selective and controlled access to data transferred.

5. Architectural components of proposed DRM Architecture

A typical DRM system comprises of different types of components and should also provide different functionalities; shortly, we identify (a) Secure Storage Containers, which are used to protect the digital content from unauthorized access. Such containers can use various security functions like cryptographic primitives, trusted computing modules, etc; (b) Rights expressions tools, referring to the ways that the usage rights over a specific digital content are expressed. Basically those tools are combinations of languages (Rights Expressions Languages - RELs) with appropriate dictionaries. We have various examples of such methods like XrML[2], MPEG REL, OMA[3] REL, etc; (c) Description and identification of digital content, (digital items and metadata) methods. For example, MPEG-21 standard provides a well defined method for describing all digital content and its relevant metadata; (d) Identification of the involved parties in the chain of a DRM system - Authentication of interacting entities with the digital content; and (e) Forensic DRM components, refer to watermarking and fingerprinting techniques for proving subsequently if any rights violation over a digital content has occurred.
The basic requirements that are expected from a DRM system are:

a. *Interoperability*, which refers to the ability of the system to operate under different types of devices, platforms and architectures,
b. *Security*, which means that the system should provide robustness against possible attacks; and,
c. *Privacy*, i.e. whatever security mechanism is employed by a system, by any means it should take care and protect user's privacy.

Especially, in the deployment of secure DRM systems, where there is an explicit relationship between end users and contents providers, with the latter ones having as their primary goal the protection of their digital assets, protection of end user's privacy is an issue that requires special treatment.
To be more specific, the core components that we identify in our architecture are, (a) the digital content creator, who is the entity that generates digital contents that are stored into the DRM server, has the copyright of those contents, triggers the license generation procedure and has the possibility to assign rights to third parties to generate licenses for his contents, (root grants); (b) the Content/Profile Server, which is the part that contains the digital contents, it is responsible for controlling the whole DRM functionality; (c) a License Server (LS), which is the part that creates valid licenses according to DRMS directives that are associated with a specific digital item - the licenses are MPEG-21 compatible; and (d) the DRM Client, which is a device that is located into the Personal Area Network of the end user

[2]XrML - The Digital Rights Language for Trusted Content and Services, http://www.xrml.org/
[3]Open Mobile Alliance, http://www.openmobilealliance.org/

and it is capable to "consume" the digital content that is provided by the Content Server according to the directives that are specified by the accompanying license. Both the Content Server and License Server are located into the content creator, while the DRM client resides into the Personal Area Network of the end user, to whom the content creator wants to deliver protected digital content. In the following scheme, the three main components can be identified. Furthermore, three communication channels can be seen, (i) end-user devices and License Server, (ii) end-user devices and Content/Profile server, and, (c) License Server and Content/Profile server.

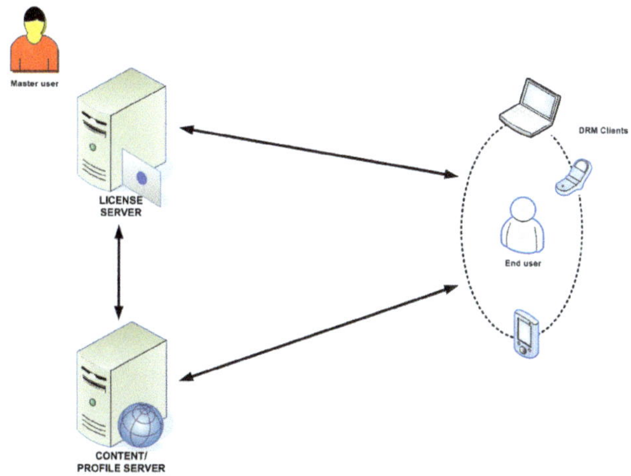

Fig. 15. Basic components and generic view of a DRM architecture with appliance to user-centric multimedia environments.

5.1 License server

The License Server (LS) accepts the request from the content's distributor, (in the general scenario), for the generation of an MPEG-21 compatible license. Initially, the LS execute the MPEG-21 authorization algorithm and in the case that its result is "yes", it creates the license and it digitally signs the license with the public key of the intended user. In a second phase the LS utilizing security mechanisms like TPM or various security libraries, it creates the packaged license, i.e. a simple concatenation of the MPEG-21 license and the content encryption key, then it encrypts the packaged license with the end users' public key and transmits the encrypted license towards the end user. In the following figures, we can see some diagrams that depict a simple scenario, where a user has acquired an encrypted digital content to his device and he requests a license from the distributor, who in turn redirects his request to the LS, (see Figures 6, 7). The basic functionalities that are implemented by the LS are,

1. Access and knowledge of the end-users' public key
2. Access to a set of root grants
3. Access to a set of MPEG-21 licenses, (optional)
4. Takes as input a request from the user that has the appropriate elements for the license creation, e.g. for which digital content, any conditions, *keyHolder* valid information, etc.

Then the LS executes the authorization algorithm, taking in care the users' request, the set of root grants and the possible set of additional licenses. If the result of the execution of the authorization algorithm is "yes" then it creates a valid MPEG-21 license, it digitally signs it with the end-users' public-key and pass it to the module for the creation of the packaged encrypted license, which is finally transmitted to the user. In case of negative response from the authorization algorithm, the user is acknowledged for the result and the LS end the process.

5.1.1 Architecture of the license server

In order to facilitate the development and implementation as well as the analysis of the needs, the software architecture of the license server is divided into the following levels as they are depicted in Figure 16.

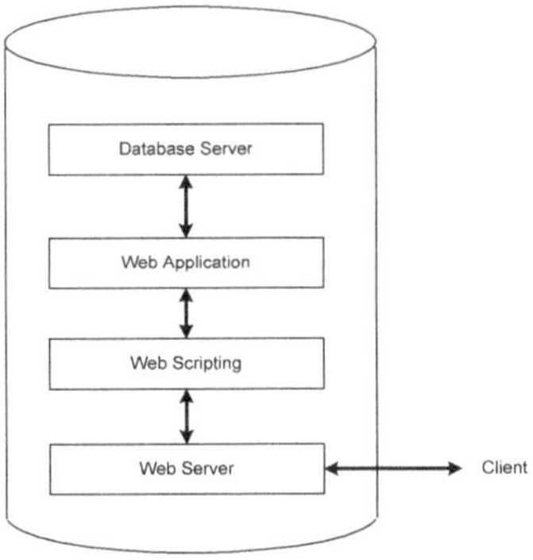

Fig. 16. Basic internal components of the License Server.

5.1.1.1 Web server

The web server is needed in order to provide the HTTP connectivity necessary to exchange the information and also provide an interface with basic messaging information such as to denote success or failure of the requests. This service is provided by the basic functions of the Internet Information Services. This allows an easy connectivity over any TCP/IP connection and instant compatibility with a variety of web-enabled clients.

5.1.1.2 Web scripting interface

A scripting interface based on the ASP scripting language is executed on the web server and it plays the role of the basic interface of the web service. This is the bridge between the HTTP requests and responses and the core functionality of the server. The ASP interface can provide simple mechanisms such as session management, user menu control and also deal with all interactions not related to the license creation mechanism itself.

5.1.1.3 Web application

The web application is a full scale (not scripting) application that is executed by the scripting interface. The web application, written in C#, executes the functionality of the main task of the application. Main tasks are as follows:

- Managing communication
- Authenticating user
- Accessing the license server application
- Encrypting license information

Communication is managed by one end via the scripting interface to the web peer and on the other application to the database server containing license information. As a result, the communication task of the web application is the bridge connecting end-to-end the communication parties. It is the communication task of the web application that translates all requests and messages. The other levels of the architecture simply adapt and transmit the information in the appropriate way. When the main application receives the requests for license creation from the user, it needs to check if the requested license can be legitimately created. In order for this to take place the user needs to be authenticated so this is an important part in the whole security of the system and it executed by the web application.
The last task of the web application involves the creation of the license for the requested content. The license is created after accessing the license server application and retrieving the necessary information. The license will be created in the format of an XML file and its content will be encrypted using the security libraries.

5.1.1.4 Database server

The database server is implemented with a Microsoft SQL server since this provides maximum compatibility with the IIS server and the C# modules. The database server is accessed by the main application and contains all the information related to users and content necessary for the creation of the license.

5.1.1.5 Communication protocol and methods

The proposed web connectivity method requires simple HTTP protocol use. It was chosen not to use security at this level, with the possible use of HTTPS protocol, since there is inherent security embedded in the content as the content of the license is encrypted. This allows avoiding the establishment of keys or the usage of certificates which would further complicate the communication.

5.1.1.6 License templates

As we have already stated the license is meta-described with the utilization of MPEG-21 standard. MPEG-21 contains a specific part, (Part 5), named MPEG REL, which provides a simple XML-based data model which allows to the content creator to meta-describe the license that describes the usage rules over a specific digital content. The use of MPEG-21, leads to a DRM scheme that is adaptive to the end-user needs, i.e. different users must have different usage rights over the same digital content, while also characterized by interoperability.
In the context of our work, the license files are not created dynamically from scratch, rather we have some pre-arranged template files that have some specific fields, which are updated with the corresponding parameters that are passed from the client's web interface towards the License Server, whenever it is requested the generation of a MPEG-21 license.
Considering the structure and creation mechanisms of licenses, as we have already stated the license is meta-described with the utilization of MPEG-21 standard. MPEG-21 contains a specific part, (Part 5), named MPEG REL, which provides a simple XML-based data model which allows to the content creator to meta-describe the license that describes the usage rules over a specific digital content. The use of MPEG-21, leads to a DRM scheme that is adaptive to the end-user needs, i.e. different users must have different usage rights over the same digital content, while also characterized by interoperability. The license files are not created dynamically from scratch; rather we have some pre-arranged template files that have some specific fields, which are updated with the corresponding parameters that are

passed from the client's web interface towards the License Server, whenever it is requested the generation of a MPEG-21 license. In Appendix (section 10.1), we present such a template for an MPEG-21 based license which sets rules and specifies rights for contents usage between two different types of mobile devices.

5.2 DRM client

One the core software components of our architecture is the DRM client application, which as it has been stated before is a software-based application that resides into the end user's device and it is responsible (a) for the enforcement of the license that accompanies a digital content, i.e. allow the unhindered and according to the license's terms "consumption" of the digital content; and, (b) for the rendering of the content.

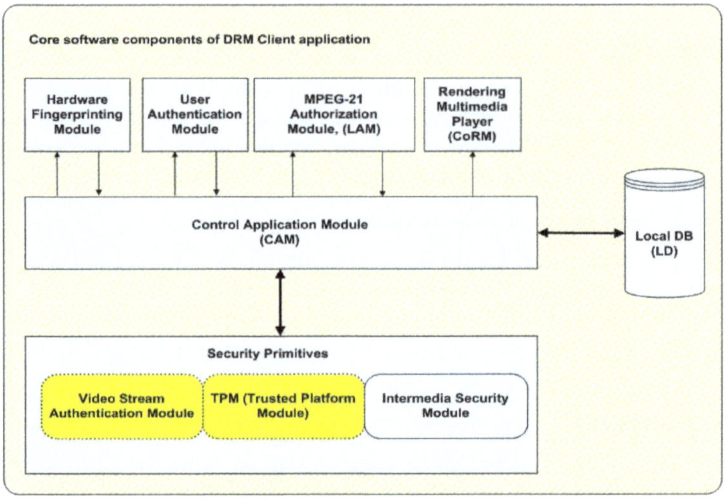

Fig. 17. The core components of the DRM Client application.

In a first approach, both the encrypted digital content and the relevant encrypted license are located in the device's file system, for example, both can feed to the user through a secureSD module. In an extended version, the DRM client is enforced with functionalities that allow to the user to access a pool of available digital contents and request the generation of valid licenses, in a near real-time manner. The basic core components of the DRM client application can be seen at the following figure,

5.2.1 User authentication

Whenever a user initiates the DRM client application, a user authentication process takes part, i.e. a user login interface pop-ups to the user, requesting some credentials (username - password), to be entered by the user, see following figure. Both the username and the password have been given to the user in a pre-arranged manner and are stored to the embedded database. When data are inserted a check is done on the fly and data are compared with those that are stored in the database; in case of erroneous situation the end-user is informed and the application terminates. Also, the user is informed in case that any other error occurs, e.g. for some reason the application cannot initiate a connection with the database.

5.2.2 Hardware fingerprinting module

The hardware fingerprinting module, mainly, is used in order to protect unauthorized delivery of both client application and transferred data (encrypted content and encrypted license) to third party devices. As an example of this, we can think the case in which a legitimate user gives both the DRM Client application and the encrypted data (license and content) to another user - to play it to his personal device, while also provides him with his credentials. To avoid such issues, in a pre-arranged manner, valid hardware fingerprints of devices are stored in the embedded database that accompanies the client application, (Figure 18 depicts Hardware Fingerprint creation and registration towards License Server).

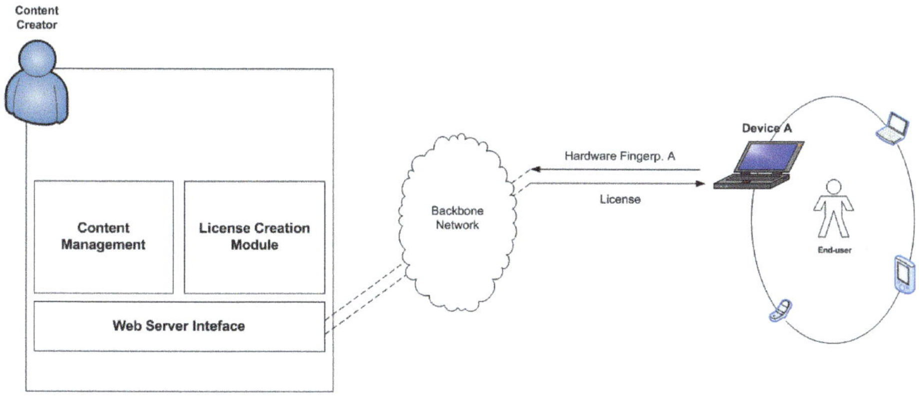

Fig. 18. Creation and registration of device's hardware fingerprint to the License Server.

The hardware fingerprint is a 16 byte unique identification code of a computer, e.g. 4876-8DB5-EE85-69D3-FE52-8CF7-395D-2EA9, and it is generated as message digest from the device's hardware components like for example CPU Id, its BIOS Id, physical hard drive information, motherboard Id, etc. Programmatically, we can get such information utilizing *System.Management* Namespace that is provided from the .net, which provides access to a rich set of management information and management events about the system, devices, and applications instrumented to the Windows Management Instrumentation (WMI) infrastructure.

The device's hardware fingerprint is generated in an a priori manner, i.e. the end-user has to inform the license/content provider about the device or devices that the user is intended to render the content. It is based upon the following device's unique characteristics,

a. CPU Id
b. BIOS Id
c. Motherboard Id
d. HDDs characteristics
e. Video controller ID
f. MAC Address of network card.

5.2.3 Web interconnectivity with license server

The DRM client needs to communicate with the License Server in order to request a license creation and then receive the license file. This communication must be protected using encryption since the license file contains sensitive information that allows decryption of

content. Moreover, the client could be simultaneously authenticating at the time of the license request so sensitive information such as a user PIN should be protected.
Since the amount of data that needs to be exchanged is very small, typically of a few kilobytes corresponding to the XML file containing the license, the methods used to communicate with the license server could be simple and straightforward. It was chosen to use web-based methods for this communication since this allows for easy development using existing web architecture frameworks and good interoperability between the distinct software components of the server.

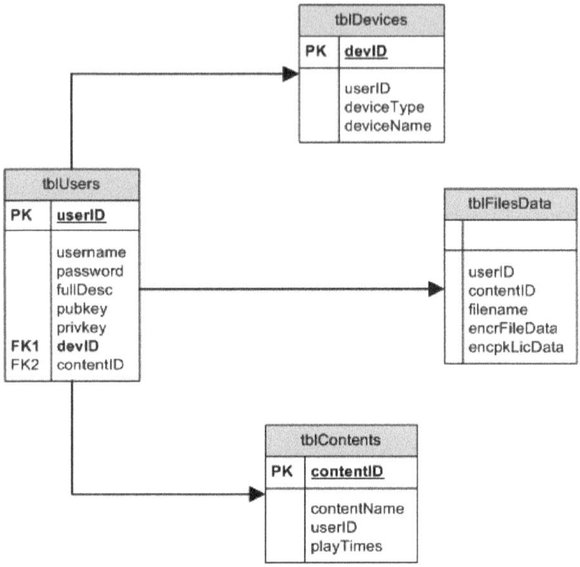

Fig. 19. DRM Client's application embedded database structure, in tables level.

5.2.4 Embedded secure database

An embedded secure database is used for storage and retrieval of data. In general, an embedded database system is a database management system (DBMS) which is tightly integrated with application software that requires access to stored data, such that the database system is "hidden" from the application's end-user and requires little or no ongoing maintenance. It is actually a broad technology category that includes database systems with differing application programming interfaces (SQL as well as proprietary, native APIs); database architectures (client/server and in-process); storage modes (on-disk, in-memory and combined); database models (relational and object-oriented); and target markets. For the purpose of our application, we have used Microsoft SQL Server Compact Edition 3.5 (MS SQL CE[4]), which is an embedded database that allows us to integrate it in our desktop and mobile applications. It takes about 1.5 MB on HDD and consumes about 5 MB of RAM. One of the biggest advantages of using MS SQL CE as the core DBMS system is

[4] MS SQL Server Compact Edition 3.5,
http://www.microsoft.com/Sqlserver/2005/en/us/compact.aspx

that it allows the encryption of the database during its creation with a master password; the password to access the database is hard coded into the application. So it provides a high level of security to the provider of the DRM client application towards the end-user. In the development level, the calls to the database are done with the utilization of C# and SQL primitives, i.e. all transactions to the database, for example data retrieval, insertion and update of data, are done through SQL queries and (INSERT, UPDATE, SELECT) statements.

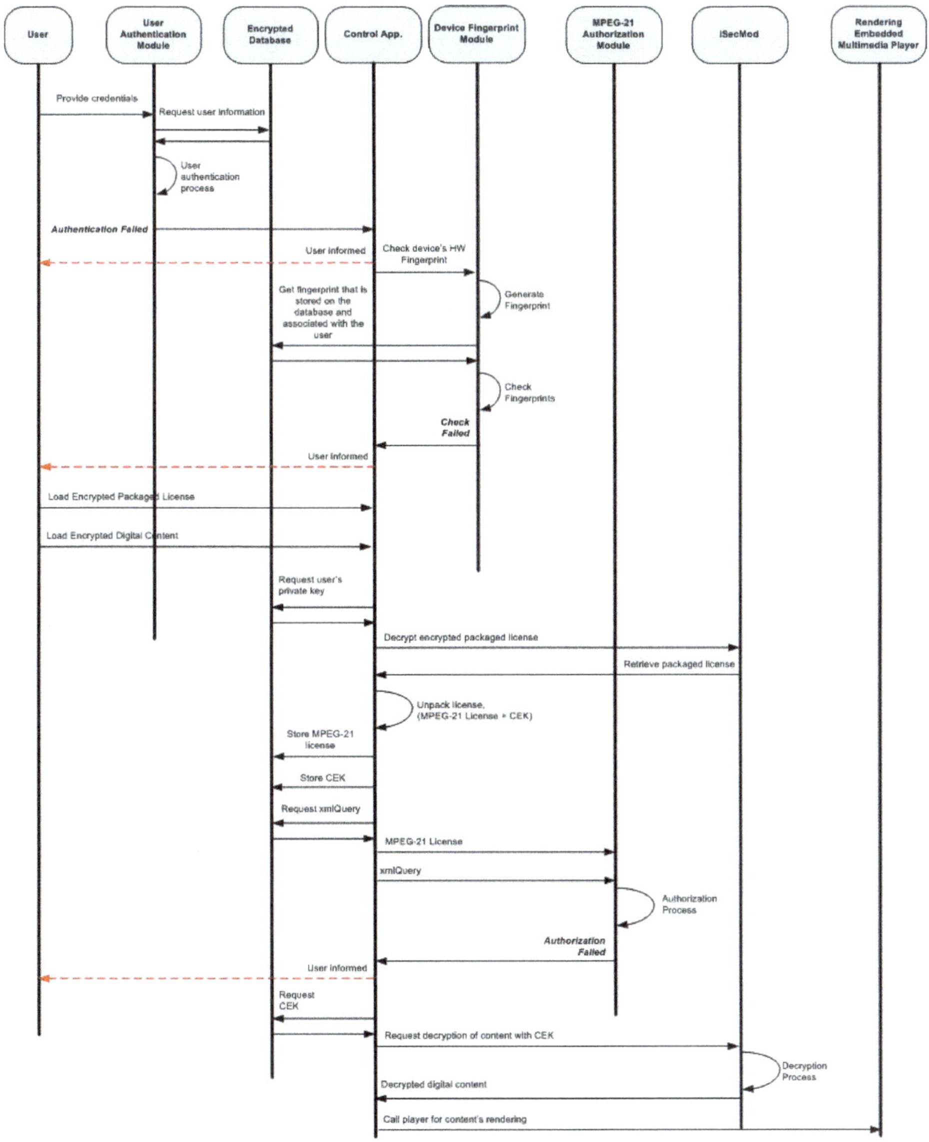

Fig. 20. Information flow between the core components of the DRM Client application.

5.2.5 Control application module

The Control Application Module is the core software component that it is responsible for the supervision, control and handling of the pre-mentioned software components. In the following figure it can be seen information flow (sequence diagram) between various internal components of DRM application.

6. DRM and security aspects in streaming data

In this section a description is done regarding, integration of pre-mentioned DRM architecture in a session migration framework. During session migration, a session, (e.g. rendering a multimedia content to device A), is migrated to device B, transparently and without any disruption to the end user. It is easily understood that such architecture, in case that it is needed to be enhanced with DRM functionalities, special methods have to be used, since DRM adds an overhead to normal operation. In (Repetto et al, 2010), it has been proposed an architecture for session migration in high mobility environments, utilizing with security and DRM aspects.

6.1 Session migration architecture integrated with security and DRM aspects

In a typical scenario for pervasive user-centric computing, the backbone architecture must provide to the end user the ability to have seamless access to digital contents in a variety of different types of devices, without loss of session; i.e. if a digital content is rendered on a PDA, when the user is located in front of another personal multimedia device, the session has to be transparently moved from the PDA to this device. This process has to be enhanced with security and DRM aspects.

As it has been stated before, the main objective is to generate a shared vision of user-centric multimedia services for modern nomadic people, who have high level of mobility and are connected to network for most of times. One of the main implications of the user-centric approach is seamless and secure access to content, regardless of its location and users' terminal device(s), which raises new concepts and expectations about Internet access, leading to the need for pervasive media architectures, which have to be aware of security and DRM aspects. Such pervasive environments rely on the presence of lots of devices in the user surrounding and on almost global network coverage, which is not an utopia ever since today, at least for what concerns 2D media: this is a key factor in fostering an interactive and continuous participation of users in the active content chain, but it is not enough to build user-centric systems. Indeed, users should interact with remote applications, content and services in a transparent, continuous and seamless way while having mechanisms that allow protecting their intellectual properties and digital rights, independently of the current device and access network.

From a networking perspective, our vision essentially targets secure user-centric media access in pervasive communication. In this scenario, users create their own content and store it at some location. That may be personal or public content made for profit or for amusement; in all cases it is likely that content has to be shared with other people. Content creators often wish to manage how other users access their content and what kind of operations those people are allowed on the content (e.g., rendering, copy, redistribution, editing), especially when they intend to get profit from it. This usually brings to the concept of license and Digital Right Management, but enforcing the observation of a license on digital media is a very hard task; currently, it mainly relies on honesty and integrity of users. Several devices may be available for users in their environment: personal or public, fixed or portable, with different capabilities and interfaces. Moreover, a large number of network connections are available: public and private, wired and wireless, covering body, local, metropolitan or wide areas, with different services and performance. In modern computing environments, Digital Right Management is already an issue in itself, but it gets even harder in pervasive communication environments, where multiple devices may be used by users to access secured DRM-enabled digital contents.

In the typical scenario, end users play a leading and twofold role (see Fig. 19): they create content and in the meantime they also are content consumers. On the one hand, content creators are mainly interested in preserving their intellectual properties. On the other hand, consumers require a mobility infrastructure to build the pervasive paradigm. Three entities are required to cope with these tasks: content providers host user's media and make them accessible by other users (content sharing), a DRM system takes the responsibility of enforcing policy rules on content access, rendering and duplication (DRM management) and a service provider must account for user mobility (mobility management). Finally, information about the user's surrounding may really enhance the system (context awareness), providing local information about the physical environment, presence of devices, available services and so on. Figure 21 also reminds the initial approach and assumptions based on the user-centric paradigm.

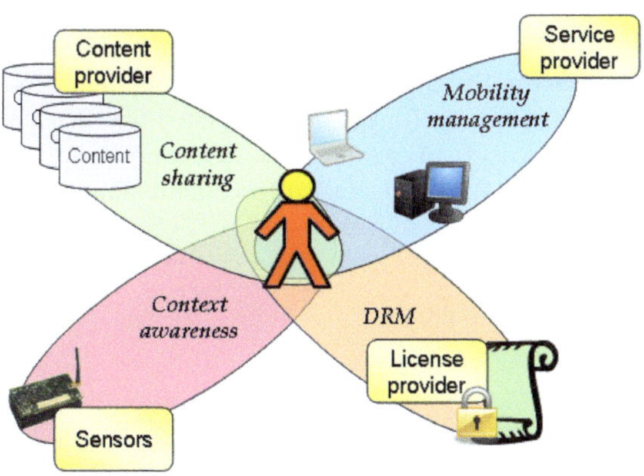

Fig. 21. Content access with DRM for mobile users in the InterMedia approach.

Figure 22 depicts the building blocks of the integrated approach. Content is available at some location (user's home or public repositories). The DRM subsystem is a trustworthy entity that knows about user's rights and policies to be applied to content; it enforces the user's client to not break the owner's agreement and terms of use (the license). As the picture shows, content does not need to be stored inside the DRM system, whilst the user's client must be part of DRM to avoid malicious users to bypass the permissions granted by the license. A mobility component integrates with the DRM system; it allows both terminal mobility (handover) and session migration (users can change their terminal without any session break). Note that no support at all is required at content location (transparency to third parties). The mobility framework includes localization facilities in order to automate the migration process; users are tracked and their position is matched against device's location to find out the most suitable one.

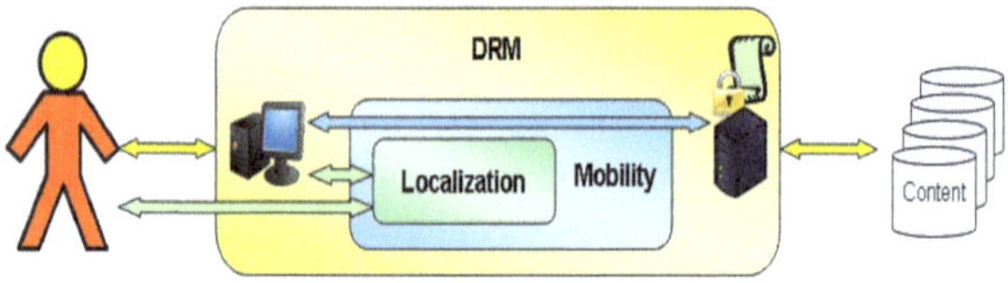

Fig. 22. Overall schematic system architecture.

In a typical DRM architecture, we can identify three main entities: (a) the digital content creator, who is the entity that generates digital contents that are stored into a server, which is located in users' premises. The user has the copyright of those contents and triggers the license generation procedure; (b) the License Server, which is the part that creates valid licenses according to user's directives that are associated with a specific digital item - the licenses are MPEG-21 compatible - and is responsible for controlling the whole DRM functionality. The license server may also contain the digital contents or it can have access to some private repositories where they lie; and, (c) the DRM Client application, which is a software application that resides into end user's device. This application is responsible for enforcing the license regarding digital content's "consumption". The basic actions, (see Figure 23, for a schematic depiction), that the content provider (the user that creates digital content) has to perform in the chain of a DRM protected media delivery towards a third party, are: (a) protect and secure the digital content; (b) generate a license that describes the usage rights over the digital content; (c) provide a secure way for contents' and licenses' transmission towards the end-user; (d) enforce the license on end-users' devices; and, (e) inform content creator/owner about the license enforcement status.

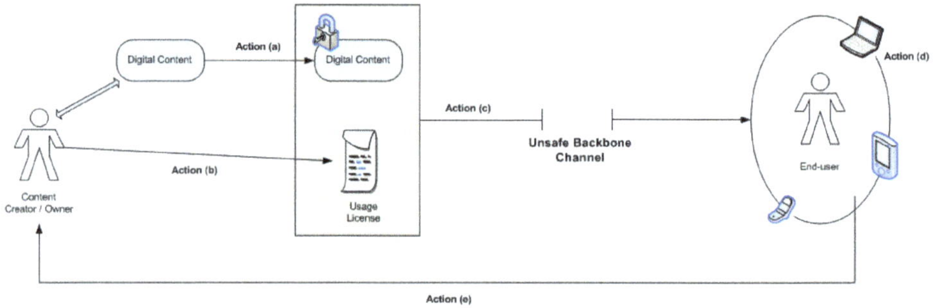

Fig. 23. Actions due to delivery of a DRM-protected digital content towards a remote entity.

6.2 Common testbed for architecture evaluation

We have implemented the whole architectural framework in a common test bed. Most of the architectural parts described in previous sections are already working; we are currently designing the internal architecture of the most complicated elements.

Our test bed specifically addresses multimedia real-time streaming, with integrated security mechanisms and DRM functionalities. It substantially implements the general scenario described in previous section with three main actors: content creators/owners, content providers and content consumers. Content providers are trustworthy parties that create licenses on behalf of content owners and give access to their media. The implementation requires a Media Repository, a Content Server, a User Client and a Context Server (see Figure 24). These elements are made up of the architectural frameworks depicted in previous section; here we discuss how the whole framework is expected to work, describing the interfaces and relationships among the above components, without going into details about their internal software architecture.

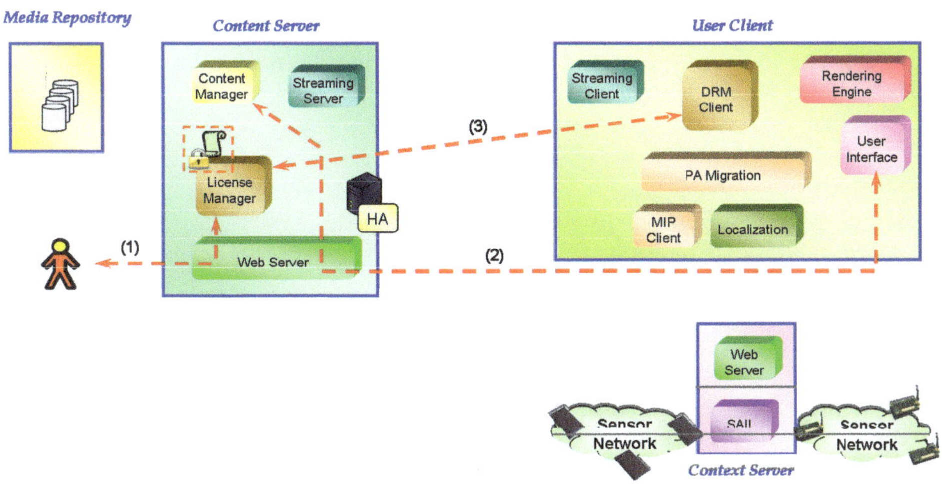

Fig. 24. Test bed for evaluating proposed architecture; initial signaling.

For the sake of generality, we assume content is stored anywhere in the Internet. Content creators register their media and the license policy by mean of some interface, for example a web interface (1). Users browse content list at the Content Server, again through the same interface (2). Once the user has selected the media, he requests a license to get it (3) and he is usually charged for this - License Manager is responsible for performing this action. User mobility implies the session may be handled by heterogeneous devices or even different implementations of the same application, and our proposed DRM system is ready to cope with this issue. Indeed, the license might be bound either to the user or to its device. Due to the pervasive nature of the scenario, the license is bound to the user, since this is a strict requirement to allow digital right management in pervasive environments. Thus, the License Manager component generates a license for that user (user-centric approach); this is possible by binding the license to a Personal Address (PA) which is assigned to the user at the same time (as well as all cryptographic material for MIP operations). Furthermore, as added security we add some device related information to the license, like for example a unique hardware fingerprint, that allows access to content only to authorized devices. This procedure makes the system dynamic. However, at the current stage of implementation we do not provide the web interface and a communication protocol between the DRM Client and Server components; instead, we get the license and the PA offline. The PA Migration adds the PA to the network interface and the MIP Client registers its current location with the HA, following the standard MIP procedure (either the client or the local FA addresses can be used.

The Content Server may also be split into two separate parts, content management (Content Manager, Streaming Server and Home Agent) and the License Server; however that would require an additional secure interface between them (the License Manager and the Home Agent (HA) have to exchange cryptographic material). Once the user requests the content, the DRM Client checks the license and enforces any restriction therein. If this check is successful, the Streaming Client requests the media by some suitable protocols; we chose RTSP for our implementation. The PA is used at this step, thus all packets are routed through the Mobile IP infrastructure (HA and eventually local FAs). This assures the request comes from an authorized client, as MIP tunnels are encrypted. At this point the Streaming Server locates the content and streams it, again through the MIP tunnel (see Figure 25). The Rendering Engine takes care of playing the media through the available interface.

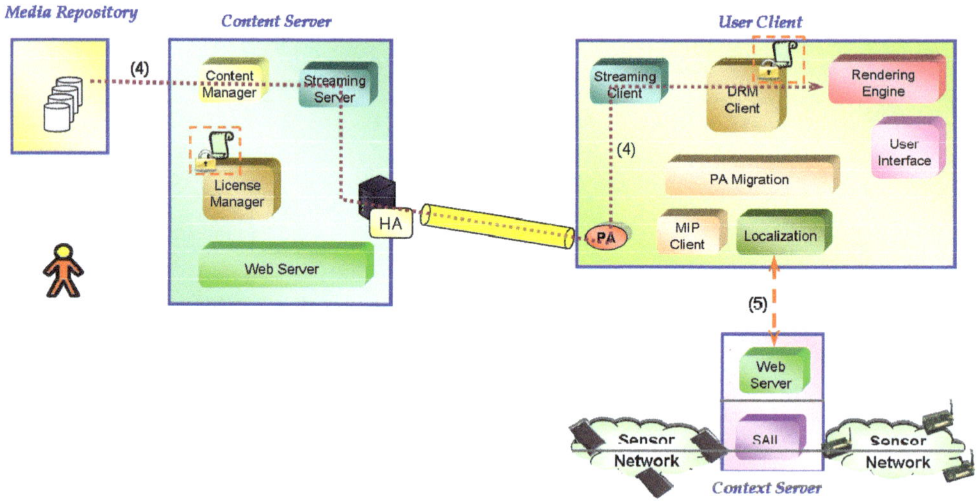

Fig. 25. The testbed architecture: media are routed through the MIP infrastructure.

At session start-up, the User Client also subscribes to the Context Server; for all session duration the Localization module polls the latter to timely detect any user movement and to discover any device close to him. Currently, we use HTTP for this interface. When the migration is needed, the PA Migration function saves the current context (license, PA, cryptographic material, media codec, transport ports, etc.) and transfers it on the new device; the MIP Client at the new device update the MIP registration and then session is restored transparently to any component behind the HA.

Furthermore, in Figure 26, we depict a sequence diagram that shows all the actions that take part, towards a session migration scenario between two or more devices. Moreover, in this figure it is depicted the request of device hardware fingerprint from the License Server, which uniquely identifies the specific device.

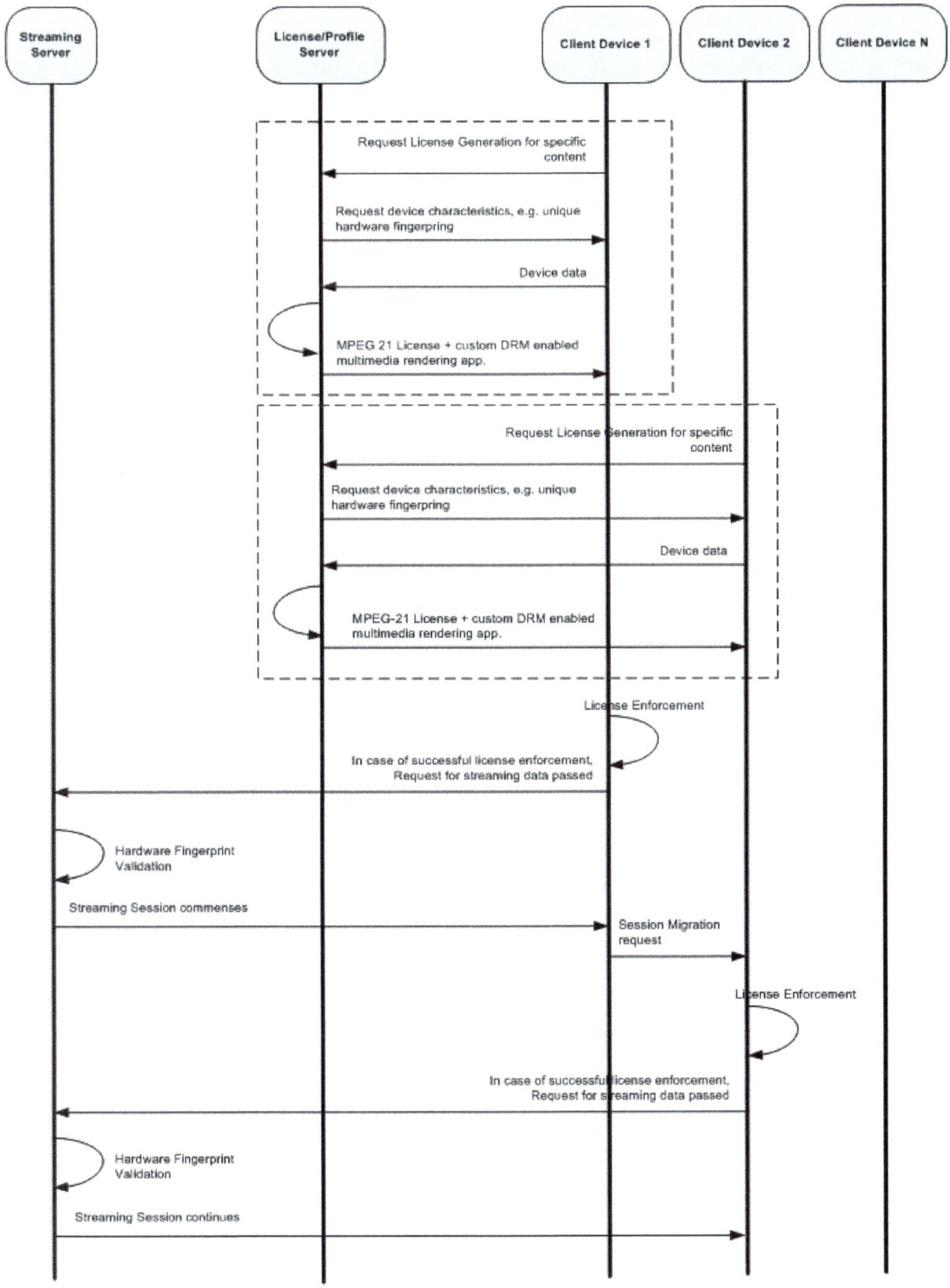

Fig. 26. Session migration scenario between two or more devices.

7. Research initiatives – related work

In this section we present, in short, some recent research initiatives and European projects, which utilized security and DRM aspects, in the context of user-centric multimedia converged nomadic environments. Mainly, we are focusing in Intermedia Network-of-Excellence, (Intermedia, 2006), and CHIRON, (Chiron, 2010).

7.1 Intermedia

There have been considerable efforts to have Audio Video systems and applications converge, in particular in home environments with homes as spaces of convergence, and for nomadic users with advanced mobile devices as points of convergence. These trends are important but also have limitations that we seek to address and overcome: home-centric systems fail to account for increased mobility and the desire to provide continuous service across spatial boundaries outside the home; device-centric convergence, e.g. in 3G phones, supports nomadic use but provides a very limited user experience as no single device and interface will fit many different applications well.

INTERMEDIA project, (Intermedia, 2006), investigated to progress beyond home and device-centric convergence toward truly user-centric convergence of multimedia. Its vision was The User as Multimedia Central: the user as the point at which services (multimedia applications) and the means for interacting with them (devices and interfaces) converge. Key to our vision is that users are provided with a personalized interface and with personalized content independently of the particular set of physical devices they have available for interaction (on the body, or in their environment), and independently of the physical space in which they are situated. Towards this vision is has been investigated a flexible wearable platform that supports dynamic composition of wearable devices, an ad-hoc connection to devices in the environment, a continuous access to multimedia networks, as well as adaptation of content to devices and user context.

The project's main objectives can be summarized to the following, (a) Vision toward device-free user-centric media environments; (b) Constructing multidisciplinary research groups; and (c) Building a common platform with a vision towards semantic convergence.

Towards the context of security and DRM, the following directions have been utilized; (a) Current applications consume a large amount of optimal quality multimedia data suited for a variety of devices, networks, and capabilities. Anywhere and anytime feed the need of content consumers to have the ability to transparently switch the consumption of their session to another (mobile) terminal without any disruption of the session. INTERMEDIA targeted to seamless continuity in multimedia sessions over heterogeneous networked devices through transparent handover of multimedia sessions and accompanied contents. This concept is called session mobility and can be realized using different MPEG-21 technologies for instance; (b) Protection of Intellectual Property (IP) through DRM solutions is a necessity in modern multimedia services. The concerns of content providers for loss of revenues constitute a strong obstacle to the wide deployment of services that involve distribution of IP protected content. Thus, it is a requirement to integrate DRM solutions in services and applications that target delivery of IP protected content to a large base of clients. Considering the technical problems that result from add-on security solutions to independently developed network services, the approach of our efforts to develop architectures with security and DRM as inherent requirements will lead to secure solutions that will increase the trust placed by content providers on the system and thus, it will lead to wider availability of services to a larger population. Such a delivery architecture needs to be scalable, in order to accommodate an increasing client population, leading to a requirement for optimized use of such resources as bandwidth and computational power, while achieving the target Quality-of-Service (QoS) characteristics for all clients of the service.

7.2 Chiron

The CHIRON Project (Chiron, 2010) intends to combine state-of-the art technologies and innovative solutions into an integrated framework designed for an effective and person-centric health management along the complete care cycle.

In this vision,

- CHIRON will address and harmonize the needs and interests of all the three main beneficiaries of the healthcare process, i.e., the citizens using the services, the medical professionals and the whole community;
- CHIRON will position the citizens at the core of the whole healthcare cycle by considering them as "persons" with specificities and identities and will empower them to manage their own health;
- CHIRON will enlarge the boundaries of healthcare by fostering a seamless integration of clinical setting, at home setting and mobile setting in a concept of a continuum of care;
- CHIRON will speed up the move from treatment of acute episodes to prevention;
- CHIRON will provide the physicians with extensive support for treatment monitoring and management, timely decisions and appropriate actions in both the clinical and home environments;

In CHIRON - rather than aiming at coming up with new security means- our activities on security will advance the SoA by ensuring seamless integration of security features in an extremely heterogeneous environment. So our contribution will be,

- Security architecture featuring:
- Strong security on resource restricted devices such as wireless sensor nodes and within the well equipped hospital infrastructure.
- Seamless integration of devices stemming from different administrative domains e.g. integration of user devices into the hospital architecture and vice versa.
- End to end security in an environment which integrates heterogeneous communications technologies
- Privacy architecture which exploits the security architecture to ensure confidentiality of data and which allows to adapt the level of confidentiality of data according to external conditions such as type of diseases, role e.g. patient versus doctors/nurses.

With regard to privacy whereas many privacy and data protection means aim at a declarative approach only, there are limited approaches for technical protection. To the best of our knowledge there are only a few approaches14 that try to make sensitive data available to a third party while ensuring secrecy of that data.

Main activities of technological innovation will be focused in the following areas:

- Authentication, Authorisation and Accounting (AAA), in order to guarantee the access, the quality of the service and the reliability of an operator.
- Security, in order to guarantee data confidentiality and protect the system through the predictive discovery of attacks.

- Privacy, in order to prevent access or processing of data for purposes other than supposed. Further to control inference of information from the aggregated medical data or even from the mere use of the protocol independent from the actual data content.

7.3 Enabling privacy in person centric e-health environments using DRM aspects

Telemetric monitoring of vital parameters of patients with chronic diseases is recognized to improve their medical condition and hence their quality of life. It also improves treatment adjustments, reaction time in acute cases and helps to reduce duration and costs of hospitalization. As a result of this, there are plenty of products and solutions for personal health monitoring available today that acquire physiological data in real-time. In order for such systems to be widely acceptable and utilized by the medical community and the patients, they must be developed satisfying the security requirements imposed by real-time data communication and protection of sensitive physiological data and measurements, data integrity and confidentiality, and protection of the monitored patient's privacy. By utilizing MPEG-21 standard's primitives, we argue that protection of transmitted medical information and enhancement of patient's privacy is accomplished, since there is selective and controlled access to medical data that sent toward the hospital's servers, (Fragopoulos et al, 2010).

In a person-centric e-health monitoring environment the primary goal is to collect, process, monitor and store medical data from different types of Wearable Embedded Monitoring Devices (WEMDs), which are located on the patient, while furthermore forward those information into general-purpose computing devices – for more sophisticated and complex processing. In such environments different security aspects may arise, so in order to be able to identify the security mechanisms that should be taken in consideration and possibly implemented into the proposed architecture for person-centric e-health monitoring infrastructure, we have to identify and classify, (i) possible attackers and malicious users of the aforementioned environment; (ii) security and privacy requirements. In general, security is a fundamental requirement in modern computing systems, but in user-centic environments focused in e-health, security is a critical and imperative requirement that needs special care in all levels, since in such systems there is flow of sensitive information between various entities.

Although DRM architectures are mainly used to the multimedia world, we argue that utilization of DRM aspects using MPEG-21 standard's primitives, can lead to protection of transmitted medical information and enhancement of patient's privacy, since there is selective and controlled access to physiological data that sent toward the hospital's servers.

A lot of research work (Fragopoulos et al, 2009), (Fragopoulos et al, 2010), (Leister et al, 2009), (Jafari et al, 2010), is towards this direction, while furthermore we are investigating issues related with the safe destruction of the medical data after their viewing; in that context the use of Trusted Platform aspects.

8. Conclusion

In the context of ubiquitous and pervasive computing environments, considerable efforts have been done in order to have audiovisual systems and applications converge, especially in home environments where homes can be considered as spaces of convergence, and for nomadic users with advanced mobile devices as points of convergence. One of the biggest challenges that we have to face in the deployment of architectures in such environments is related to, on the one hand, with the security and protection of digital contents that interchanged between users of such architectures and on the other hand with the provision to the users with security mechanisms that allow them to perform secure transactions (e.g. authentication, privacy protection, secure data transfer, etc.) in those environments. Moreover intellectual property protection is a mandatory request in modern multimedia environments like the ones that are going to be deployed in the InterMedia context. Today the end-users are equipped with different types of small devices that allow them to be the digital contents creators, thus creating digital content that wish to share with third parties. In most cases, the end-users would like to have mechanisms which would give them the possibility to protect the content which have created and possibly to set their own usage rights over it, thus specifying towards third users how their digital content shall be used.

9. Acknowledgment

The work that has been presented in this book chapter comprises part of extensive research work that has been done to the EU Network of Excellence "INTERMEDIA", FP6 - IST-38419. Also, this work has been funded by CHIRON Project, (Chiron, 2010).

10. Appendix

10.1 MPEG-21 license template

Following, we present a detailed XML view of an MPEG-21 license, generated by the License Server, which allows streaming of a specific content to a specific user on two pre-defined devices.


```xml
<?xml version="1.0" encoding="utf-8" ?>
<r:license licenseID = "0001"
 xmlns="urn:mpeg:mpeg21:2003:01-REL-R-NS"
 xmlns:mx="urn:mpeg:mpeg21:2003:01-REL-MX-NS"
 xmlns:dsig="http://www.w3.org/2000/09/xmldsig#"
 xmlns:xenc="http://www.w3.org/2001/04/xmlenc#"
 xmlns:r="urn:mpeg:mpeg21:2003:01-REL-R-NS"
 xmlns:sx="urn:mpeg:mpeg21:2003:01-REL-SX-NS"
 xmlns:m1x="urn:mpeg:mpeg21:2005:01-REL-M1X-NS"
 xmlns:m2x="urn:mpeg:mpeg21:2006:01-REL-M2X-NS"
 xmlns:xsi="http://www.w3.org/2001/XMLSchema-instance">

 <!--Grant for protected resource with ID=protectedResource1, for -->
 <!--Target Device with identifierm DEV_A -->
 <r:grant>
  <r:keyHolder>
   <r:info>
    <dsig:KeyValue>
     <dsig:RSAKeyValue>
      <dsig:Modulus>
                 base64_encoded_modulus_of_user's_public_key
          </dsig:Modulus>
      <dsig:Exponent>
                 base64_encoded_exponent_of_user's_public_key
          </dsig:Exponent>
     </dsig:RSAKeyValue>
    </dsig:KeyValue>
   </r:info>
  </r:keyHolder>
```

```
    <!--Device's Domain Indentifier-->
    <m1x:identityHolder licensePartId="domain_A">

      <m1x:idSystem>
       urn:mpeg:mpeg21:2006-01-REL-M2X-NS:DM-1000 <!--Domain Manager URI-->
      </m1x:idSystem>
      <m1x:idValue>
       DO0001 <!--Domain Identifier-->
      </m1x:idValue>

    </m1x:identityHolder>

    <!--Digital Resource Details-->
    <!--Content is protected, i.e. encrypted, and encryption key is encrypted-->
    <!--and it is embedded into the protectedResource element-->
    <m1x:protectedResource licensePartId="protectedResource1">

     <digitalResource>
      <nonSecureIndirect URI="URI_of_digital_content"/>
     </digitalResource>

     <xenc:EncryptedKey>
      <xenc:CipherData>
       <xenc:CipherValue>
       <!--Digital Content, BASE64 encoded, -->
               <!--encrypted key with user's pub. key-->
                         Base64_encrypted_key_with_user_public_key
       </xenc:CipherValue>
      </xenc:CipherData>
     </xenc:EncryptedKey>

    </m1x:protectedResource>

    <!--Conditions that has to be apply for the Target Device-->
    <r:allConditions>

     <m2x:destinationPrincipal> <!--Allowed Target Device-->
      <m1x:identityHolder>
       <m1x:idSystem>
        urn:mpeg:mpeg21:2006-01-REL-M2X-NS:DM-1000:DO-0001
       </m1x:idSystem>

       <!--Target Device A-->
       <m1x:idValue>
        DEV_A, (Target_Device_ID_Unique_HW_Fingerprint)
       </m1x:idValue>
      </m1x:identityHolder>

     </m2x:destinationPrincipal>

     <sx:exerciseLimit>
      <sx:count>Number of Times that content is allowed to rendered</sx:count>
     </sx:exerciseLimit>

    </r:allConditions>

   </r:grant>

   <!--Grant for protected resource with ID=protectedResource1, for -->
   <!--Target Device with identifierm DEV_B -->
   <r:grant>

    <m1x:identityHolder licensePartIdRef="domain_A"/>
    <m1x:protectedResource licensePartIdRef="protectedResource1"/>
    <allConditions>
```

```
    <!--TARGET DEVICE B-->
    <m2x:destinationPrincipal>

      <m1x:identityHolder>
        <m1x:idSystem>
          urn:mpeg:mpeg21:2006-01-REL-M2X-NS:DM-1000:DO-0001
        </m1x:idSystem>

        <!--Target Device B-->
        <m1x:idValue>
          DEV_B, (Target_Device_ID_Unique_HW_Fingerprint)
        </m1x:idValue>
      </m1x:identityHolder>

    </m2x:destinationPrincipal>

    <!--Maximum number of play times-->
    <sx:exerciseLimit>
      <sx:count>Number of Times that content is allowed to rendered</sx:count>
    </sx:exerciseLimit>
  </allConditions>
</r:grant>

<r:issuer>
  <keyHolder>
    <info>
      <dsig:KeyName>Rights Issuer Public Key Name</dsig:KeyName>
    </info>
  </keyHolder>
</r:issuer>

</r:license>
```

11. References

Adams, Carlisle; (2006) *A Classification of Privacy Techniques*, UOLT Journal, 2006, http://www.uoltj.ca/articles/vol3.1/2006.3.1.uoltj.Adams.35-52.pdf

Anderson, Ross J.; (2001) *Security engineering: a guide to building dependable distributed systems*, Wiley, 2001, ISBN. 0471389226

Burnett, Ian (Ed(s)). 2006. *The MPEG-21 Book*, Willey, ISBN 0470010118, England

CHIRON European Project, (2010), Cyclic and person-centric Health management, JU ARTEMIS Grant Agreement # 2009-1-100228, 2010-2012

Delgado. J.; Prados, J.; Rodríguez, E.; 2005, *An MPEG-21 REL mobile profile*, ISO/IEC JTC 1/SC 29/WG11/M12229, July 2005, Poznan (Poland).

Fragopoulos, A. & Serpanos D. N. (2005). *Intellectual Property Protection Using Embedded Systems*, in Security & Embedded Systems, Vol. 2, IOS Press (Amsterdam, The Netherlands, pp. 44-56, 2005.

Fragopoulos, A.; Serpanos, D. & Voyiatzis, (2009). *Design Issues in Secure Embedded Systems*, book chapter in Embedded Systems Handbook, 2nd ed., ISBN 9781420074109.

Fragopoulos, Anastasios; Gialelis, John; Serpanos, Dimitrios; (2009), *Security Framework for Pervasive Healthcare Architectures Utilizing MPEG-21 IPMP Components*, International Journal of Telemedicine and Applications 2009: 9

Fragopoulos, Anastasios; Gialelis, John; Serpanos, Dimitrios; (2010), *Imposing Holistic Privacy and Data Security on Person Centric eHealth Monitoring Infrastructures*, in 12th International Conference in e-Health Networking, Application & Services IEEE

Gasser, Urs; Palfrey, John, *DRM-protected Music Interoperability and e-Innovation*, November 2007, Berkmann Publication Series, http://cyber.law.harvard.edu/interop/downloads.html
Hankerson, D.R.; Vanstone S. A.; and Menezes, A. J.; (2004), *Guide to Elliptic Curve Cryptography*. New York: Springer, 2004, pp. 311.
Heileman, Gregory L.; Jamkhedkar, Pramod A., (2005), *DRM Interoperability Analysis from the Perspective of a Layered Framework*, in: Proceedings of the Fifth ACM Workshop on Digital Rights Management, 17-26, Alexandria, Nov. 2005, p. 20.
Henk C.; van Tilborg, A. (Eds), *Encyclopedia of cryptography and security*, Springer, 2005, ISBN. 038723473X.
INTERMEDIA Network-of-Excellence, (2006), *Interactive Media with Personal Networked Devices*, http://intermedia.miralab.unige.ch:80/, FP6 – IST- 38419, 2006-2010
International Standards Organization (ISO), (2004), *Information technology -- Multimedia framework (MPEG-21) -- Part 1: Vision, Technologies and Strategy*, ISO/IEC TR 21000-1:2004
International Standards Organization (ISO), (2004), *Information technology -- Multimedia framework (MPEG-21) -- Part 5: Rights Expression Language*, ISO/IEC 21000-5:2004
International Standards Organization (ISO), (2006), *Information technology -- Multimedia framework (MPEG-21) -- Part 4: Intellectual Property Management and Protection Components*, ISO/IEC 21000-4:2006
Jafari, Mohammad ; Safavi-Naini, Reihaneh ; Saunders, Chad ; and Sheppard, Nicholas Paul; (2010) *Using digital rights management for securing data in a medical research environment*, In Proceedings of the tenth annual ACM workshop on Digital rights management (DRM '10). ACM, New York, NY, USA, 55-60
Leister, Wolfgang; Fretland, Truls ; Balasingham, Ilangko; (2009), *Security and Authentication Architecture Using MPEG-21 for Wireless Patient Monitoring Systems*, International Journal in Advances in Security, vol. 2, no. 1, 2009, http://www.iariajournals.org/security/tocv2n1.html
Lipton, R.J.; Rajagopalan, S.; and Serpanos, D.N.; (2002) *Spy: A Method to Secure Clients for Network Services*, in ICDCS Workshops, 2002, pp. 23-28, http://csdl.omputer.org/omp/proeedngs/dsw/2002/1588/00/15880023abs.htm
Menezes, A.J.; van Oorschot, P. C.; and Vanstone, S. A.; (1996), *Handbook of Applied Cryptography*, CRC Press Inc., 1996
Messerges, T.S.; Dabbish, E.A.; (2003), *Digital rights management in a 3G mobile phone and beyond*, in Proceedings of the 2003 ACM workshop on Digital rights management, 2003, pp. 27-38.
Repetto, Matteo; Rapuzzi, Riccardo; Chessa, Stefano; Lenzi, Stefano; Gialelis, John and Fragopoulos, Tasos; (2010), *The InterMedia Networking and Security Architecture for User Centric Multimedia Convergence*, International Symposium on Performance Evaluation of Computer and Telecommunication Systems (SPECTS), 2010, Ottawa, Canada.
Roush, W.; (2006), Inside the Spyware Scandal, MIT Technology Review, May-June 2006
Serpanos, D.N.; and Lipton, R.J.; (2001), *Defense Against Man-in-the-Middle Attack in Client-Server Systems*, in ISCC, 2001, pp. 9-14, http://csdl.omputer.org/omp/proceedings/s/2001/1177/00/11770009abs.htm
Schneier, B. ; (1996), *Applied Cryptography*, (Second Edition), John Wiley & Sons, 1996, ISBN. 0-471-11709-9

Part 4

Wireless Sensor Networks

12

Effect of Decentralized Clustering Algorithm and Hamming Coding on WSN Lifetime and Throughput

Nora Ali[1], Hany ElSayed[1], Magdy El-Soudani[1], Hassanein Amer[2] and Ramez Daoud[3]
[1]*Cairo University,*
[2]*American University in Cairo,*
[3]*KAMA Trading,*
Egypt

1. Introduction

Wireless Sensor Networks (WSN) has become an interesting field of research because of its wide range of applications such as environmental monitoring, electromagnetic pollution monitoring, medical applications and industrial applications (Teo et al., 2007; Margi et al., 2009; Castelluccia et al., 2005; AbouElSeoud et al., 2010; Tavares et al., 2008). WSN consists of multi-functioning sensor nodes with limited power capacity, so prolonging the lifetime is essential and is one of the main concerns (Castelluccia et al., 2005; Schmidt et al., 2009; Karlsson et al., 2005).

For this reason different routing protocols are obtained to increase network lifetime. The clustering routing protocol is one of the most commonly routing protocols because it is energy efficient (Heinzelman et al., 2000, 2002). In any clustering protocol, the network is divided into clusters where some nodes are responsible for others. These nodes are called cluster heads (CHs) or network masters (NMs). There are different algorithms and different methods of choosing the CHs. For example, LEACH (Heinzelman et al., 2000) used the randomized rotation to choose CH nodes. This randomized rotation allows some nodes to act as CHs and the others cannot. Therefore LEACH was improved to be LEACH-C (Heinzelman et al., 2002) that uses central algorithm to choose the CHs and allows only the nodes in the center of each cluster to act as CHs.

Also two different algorithms of choosing the NMs are considered in (Botros et al., 2009). The network is considered as one cluster; therefore the CH node that is responsible for collecting data from other nodes is called NM. In the first algorithm, the sensor could become NM more than once for a fixed number of cycles. It was proven that this algorithm provided a lifetime longer than the lifetime obtained by LEACH and LEACH-C algorithms (Heinzelman et al., 2000, 2002). However, this algorithm has some residual energy after the network failure and this energy cannot be used anymore. Therefore, the second algorithm is obtained to improve the first one by allowing each sensor to become NM once with a different number of cycles and acts as an active node or ordinary node (that senses the surroundings and sends the sensed data to the NM node) till network failure. Using this algorithm increased the lifetime by approximately 5% compared to the first algorithm.

In this chapter, the network is divided into clusters and all the nodes inside each cluster will act as CHs once with a different number of cycles. The optimum number of clusters is obtained taking into account that the algorithm is decentralized, i.e., some nodes cannot reach the sink because their initial energy is too low. Also, the effect of clustering on networks covering large areas and on the applications that do not need data aggregation is examined.

On the other hand, in some important applications such as industrial applications and critical applications such as medical applications (Tavares et al., 2008), lifetime is not the only important factor; receiving all sensor data correctly at the sink might be more important to prevent taking wrong decisions (Margi et al., 2009). But, this is very difficult in noisy environments. Hence Error Correcting Codes (ECC) must be used to improve data integrity (Schmidt et al., 2009). ECC can have an adverse effect on network lifetime due to the processing energy consumed for encoding and decoding.

Therefore, in this work, the Hamming code will be introduced as an example of ECC due to its wide use in sensor networks (Karyonen & Pomalaza-Ráez, 2004; Sadeghi et al., 2006). It will be compared to the Cyclic Redundancy Check (CRC) as an example of a very widely used error detecting technique (Nguyen, 2005). In CRC, the error is only detected and the correction is done by retransmitting data. In contrast, in the Hamming code, the error can be detected and corrected without retransmission. The processing energy for coding will be investigated based on hardware implementations of encoding/decoding circuits. Moreover, a metric that represents a compromise between the lifetime and the amount of correct received data (throughput) will be introduced. The effect of using NM as a repeater is examined to improve network performance. Finally, the effect of using coding in case of fixed data length is investigated. In this research, network lifetime is defined as the time to the first node failure due to battery outage (Botros et al., 2009; Sadeghi et al., 2006).

This chapter is organized as follows. Section 2 describes the decentralized algorithm, the optimum number of clusters using this algorithm and the effect of clustering on networks covering large areas and on the applications that do not need data aggregation. Section 3, describes coding in sensor networks and the processing energy for coding for both Hamming code and CRC. Section 4 describes the simulation results. Section 5 describes the fixed data length scheme and its simulation results. Finally, section 6 concludes the chapter.

2. Decentralized algorithm

In many WSN applications that cover large areas, some nodes may be out of the sink's range and cannot reach it because the required energy is higher than their initial energy, i.e., a more powerful battery would be needed to reach the sink. According to the technology or the cost, there may be some constrains which prevent increasing the power of the node battery. In such cases, some nodes cannot reach the sink and the sink does not have any information about these nodes. Consequently, these nodes are considered dead and out of range throughout network lifetime and hence the area in question is not fully covered.

Therefore, a new algorithm which is called the *decentralized algorithm* is developed. When applying this algorithm, the nodes that are out of the sink's range, will be in range and operate as active nodes. In the set up phase, the nodes that are in range send their information to the sink that will divide the network into clusters and compute the number of cycles of the CHs in the first rotation. After this first rotation, the role of the sink is over and the responsibility of choosing the next CHs is transferred to the current CHs. This means that the CHs in any rotation are responsible for choosing the CHs of the next rotation and computing their number of cycles.

In order to permit the CH to undertake the new responsibilities, some information about the nodes in its cluster (such as the IDs and the remaining energy of each node) must be known. Therefore it will send a broadcast message to all nodes in its cluster. Since the distances between nodes and CHs within a cluster are less than the distances between nodes and the sink, the nodes that are out of the sink range will be within the CH range. All nodes in each cluster will send their information such as IDs and energy levels to the CH. Also, the sink will send the IDs of all nodes that can work as CHs at the end of the first rotation. By using this algorithm, the nodes that were out of the sink range will be active nodes, sense the surroundings and communicate with the CHs which makes the network fully covered. On the other hand, some overhead energy will be consumed by the CHs due the additional responsibility. This overhead energy is explained next according to the network parameters and variables shown in Table 1:

Parameter	Value
Number of Sensors (N)	100
Initial Energy	2J
Transmitter/Receiver Electronics (E_{elec})	50nJ/bit
Transmitter Amplifier for multi path model (E_{mp})	0.0013 pJ/bit/ m^4
Transmitter Amplifier for free space model (E_{fs})	10 pJ/bit/ m^2
Aggregation Energy (E_{DA})	5 nJ/bit/signal
Transmitted Frame Length (L)	4000 bits
CRC Generator Polynomial Length (h)	12 bits

Table 1. Network parameters and variables

According to the above network paramters and variables, the overhead energy is as follows:

1. Broadcast energy (E_{bc}):

The energy dissipated by the CH in order to inform all the nodes in its cluster about the next CH and to activate the nodes that are out of the sink's range. It is calculated as follows, assuming that the network is divided into K clusters.

$$E_{bs} = \frac{N}{K} l E_{fs} d_{CH\_N}^2 \qquad (1)$$

where l is the number of bits that is transmitted by the CH to declare the next CH and is considered to be equal to 8 bits and d_{CH_N} is the distance from the CH node to any sensor node.

2. Processing energy for ID comparison (E_{ID}):

The processing energy that is dissipated by the CH to compare between the transmitted nodes IDs and the IDs stored in its data base to choose the next CH. It is calculated according to the following equation:

$$E_{ID} = N_{op} \times E_{oper} \qquad (2)$$

where N_{op} is the number of binary operations and E_{oper} is the energy per binary operation and is equal to 10^{-14}J according to the new technology (Ali et al., 2011).

3. Processing energy for computations (E_{Cy}):

The processing energy that is dissipated by the CH in order to calculate the number of cycles that will be allocated to the next CH. It is calculated in the same manner as the processing energy for ID comparison (E_{ID}).

4. Announcement energy (E_{an}):

The energy dissipated by the CH in order to announce the next CH node about its number of cycles. It is calculated as follows:

$$E_{an} = lE_{fs} d^2_{CH-N} \tag{3}$$

2.1 Optimum number of clusters for the decentralized algorithm

The optimum number of clusters is obtained by minimizing the total energy consumed per cycle. This is because the total energy consumed by the sensor node is the energy consumed per one cycle multiplied by the total number of cycles; this is considered to be the network lifetime. Note that the total energy consumed by the sensor node is almost equal to its initial energy because the remaining energy is very small (close to zero) using the algorithm of (Botors et al., 2009). Therefore, the lifetime is maximized by minimizing the total energy consumed per cycle. This energy depends on the energy consumed by the node when it acts as a CH and when it acts as an active node. Since the nodes cover the entire area under study, the energy consumed by the node when it acts as an active node and CH will be as follows:

$$E_{active} = E_{Rx} + LE_{elec} + LE_{fs} \frac{M^2}{\pi K} \tag{4}$$

$$E_{CH} = LE_{elec}(\frac{N}{K} - 1) + LE_{DA} \frac{N}{K} + E_{ID} + E_{Cy} + E_{bs} + E_{an} + LE_{mp} d^4_{\text{Sink}} \tag{5}$$

where E_{Rx} is the energy dissipated by the active node in order to receive an announcement from the CH.

It is assumed that the network consists of N nodes and is divided into K clusters with approximately (N/K) sensors per cluster; therefore, the energy dissipated inside a single cluster during one complete cycle is as follows:

$$E_{Cluster} = E_{CH} + (\frac{N}{K} - 1)E_{active} \tag{6}$$

Consequently, the total energy consumed during a single cycle or during transmitting single frame.

$$E_{Cycle} = K \times E_{Cluster} \tag{7}$$

By substituting from equations (4), (5) and (6) into (7), the energy consumed in the entire network during a single cycle is as follows:

$$E_{Cycle} = K(LE_{elec}(\frac{N}{K}-1) + LE_{DA}\frac{N}{K} + E_{ID} + E_{Cy} + lE_{fs}\frac{M^2}{\pi K} + LE_{fs}\frac{N}{K}\frac{M^2}{\pi K} + LE_{mp}d_{Sink}^4 + \frac{N}{K}LE_{elec} + l\frac{N}{K}E_{elec} + LE_{fs}\frac{M^2}{\pi K^2}) \quad (8)$$

The optimum number of clusters K_{opt} is obtained by setting the derivative of E_{Cycle} with respect to K to zero and K_{opt} will be as follows:

$$K_{opt} = \sqrt{\frac{E_{fs}NM^2(L+l)}{\pi(E_{ID} + E_{Cy} + E_{mp}Ld_{Sink}^4)}} \quad (9)$$

By ignoring the processing energies E_{ID} and E_{Cy} (because they are in the order of 10^{-14} J (Ali et al., 2011)), the optimum number of clusters will be simplified as follows:

$$K_{opt} = \sqrt{\frac{E_{fs}NM^2(L+l)}{\pi E_{mp}Ld_{Sink}^4}} \quad (10)$$

It is obvious from the above equation that the optimum number of clusters depends on the network parameters and the network area. Therefore it is more reasonable to investigate the effect of the decentralized algorithm on the networks that cover large areas to see its effect on lifetime and network coverage.

2.2 Effect of decentralized algorithm on networks covering large areas

Lifetime is expected to decrease with increasing the network area. This is because of the large distances between nodes and the sink that prevents a lot of nodes from reaching the sink (require energy more than its initial energy). Consequently, these nodes will be considered as dead nodes with respect to the sink. Therefore, applying the decentralized algorithm on the applications that cover large areas is more reasonable. A MATLAB (Matlab) simulation model is built to study the effect of the decentralized algorithm on networks covering large areas. The lifetime for different network areas is shown in Table 2. Assuming that the number of sensor nodes and the sink location vary according to the network area (For example, the sink location is (0, -125) and the number of sensor nodes equals 100 nodes for a network of 100m×100m. For a network of 200m×200m, sink location is (0, -250) and the number of sensor nodes equals 400 nodes and so on)

The area	Sink location	Lifetime without clustering	Optimum number of clusters	Lifetime	Percentage of increase
200m×200m	(0, -250)	2600	3	2900	11.5%
300m×300m	(0, -375)	1800	3	2200	22.2%
400m×400m	(0, -500)	1000	4	1290	29%

Table 2. Lifetime of different network areas using the decentralized algorithm

The table shows that the clustering technique improves the lifetime for the different network areas especially in larger network areas. This is because, as the network area increases, the number of sensors that cannot reach the sink increases; this increases the efficiency of the decentralized algorithm. On the other hand, a large increase in the network area will lead to a slight increase in lifetime (for example for 500m×500m area the lifetime increased by approximately 17.5%) because some clusters will have a lot of nodes that cannot act as CHs and the nodes that will act as CHs will operate for a small number of cycles due to the large communication distance from nodes to the sink. The simulation is also run for a small network area (100m×100m) and it is found that the lifetime is increased by only 5%. This means that the clustering with decentralized algorithm is not efficient on small network area because all nodes can reach the sink.

2.3 Effect of decentralized algorithm in case of no data aggregation

In some applications such as environmental mentoring and industrial applications, data of each and every sensor is important. This means that data cannot be aggregated and the CH collects data from the nodes and sends it as is to the sink without aggregation. Therefore, the decentralized clustering algorithm is investigated in case of no data aggregation and the energy dissipated by the CH and the total energy dissipated inside the network during single cycle will be as follows:

$$E_{CH} = 2LE_{elec}(\frac{N}{K}-1) + E_{ID} + E_{Cy} + E_{bs} + E_{an} + LE_{mp}(\frac{N}{K}-1)d_{Sink}^4 \tag{11}$$

Keeping the energy of a node when it acts as an active node the same as in the case of data aggregation (because in both cases, it senses the surroundings and sends the data to the CH node), the energy dissipated in one cluster will be as follows:

$$\begin{aligned} E_{Cluster} &= 2LE_{elec}(\frac{N}{K}-1) + E_{ID} + E_{Cy} + lE_{fs}\frac{M^2}{\pi K} + lE_{fs}\frac{N}{K}\frac{M^2}{\pi K} \\ &+LE_{mp}(\frac{N}{K}-1)d_{Sink}^4 + \frac{N}{K}LE_{elec} + l\frac{N}{K}E_{elec} + LE_{fs}N\frac{M^2}{\pi K^2} \end{aligned} \tag{12}$$

And the total energy dissipated inside the network during one complete cycle will be as follows:

$$\begin{aligned} E_{Cycle} &= 2LE_{elec}(N-K) + LE_{mp}(N-K)d_{Sink}^4 + KE_{ID} + KE_{Cy} \\ &+lE_{fs}\frac{M^2}{\pi}(1+\frac{N}{K}) + E_{elec}N(L+l) + LE_{fs}N\frac{M^2}{\pi K} \end{aligned} \tag{13}$$

By setting the derivative of E_{Cycle} with respect to K to zero to obtain the optimum number of clusters, K^2 will be as follows:

$$K^2 = \frac{(\frac{N}{\pi}E_{fs}M^2(l+L))}{E_{ID} + E_{Cy} - (2LE_{elec} + LE_{mp}d_{Sink}^4)} \tag{14}$$

This equation has no solution since the denominator will always be negative (because the values of the processing energy are very small compared to the energy consumed in reception and transmission, i.e., $E_{ID} + E_{Cy}$ is always less than $2LE_{elec} + LE_{mp}d_{Sink}^{4}$). Therefore, using clustering technique is not efficient and only one cluster is preferred in case of no data aggregation using decentralized algorithm.

3. Coding in sensor network

In the recent years a significant amount of research has focused on the lifetime prolongation which is the main concern in different WSN applications. But, in important and critical applications such as industrial and medical applications, the throughput may be more important (Margi et al., 2009). Therefore, in this part, it is assumed that the CH collects data from all sensors and sends it to the sink and only one cluster is considered because it was proved in the previous section that the clustering technique was not efficient in case of no data aggregation. Therefore, the CH will be called the network master (NM). But in noisy environments, data received at the sink may be corrupted by noise and wrong decisions may be made. In order to guarantee data integrity at the sink, Error Correcting Codes (ECC) are used. In this work, the use of the Hamming code is considered with different rates. Also, the Hamming code is compared to the CRC which is the most commonly used error detecting technique (Nguyen, 2005). With CRC, correction is attempted by retransmission of the data once.

Also, the metric that compromises between the throughput and the lifetime is introduced. The metric is called Information per Joule (*IPJ*). It is defined as the overall throughput during network lifetime (*Tp*) divided by the total energy consumption (E_{all}) as explained in the following equation:

$$IPJ = \frac{Tp}{E_{all}} \tag{15}$$

The throughput represents the amount of information that can correctly reach the sink during the lifetime. It is calculated according to the following equation:

$$Tp = \sum_{j=1}^{N} \sum_{\substack{i=1 \\ i \neq j}}^{N} (1 - BER_i) \times L \times r \times C_j \tag{16}$$

where r is the code rate, C_j is the number of cycles allocated to NM and *BER* is the Bit Error Rate and is defined as the amount of error in the received frames divided by the total frame length.

3.1 Processing energy for coding

It is assumed in this work that the coding and decoding processes are part of the sensor hardware architecture. The processing energy is assumed to be the number of binary operations multiplied by the energy per binary operation (Karlsson et al., 2005). The binary operation is defined as the exclusive-or of two bits and the energy per binary operation is the energy consumed in processing of a two-input Exclusive-Or (XOR) gate. It is assumed in this work that the design of the XOR gate is based on static CMOS which is commonly used in sensor networks (Teo et al., 2007, Enz et al., 2004). The energy per binary operation (E_{oper}) depends on the fabrication technology (Hempstead et al., 2006, Ragini et al., 2009). The following values: 10^{-10}, 10^{-12} and 10^{-14}J respectively will be used (from older to newer technology).

3.1.1 Hamming code processing energy

The number of binary operations in the encoder circuit is equivalent to the number of two-input XOR gates in the parity tree. The decoder contains two circuits, one for error detection and the other for error correction (Fu & Ampadu, 2010). The processing energy for encoding (E_{enc}) and decoding (E_{dec}) will be as follows:

$$E_{enc} = [(k-2) \times (n-k) \times N_c] \times E_{oper} + E_{oh} \tag{17}$$

$$E_{dec} = E_{syndrome} + E_{correction} + E_{oh} \tag{18}$$

$$E_{syndrome} = [(k-2) \times (n-k) \times N_c + (n-k) \times N_c] \times E_{oper} \tag{19}$$

$$E_{correction} = [k \times (n-k-1) \times N_c + k \times N_c] \times E_{oper} \tag{20}$$

where k is the information block length, n is the codeword length, N_c is the number of codewords in the transmitted frame and E_{oh} is the overhead energy due to the sensor's microprocessor instructions execution.

3.1.2 CRC processing energy

A 12-bit CRC is used since it is more suitable for the assumed frame length (for simplicity, it is assumed to be 2048 bits instead of 4000 bits) (Nguyen, 2005). The implementation of CRC using XOR operations is obtained. The processing energy for CRC encoding or decoding is the same and consists of the following steps:

5. Zero padding the polynomial vector to have the same length of data.
6. Exclusive-ORing the output vector and the data vector.
7. Ignoring the most significant zeros in the output vector because they represent the quotient.
8. Exclusive-ORing the remainder and the polynomial vector.

These steps are repeated until a remainder of length equal to the length of the used polynomial (12 bits) is reached. This final remainder represents the CRC check bits that are included in the transmitted frame. Consequently, the processing energy for CRC encoding or decoding according to these steps is as follows:

$$E_{CRC} = \left(\sum_{i=0}^{L-2h} (L-h-i)\right) \times E_{oper} + E_{oh} \tag{21}$$

where h is the generator polynomial length.

3.2 Network assumptions

At any time instant, the network consists of sensors that sense the surroundings and an NM that collects the sensed data and forwards it to the sink. This means that there are two noisy communication channels: one from sensors to the NM and the other from the NM to the sink. The following assumptions are made regarding applying error correcting or detecting techniques in the sensor network. For the Hamming code, sensors encode the data and send it to the NM that decodes then re-encodes the data again before sending it to the sink. For the CRC, sensors compute the CRC check bits and send the coded data to the NM that decodes it; if an error is detected, the NM will request retransmission. The NM decodes the retransmitted data and if no errors are detected it sends it to the sink. However, if any frame error is detected, the NM will discard the frame because, it is assumed that retransmission only occurs once.

According to these assumptions, the energy consumed by the sensor node for both Hamming code and CRC is as follows:

Hamming Code:

$$E_{sensor_j} = C_j \times E_{NM_j} + \sum_{i=1}^{N-1} C_i \times (E_{enc} + E_{tx_j - NM_i}) \tag{22}$$

$$E_{NM_j} = \sum_{i=1}^{N-1} E_{rx} + \sum_{i=1}^{N-1} E_{enc} + \sum_{i=1}^{N-1} E_{dec} + \sum_{i=1}^{N-1} E_{tx} \tag{23}$$

for $j = 1, 2, \cdots, N$

CRC:

$$E_{sensor_j} = C_j \times E_{NM_j} + \sum_{i=1}^{N-1} C_i \times (E_{CRC} + mE_{tx_j - NM_i}) \tag{24}$$

$$E_{NM_j} = m\sum_{j=1}^{N-1} E_{rx} + m\sum_{j=1}^{N-1} E_{CRC} + \sum_{j=1}^{S1} E_{tx} \tag{25}$$

for $j = 1, 2, \cdots, N$

where

$$E_{tx} = E_{elec} \times L + E_{amp} \times L \times d^{p}_{\mathrm{sink}_j} \tag{26}$$

$$E_{rx} = E_{elec} \times L \tag{27}$$

$$E_{tx_j - NM_i} = E_{elec} \times L + E_{amp} \times L \times d^{p}_{tx_j - NM_i} \tag{28}$$

where p is the path loss factor (for example, it equals 2 for the free space model), $S1$ is the number of corrected frames received by the NM, d_{sink_j} is the distance from NM_j to the sink, $d_{tx_j - NM_i}$ is the distance between a sensor and the NM and C_j is the number of Cycles allocated to NM_j and it is calculated according to the following relation:

$$E_{sensor_i} \le E_{initial} \tag{29}$$

for $i = 1,2,\cdots,N$

where $E_{initial}$ is the sensor node initial energy.

4. Simulation results and analysis

Simulations are run using MATLAB to study the effect of using error correction/detection techniques on the lifetime and the IPJ in case of additive white Gaussian noise (AWGN) channel. Assuming an area of 100m×100m for simplicity, Table 3 shows the values of the lifetime in cycles for CRC and Hamming code at different values of E_{oper} and SNR. The values prove that the Hamming code has a very low effect on network lifetime. It produces almost the same lifetime of a system without coding (uncoded system) which is equal to 2950 cycles (the lifetime is reduced in case of no data aggregation). In contrast, CRC decreases the lifetime by 37.3% compared to the uncoded system at $E_{oper} = 10^{-10}J$ due to its high processing energy.

The other important metric is the IPJ. Fig. 1 shows the IPJ for the used Hamming codes and the CRC at different energy per binary operation. It is found that the IPJ of Hamming (63, 57) outperforms the IPJ of both Hamming (7, 4) and CRC. It is also noticed that the IPJ of Hamming (63, 57) at the lowest SNR and highest E_{oper} is higher than the IPJ of CRC and Hamming (7, 4) at highest SNR and lowest E_{oper}. Therefore, it is better to use the long Hamming code.

To improve network performance, another method is investigated in which the NM acts as a repeater or a relay that collects data from sensors and forwards it to the sink without decoding or encoding to reduce the processing energy at the NM. MATLAB simulations indicate that the lifetime for both Hamming lengths will increase and become almost equal to the lifetime of the uncoded system. Also the IPJ is slightly improved as shown in Fig. 2. The figure shows that using the NM as a repeater is more suitable for high rate Hamming codes lengths especially at low SNR.

Error Detection & Correction Techniques	Energy per binary operation (E_{oper})		
	10^{-10}J	10^{-12}J	10^{-14}J
CRC (low SNR)	1250	2080	2110
CRC (high SNR)	1850	2900	2930
(7, 4) Hamming	2940	2945	2945
(63, 57) Hamming	2925	2945	2945

Table 3. Lifetime at different E_{oper}

On the other hand, it is more reasonable to consider Rayleigh fading channel in the wireless communications channels (Karyonen & Pomalaza-Ráez, 2004). Slow fading is considered in this work due to the small size of the area under study (100m×100m). The same two lengths of Hamming are examined in case of AWGN channel with Rayleigh fading (Rayleigh fading channel). The fading channel adds a large amount of errors to the data which decreases the probability of finding a single error in long codeword lengths such as a codeword of length 63. However, the overall IPJ of Hamming (63, 57) is higher than the IPJ of Hamming (7, 4) as shown in Fig. 3, due to its higher code rate. The figure shows the IPJ over the AWGN channel and the Rayleigh fading channel of the used low rate and high rate Hamming code.

The IPJ is obtained at E_{oper} = 10^{-10} J, because at this value, the processing of coding has a noticeable effect on the lifetime and the IPJ. It is observed from the figure that some degradation of the IPJ in case of Rayleigh fading channel occurs in both lengths of Hamming as a result of the large number of errors added from the fading, which has an adverse effect on network performance.

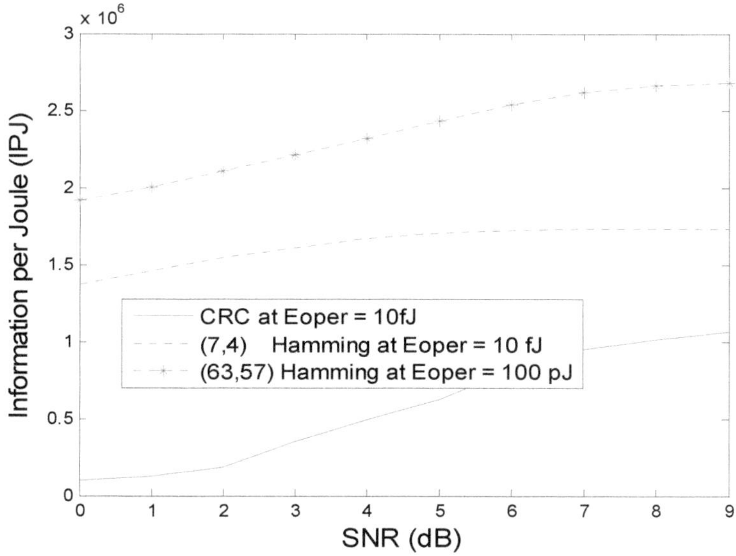

Fig. 1. IPJ of Hamming code and CRC at different E_{oper}

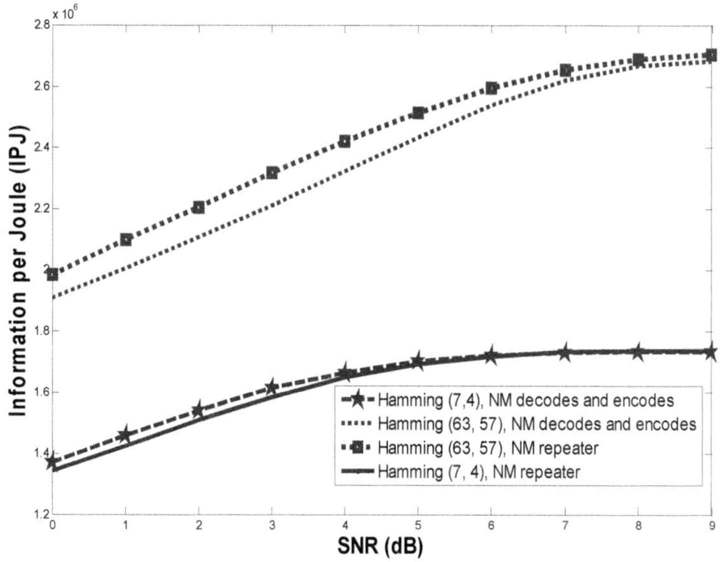

Fig. 2. IPJ of Hamming code in case of NM as a repeater

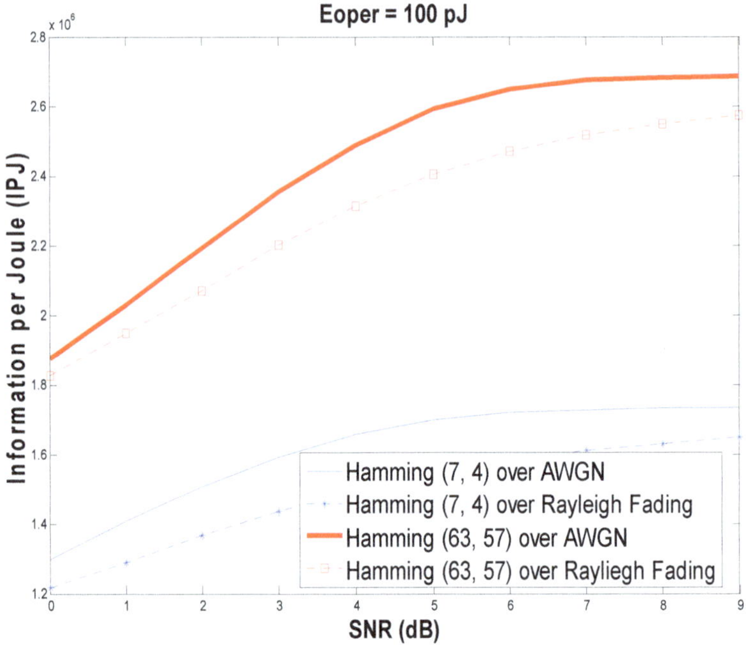

Fig. 3. IPJ of Hamming code over AWGN and Rayliegh fading channel

5. Fixed data length scheme

All the previous results were based on transmitting a frame of fixed length by all sensors with an amount of data that varies according the code rate. Consequently, all the sensor nodes consume the same amount of transmitted energy and have approximately the same lifetime. On the other hand, in some applications such as environmental monitoring, the sensor collects a fixed amount of data. The sensor can have a fixed amount of data and sends a frame of variable length according to the coding technique used. This length will depend on the amount of added parity by the coding technique. Therefore, in this section, it is assumed that all sensors have a fixed amount of data of length 2048 bits and transmit a frame of length that varies according to the amount of added parity. Consequently, the amount of energy consumed by any sensor node and an NM will change due to variations in transmitting and receiving energy as follows:

$$E_{rx} = E_{elec} \times K_{tx} \tag{30}$$

$$E_{tx} = E_{elec} \times K_{tx} + E_{amp} \times K_{tx} \times d^{p}_{\text{sink}_j} \tag{31}$$

$$E_{tx_j-NM_i} = E_{elec} \times K_1 + E_{amp} \times K_{tx} \times d^{p}_{tx_j-NM_i} \tag{32}$$

where K_1 is the data length that equals to 2048 bit and K_{tx} is the frame length and is calculated according to the following equation:

$$K_{tx} = \frac{K_1}{k} \times n \tag{33}$$

Simulations are run to study the effect of using this scheme on the overall network performance. The IPJ for Hamming (7, 4) and (63, 57) over AWGN at $E_{oper} = 10^{-10}$J are shown in Fig. 4.

The figure shows that the IPJ of Hamming (63, 57) outperforms the IPJ of Hamming (7, 4).

The rationale behind this result is investigated and it is found that the low rate code such as the Hamming (7, 4) adds a large amount of parity which increases the transmitted energy and has an adverse effect on the lifetime. In contrast, the high rate code such as the Hamming (63, 57) adds a small amount of parity which does not strongly affect the transmitted energy. Consequently it does not affect the lifetime.

Table 4 shows the values of lifetime (in cycles) at different energy per operation. It is found that the lifetime of the Hamming (7, 4) is lower than the lifetime of the Hamming (63, 57) by about 33% at E_{oper} =10^{-10}J. This because of the difference in the amount of energy consumed for transmission for both Hamming lengths.

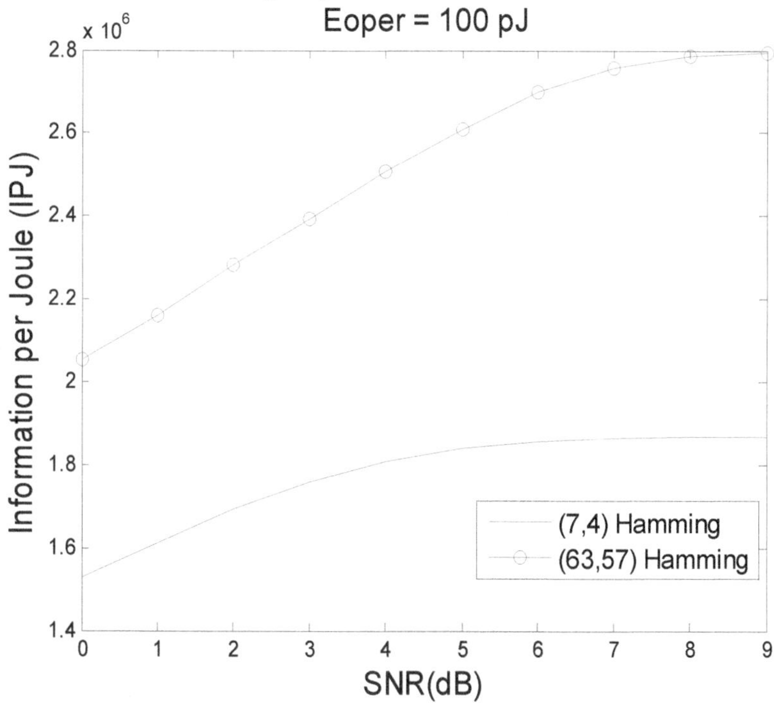

Fig. 4. Lifetime at different energy per binary operation (E_{oper})

This difference in lifetime causes the IPJ over the lifetime of the Hamming (63, 57) to outperform the IPJ of the Hamming (7, 4). Therefore, the high rate Hamming is more suitable in sensor networks than the low rate Hamming irrespective of the application and the transmitting scheme used by the sensor node (transmitting fixed frame or fixed data). Simulations show that this scheme will not change the result in the case of Rayleigh channel.

Hamming length	Energy per Binary Operation E_{oper}		
	10^{-14}J	10^{-12}J	10^{-10}J
(7, 4)	1780	1779	1775
(63, 57)	2695	2688	2675

Table 4. Lifetime at Different E_{oper} for fixed data length scheme

6. Conclusion

Different clustering algorithms and routing protocols were examined for prolonging the lifetime such as LEACH and LEACH-C. This chapter focuses on increasing network lifetime by dividing the network into clusters and making each node inside the cluster acts as a Cluster Head (CH) only once. The Decentralized algorithm is developed and studied in this chapter and it is found that the optimum number of clusters will depend on the algorithm of choosing the CHs. The effect of the decentralized algorithm on networks covering large areas is investigated. It is found that clustering is more efficient for large networks. Also, the proposed clustering algorithm is examined for applications that do not need data aggregation. Results prove that the clustering will be inefficient in case of no data aggregation and only one cluster is preferred. Also, network throughput is an important factor in case of no data aggregation; therefore error detecting and correcting codes are used to improve data integrity and the whole network is considered as one cluster. The Hamming code with different rates is used to improve network throughput and compared to CRC. It is found that the Hamming code with different lengths provides longer lifetime than CRC due to its lower processing and higher IPJ due to its higher throughput. It is also observed that the Hamming code has a negligible effect on lifetime compared to the uncoded system.

These results are taken a step further by examining different lengths of Hamming codes. It is observed that a Hamming code of length 63 is more suitable in sensor networks than that of length 7. This means that the high rate Hamming can provide a higher IPJ at low SNR than the low rate Hamming. The system is also investigated when the NM acts as a repeater or relay that collects data from sensors and forwards it to the sink without decoding or encoding. It is observed that this technique increases the IPJ for high rate code at low SNR. It also increases the lifetime because of the reduction in processing at the NM.

The effect of Rayleigh fading channel was also investigated. The results showed that the IPJ of the high rate Hamming is still higher than the low rate Hamming; even though the low rate Hamming improves the BER and makes the IPJ of the fading channel close to the IPJ of the AWGN channel.

Finally, a fixed data length scheme is examined to generalize the results for different applications that do not require data aggregation. The results of using this scheme show that the lifetime and the IPJ of the high rate Hamming codes are higher than the lifetime and the IPJ of the low rate Hamming codes. Therefore, the proposed hardware implementation of the high rate Hamming code will be one of the preferred solutions in sensor networks with different transmitting schemes and applications.

7. References

AbouElSeoud, D.; Nouh, S.; Abbas, R.; Ali, N.; Daoud, R.; Amer, H. & ElSayed, H. (2010). Monitoring Electromagnetic Pollution using Wireless Sensor Networks, *Proceedings of the 15th International Conference on Emerging Technologies and Factory Automation ETFA*, Bilbao-Spain, September 2010.

Ali, N.; ElSayed, H.; El-Soudani, M. & Amer, H. (2010). Effect of Hamming Coding on WSN Lifetime and Throughput, *Proceedings of the IEEE International Conference on Mechatronics ICM*, Istanbul, Turkey, April 2011.

Botros, S.; ElSayed, H.; Amer, H. & El-Soudani, M. (2009). Lifetime Optimization in Hierarchical Wireless Sensor Networks, *Proceedings of the 14th International Conference on Emerging Technologies and Factory Automation ETFA*, Mallorca-Spain, September 2009.

Castelluccia, C.; Mykletun, E. & and Tsudik, G. (2005). Efficient Aggregation of Encrypted Data in Wireless Sensor Networks, *Proccedings of ACM /IEEE Mobile and Ubiquitous Systems Conference MOBIQUITOUS*, San Diego-CA-USA, July 2005.

Enz, C.; El-Hoiydi, A.; Decotignie, J. & Peiris, V. (2004). WiseNET: An Ultra-Low-Power Wireless Sensor Network Solution, *IEEE Computer Society Press*, USA, August 2004.

Fu, B. & Ampadu, P. (2010). Error Control Combining Hamming and Product Codes for Energy Efficient Nanoscale on-chip Interconnects, *IET Computers & Digital Techniques*, vol. 4, no. 3, May 2010.

Heinzelman, W.; Chandrakasan, A. & Balakrishnan, H. (2000). Energy- Efficient Routing Protocols for Wireless Microsensor Networks, *Proceedings of the 33rd Hawaii International Conference on System Sciences HICSS*, Maui, HI, USA, January 2000.

Heinzelman, W.; Chandrakasan, A. & Balakrishnan, H. (2002). An Application Specific Protocol Architecture for Wireless Micro Sensor Networks, *IEEE Transactions on Wireless Communications*, vol. 1, no.4, October 2002.

Hempstead, M.; Wei, G. & Brooks, D. (2006). Architecture and Circuit Techniques for Low Throughput, Energy Constrained Systems Across Technology Generations, *Proceedings of the International Conference on Compilers, Architecture, and Synthesis for Embedded Systems CASES*, pp. 368-378, Seoul-Korea, October 2006.

Karlsson, P.; Oberg, L. & Xu, Y. (2005). An Address Coding Scheme for Wireless Sensor Networks, *Proceedings of the 5th Scandinavian Workshop on Wireless Ad-hoc Networks*, Stockholm-Sweden, May 2005.

Karyonen, H. & Pomalaza-Ráez, C. (2004). Coding for Energy Efficient Multihop Wireless Networks, *Proceedings of the Nordic Radio Symposium*, Oulu-Finland, August 2004.

Margi, B; de Oliveira, B.; de Sousa, G.; Simplicio, M.; Freitasy, F.; Barretoy, P.; Carvalhoy, T.; Näslundz, M. & Goldz, R. (2009). Demo: Security Mechanisms Impact and Feasibility on Wireless Sensor Networks Applications, *Proceedings of the IEEE International Conference on Computer Communications INFOCOM*, Rio de Janeiro-Brazil, April 2009.

MatLab, Official Site of MatLab: www.mathworks.com

Nguyen, G. (2005). Error-Detection Codes: Algorithms and Fast Implementation, *IEEE Transactions on Computers*, pp. 1-11, vol. 54, no. 1, January 2005.

Ragini, K. & Madhavi, D. (2009). Ultra-Low-Power digital Logic Circuits in Sub-threshold for Biomedical Applications, *Journal of Theoretical and Applied Information Technology*, 2009.

Sadeghi, N.; Howard, S. & Kasnavi, S. (2006). Analysis of Error Control Code use in Ultra–Low–Power Wireless Sensor Networks, *Proceedings of the International Symposium on Circuits and Systems ISCAS*, pp. 3558-3561, Island of Kos-Greece, September 2006.

Schmidt, D.; Berning, M. & When. N. (2009). Error Correction in Single-Hop Wireless Sensor Networks - A case study, *Proceedings of the Design, Automation and Test in Europe Conference DATE*, Nice-France, April 2009.

Tavares, J.; Velez, F. & Ferro, J. (2008). Application of Wireless Sensor Networks to Automobiles", *Measurement Science Review*, pp. 65-70, vol. 8, no. 3, 2008.

Teo, T.; Lim, G.; David, D.; Tan, K.; Gopalakrishnan, P. & Singh, R. (2007). Ultra Low-Power Sensor Node for Wireless Health Monitoring System, *Proceedings of the International Symposium on Circuits and Systems ISCAS*, New Orleans-LA-USA, May 2007.

Part 5

Neural Networks

13

Implementation of Massive Artificial Neural Networks with CUDA

Domen Verber
University of Maribor
Slovenia

1. Introduction

People have always been amazed with the inner-workings of the human brain. The brain is capable of solving variety of problems that are unsolvable by any computers. Is capable of detecting minute changes of light, sound or smell. It is capable of instantly recognizing a face, to accurately read the handwritten text, etc. The brain is the centre of what we call human intelligence and self-awareness. This is not limited only to the human brain. A bee, for example, has a brain that is only a fraction the size compared to the human brain. Nevertheless, the bee able of detecting nectar over long distances; it is capable to orient itself in space and find its way back to the beehive, and it is capable of transferring the information about nectar locations to other bees though a well-choreographed dance.

The basic unit of the nervous system is the neuron. A group of neurons build a neuronal network. In general, a neural network is a parallel system, capable of resolving problems that linear-computing cannot. Neural nets are used for signal processing, pattern recognition, visual and speech processing, in medicine, in business, etc.

The techniques of the neural networks are a part of a machine-learning paradigm. Using this, a system should find solutions for certain problems based only on empirical data, using unknown underlying probability distribution. In addition to this, a vast number of research has been done in the field of artificial neural networks, in order to better understand the human brain, itself. For example, in the Blue Brain Project, the goal is to reconstruct the brain piece by piece and build a virtual brain within supercomputer (BBP, 2011). This approach tries to emulate the human brain very accurately, and requires considerable computing power. Each simulated neuron requires the equivalent of a laptop computer. Several programming libraries and tools exists, which allow for building artificial neural networks of moderate sizes. In addition, several experiments have been where the neurons are emulated within hardware.

This exposure presents a study how to use massive parallel programming on general PCs for artificial neural networks (ANN), which utilizes the processing power and highly parallel computer architectures of graphic processor units (GPU). GPUs on mass-market graphical cards may greatly outperform general processors for some type of applications, both in computation power and in memory bandwidth. The graphic processor consists of a large number of processing cores that may perform a large number of tasks, in parallel. The execution of artificial neural networks is an intrinsically parallel problem. Therefore, parallel

computational architectures, such as GPUs, lead to a great improvement in speed. Until recently, the programmers of ANN could only harness this processing power with especially prepared graphical applications. What is new is that the newest GPU architectures allow for a more general approach to ANN programming, without taking into consideration the graphical aspects of GPUs.
One general-purpose parallel computing architecture is CUDA (Compute Unified Device Architecture), as developed by the Nvidia GPU manufacturer. Different aspects of ANN implementation using CUDA are discussed later. A much greater performance of ANN can be achieved by better understanding the particularities and limitations of CUDA.
The next section presents some biological background of neurons and neural network. Later, different implementation techniques are identified for artificial neural networks. The main section explains section, the implementation of ANN with the CUDA development toll. In conclusion, several experiments are demonstrated and several implementation techniques for large ANN are compared.

2. Biological background of neurons and neural networks

The biological aspects of the nervous system are studied since ancient times. Modern understanding of the neurons started toward the end of the 19th century. Advances in technology, especially regarding brain scanning techniques, the in-vivo observation of mammalian and human brains, etc., allows scientists to determine a detailed understanding of neurons and biological neural networks. In this section, only the brief introduction is given, as relevant to the rest of the matter, is given. For a detailed description regarding biological features of neurons and the nervous system, see (Nicholls et al., 2001). For a detailed survey of the theoretical neuroscience, see (Dayan & Abbot, 2001).

2.1 Neurons

The basic element of the nervous system is a neuron. Neurons are specialized cells that are capable of transmitting and processing information by electrical and chemical means. The human brain consists of roughly 10^{12} neurons. A picture of a typical neuron is presented in Fig. 1.

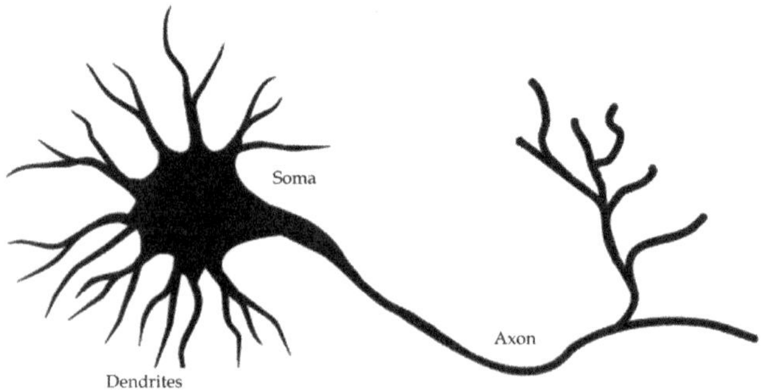

Fig. 1. The conceptual diagram of a biological neuron

Each neuron consists of a number of dendrites that receive signals from the other neuronal or sensory cells. The number of dendrite inputs varies from several thousand, in the case of typical neurons, to 100,000 inputs for some specialized cells. The signals are processed by the cell body (the soma). Under resting conditions, there is a constant electrical potential between the interior and exterior of the cell. The cell is said to be polarized. The electrical potential is maintained with the steadily flow of ions through the cell membrane. Typically, electrical potential of a neuron cell is -70mV, where the surrounding potential is set to 0mV. The electric current from the dendrites changes the electrical potential of the cell. This process is called a depolarization if the potential is increased, or hyperpolarisation, if the potential is decreased. If a neuron is depolarized enough by raising the membrane potential over a certain threshold, the neuron generates an action potential. This is a change in the cell's potential with typical amplitude of 100 mV, and duration of 1ms. The neuron generates an electric spike that is transmitted over the axons to other cells. We say that the action potential (or the spike or the cell) is fired. After an action potential has been initiated, there is a small period of time (a few milliseconds) when another spike cannot be fired. This is a so-called absolute refractory time. After that, there is a relative refractory period (up to 10ms long), where firing is more difficult. The process of neuron firing is depicted in Fig. 2.

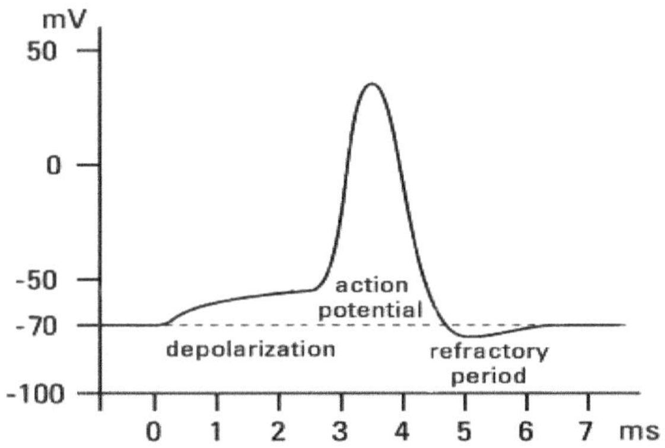

Fig. 2. A time diagram of neuron activation

Except for a few exceptions, the cells are almost never connected to each other directly. The axon of a cell is connected to a dendrite of another cell through a small insulating gap, named a synapse. The transmission of the information within the synapses is carried out by chemical processes. When an action potential is initiated, a chemical compound (a neurotransmitter) is released into the gap. An action potential signal always has the same magnitude. As a consequence, the amount of released neurotransmitters from within the synapse is also the same. The neurotransmitters bind on the receptors on the other side of the synaptic gap, and induce an electric current. Depending on the type of synapse, this currents may be either excitatory (i.e., supporting the polarization) or inhibitory (i.e., preventing the depolarization). The neurotransmitters produce the current over only a short time-interval. Therefore, a neuron is fired over only a short time-period, if the sum of excitatory and inhibitory currents depolarizes the cell membrane over the threshold. The strength of the current of a synapse depends on its structure and may change during the time. In this way, a synapse may have lesser or greater influence on the depolarization of a cell. This is the origin of learning within neural systems.

There are also other kinds of neuronal cells, which are specialized for different tasks. For example, scientists have identified over 80 different classes of neuronal cells in a human eye. The purpose of some of them are the subjects of on-going studies.

2.2 How biological neurons encode the information?

The information in biological neurons is typically encoded as a series of spikes. Although the amplitude and the duration of a spike do not change, the interval between the spikes do change. For example, photosensitive cells in the retina produce a steady sequence of spikes with a frequency that depends on the strength of the light.
There is also strong evidence that some information is encoded as a correlation between the signals of different cells. Cells within a group may fire simultaneously or with a delay, between themselves.

2.3 How biological neurons process the information?

As mentioned previously, the neuron fires only if the sum of the currents produced by synapses depolarizes the cell over a certain threshold. Therefore, a neuron may be observed as an integration function of input signals. However, because the current generated by a synapse quickly dissipates, the model is somehow more complicated, as the sum of currents also changes over the time. A synapse will generate a steady current as long as the series of spikes is present. The average amplitude of this current depends on the frequencies of the spikes, and on the characteristic of a synapse.

2.4 Biological neural networks

A single neuron can only process a small amount of information. In general, several neurons must cooperate with each other. The number of neurons involved depends on the complexity of the task. An interconnected collection of neurons is called a neuronal network. Usually the information within the neuronal network is processed in phases. Each step is performed with a group of neurons that perform similar tasks. This grouping is usually called a neuron layer. The data from one layer is then mediated to another, where more processing is done.
A typical example of this is how the visual information is handled in the mammalian brain. Firstly, the light stimuli are processed by the layer of photosensitive cells in the retina of the eye. Before this information leaves the eye, it is pre-processed by the layer of specialized cells (named ganglion cells). Each of these cells receives signals from photoreceptors within a small region of the receptive field. As a briefly simplified representation: there are two kinds of ganglion cells. The first ones, the on-centre ganglion cells, reacts most favourably to the situations where there is a light stimulus within the middle of the receptive field, and no light in the surrounding area. In a reverse situation, the cells fire almost no spikes. With the off-centre ganglion cell, the reaction is strongest if the surrounding area is lit and there is no light in the middle of the receptive field. Both kinds of cells produce moderate steady flow of spikes when the whole area of the receptive field is lit with the same intensity. By such pre-processing, the amount of data is greatly reduced. There are 130 million photoreceptors in the eye, but only about 1.2 million ganglion cells convey the information from the eye to the brain. This means that each ganglion cell processes the information from about one hundred photosensitive cells. There is another reason why nature's evolution yields such organization in an eye. Such pre-processing is a very efficient way of detecting and isolating the edges and other features within the image.

The signals from both eyes are then combined within the called lateral geniculate nucleus or LNG, from where visual information travels to the visual cortex. The visual cortex is the largest part of cerebral cortex, which is a grey-matter of the brain. Here, in a series of regions, more complex features of visual input are recognized. For example, the first region is responsible for detecting lines of different lengths and different orientations. Later on, the combination of lines are recognized and processed. In the latest stages, the visual information is combined with the signals from other senses, and associated with the memory.

2.5 Learning within biological neural networks

The strengths of the synapses are flexible. On the long term, the amount of the neurotransmitters released into the synaptic gap changes due to changes in neuronal activities. By so-called Hebb's proposal (Hebb, 1946), which started the evolution of neural networks, when an axon of one cell repeatedly excites another cell, the efficiency of the synapse is increased. At the beginning, this idea was largely speculative. Later, however, more and more biological evidence supports it (Fiete, 2003). Another basis for the learning should be the changes in the dendrites themselves. A dendrite may grow and establish a new connection with other cells. However, such growing of dendrites within an adult animal's brain is not confirmed or is extremely rare. Some neuronal cells are interconnected with each other directly. The learning process in this case is rather unclear.

3. Artificial neural networks

Digital computers are capable of performing millions of mathematical operations within a very short amount of time. However, they are less successful at solving some problems, which are effortlessly carried out by even the simplest biological systems. Development of the first artificial neural networks began in the 1940s. It was motivated by the human desire to understand the brain and to emulate some of its strengths (Fausett, 1994). The first ANNs were very simple and had only a few neurons. In the 1980s, when the computers became available to everyone, an enthusiasm for ANN was renewed. Different training strategies for ANNs were developed and more complex problems solved. Advances in hardware over recent years created a new renaissance. Using capable computers, it is now feasible to build massive neural networks.

3.1 An artificial neuron

Several models exist for inner neuron behaviour that emulates the spiking nature of the biological neuron (Gerstner & Werner, 2005). However, the representation of information and the information processing within a biological neuron are very complex and the implementations of those models can be very inefficient. As a consequence, in early artificial neuronal networks, a simplified model of a neuron was used, as shown in Fig. 3.

Instead of using series of spikes, this model uses numerical quantities (either integer or floating-point numbers). The magnitude of inputs and outputs corresponds to the frequency of spikes within a real neuron. The values may be observed either continuously or at discrete points in time.

A synapse is modelled by a multiplier with a constant. The input to the synapse is multiplied by a weight that corresponds to different synaptic behaviour. Larger values of the weights are associated with a large amount of neurotransmitters released, and vice-versa. The negative weights match the inhibitory synapses. The weights are set during the training process of ANN, and usually remain fixed after that.

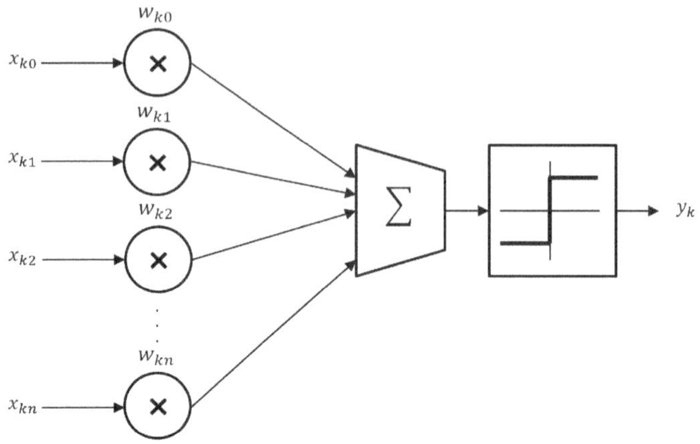

Fig. 3. Simplified model of an artificial neuron

Next, the weighted inputs are summarized by mimicking the integration function of the soma. In some neuron models, an additional correction factor or bias is also applied.

After that, the sum is processed using the so-called activation function. This activation function models the nonlinear characteristic of the action potential generation within the biological neurons. The most common activation functions are the identity function (Eq. 1), the step function (Eq. 2) and the sigmoid function (Eq. 3). The latter is frequently replaced by the hyperbolic tangent function (Eq. 4), which is a good approximation of the sigmoid function with σ=1.

$$i(x) = x \tag{1}$$

$$(x) = \begin{cases} -1, & x < -\theta \\ 0, & -\theta \le x \le \theta \\ 1, & x > \theta \end{cases} \tag{2}$$

$$g(x) = \frac{1 - e^{-\sigma x}}{1 + e^{-\sigma x}} \tag{3}$$

$$h(x) = \frac{e^{x} - e^{-x}}{e^{x} + e^{-x}} \tag{4}$$

The values of threshold θ and the steepness parameter σ are usually set accordingly to a specific problem. The represented functions (with the exception of identity) are a bipolar function with the output-value interval [-1, 1]. Unipolar versions of these functions are also common, some other output ranges may be used.

The output value of an artificial neuron represents the axon within the biological neuron. The operation of an artificial neuron can be embodied within Eq. 5.

$$y_i = f\left(\sum_{j=1}^{n} x_{ij} \cdot w_{ij}\right) = f(x_j \cdot w_j) \tag{5}$$

As seen from the equation, the weighted sum can be evaluated as the dot product between vector of the input values and vector of the corresponding synaptic weights. This can be done very efficiently using digital computers.

3.2 Topologies of the artificial neural networks

The ANN is usually represented by layered sets of neurons or nodes, interconnected by communication links. Each node incorporates an axonic value and each link is associated with a synaptic weight. With the more common ANN net architectures, the nodes of a single layer are connected only with the nodes on the prior one. However, other topologies are also used in practice. A simple multilayer topology of ANN is show in Fig. 4.

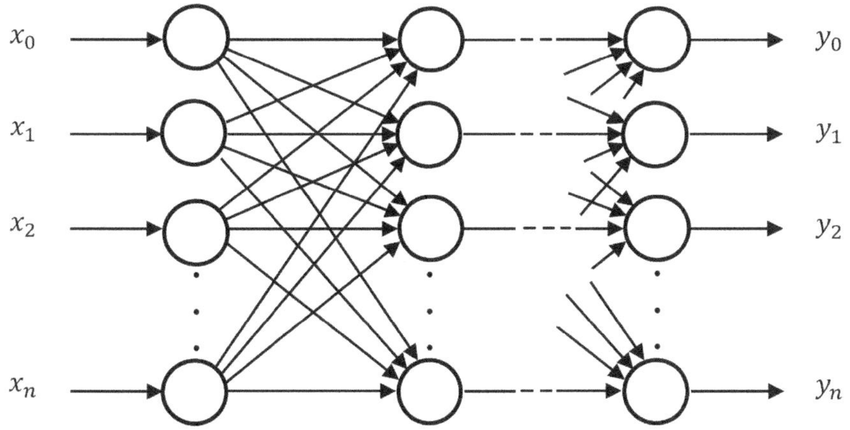

Fig. 4. Typical topology of a multilayer artificial neural network

It is unnecessary for each neuron within a layer to be connected to every neuron in the previous one. With large number of neurons within a layer, the number of connections would be enormous. As was mentioned in section 2.4, each retinal ganglion cell is connected to only one hundred photoreceptors. For the same reasons, in practice, processing a rectangular sub-region of an image is often more common than processing complete images. Such sub regions are usually referred to as regions-of-interest or ROIs.

The opposite may also be true; a single layer of neurons may be processed by several subsequent layers, simultaneously. As will be shown later in the experiments, different features may be extracted from a single picture at the same time.

3.3 Training of artificial neural networks

Before the ANN can be executed, the proper weights must be set. In addition to the architecture, the method for setting the weights' values (called training or learning) is an important characteristic of different neural nets. The training can be supervised. In this case, the neural net is exposed to a set of training patterns for which the desired outputs are known in advanced. The expected outputs are compared with the actual ones, and the weights are adjusted accordingly to minimize any error. In unsupervised learning algorithms, the weighs are adjusted without training patterns. The neural net is self-adapting. In this case, the learning process tries to groups the input patterns into clusters with similar intrinsic characteristics.

For a simple, single-layer networks, the learning algorithm is straightforward and can be achieved mathematically. The most common algorithm for the training of multi-layer artificial networks is the so-called back-propagation algorithm. This is the supervised algorithm. For each training pattern, the difference (error) between the expected and actual results is propagated back from the output layer to the input one. This process is repeated until the error for each training pattern drops under a certain accepted level. The rate of learning can be rather slow. For the back-propagation algorithms, the first derivative of the activation function must be known. The detail description of back-propagation and other multi-layer training algorithms can be found in (Fausett, 1994). In the experiments presented in this work, the focus was on the ANN execution only.

In some cases, no extra training is required. The weighs for some neurons can be determined from the nature of the neuron itself. For example, the retinal ganglion cell has a specific functionality and the weight can be determined analytically. This will be demonstrated in Fig. 17. The weights are adjusted in such a way that the output is most excitatory when a proper combination of on-centre/off-centre light shines on the visual field of the neuron (the output is +1). The opposite combination yield to the inhibitory response (the output is -1). When the entire receptive field is illuminated with the same light intensity, no response is generated (the output value is zero). With these rules and by considering the number of "black" and "white" weights, the weight values can be calculated analytically.

$$w_I = -1/(2 * N_I)$$

$$w_E = +1/(2 * N_E)$$

The w_I and w_E represent the values of inhibitory end excitatory weights, respectively. Similarly, the N_I and N_E corresponds to the number of these weights.

4. Implementation of artificial neural networks

The basic algorithm for a single neuron execution can be derived from Fig. 3 and Eq. 5. It is shown in Fig. 5.

```
sum = 0
for i = 1 to NO_OF_INPUTS do
  sum = sum+inputs[i]*weights[i]
endfor
output = activationFun(sum)
```

Fig. 5. Basic algorithm for simple artificial neuron activation

Each neuronal input is multiplied by the synaptic weight and added to the total sum. At the end, activation function is executed.

In order to execute a layer of neurons, the same algorithm must be completed for each neuron. The execution of the ANN is usually performed in a layer-by-layer fashion. Firstly, the input nodes are processed, and the nodes from the first intermediate layer, etc. This is shown in Fig. 6.

```
for i = 1 to NO_OF_LAYERS do
  for j = 1 to NO_OF_NEURONS_IN_LAYER do
    evaluateNeuron(i,j)
```

Fig. 6. Elementary algorithm for ANN execution

The first layer is usually the input layer and contains the input data. Commonly, no processing is done in this layer. Similarly, the evaluated neurons in the last layer usually represent the outputs of a network. For a massive ANN, the neurons in a layer are typically organized within two or three-dimensional arrays.

ANN can be implemented either in software or in hardware. The implementation with CUDA, which is the topic of a further section, can be observed as a hybrid between software and hardware implementation. A piece of software is executed several times on massively parallel multiprocessor architecture.

4.1 Software implementation

Most often, the ANN is implemented in software. The algorithms are simple enough for any computer language. In addition, there are numerous programming libraries and frameworks that simplify ANN construction. A very popular tool for ANN is the adaptive Neural Network Library included as an add-on in Matlab 5.3.1, and later. It is a collection of blocks that implement several Adaptive Neural Networks featuring different adaptation algorithms (Matlab, 2011). With this toll, only the simulation of a small ANNs is feasible. A good starting point for different software libraries and other resources for ANN can be found at (DMOZ, 2011).

The main advantage of software implementation is that the programmer may utilize all resources of the computer. The application has access to large amount of memory, mass storages, input/output devices, etc. For example, the processing of visual data in real-time requires a direct connection to the camera; pictures may be loaded from a databases or a disk, etc. With other kinds of ANN implementation, this is not achieved that easily; usually the data must be prepared and send to the device where it will be processed.

For massive ANNs, a large amount of storage is required to store values of weights and the intermediate results of neurons. Nowadays, this is no longer an issue. Even modest personal computers have enough computer memory for this. With other kinds of implementations, the limited memory may present a problem.

In our experiments, dynamic memory regions organized as two-dimensional arrays were used to represent the inputs, the weights and the intermediate values. The class definition of such a memory region is presented in Fig. 7.

The data type determines the data domain of those values kept in the memory. This is usually a floating-point number, but can also be an integer or a byte. For convenience, regardless of the actual data type, the values are always processed as floats. The stride represents the actual data size of bytes on each data row. For some configurations, the data is aligned within the memory for more efficient access. There are also a number of routines (not shown in the code) that allow for direct loading or storing the data in several image formats. In this way, the data can be efficiently read from or write to several image formats from a disk or a memory stream.

```
class MRegion
{
public:
        int dataType; // element data type
        int width;  // width, height in data type units
        int height;
        int stride;  // stride of data line in bytes
        void *memory;

        MRegion(...) {};

        float getData(int row,int col);
        void setData(int row,int col,float value);
        ...
}
```

Fig. 7. Data structure for abstract memory region

Another piece of information is the configuration as to how the weights are determined for each neuron. This is called weight evaluation parameters or WEP. The parameters of WEP are depicted in Fig. 8, and the The data structure is given in Fig. 9.

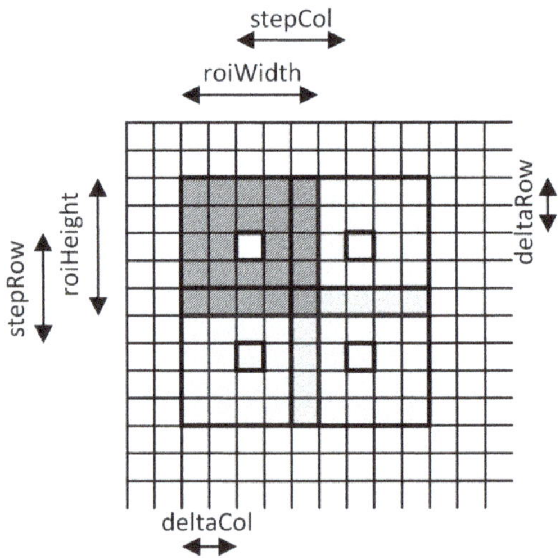

Fig. 8. Conceptual diagram of weight evaluation parameters

Each neuron within a layer has its ROI that determines a region of inputs that must be evaluated. The position of the weight row and column indexes relative to the row and column indexes of the input data is determined by deltaCol and deltaRow members. It is unnecessary for the neuron to be positioned in the centre of the ROI. The data densities of the outputs are determined by the stepCol and stepRow parameters. As in the case of the ganglion cells, the number of output neurons can be smaller than in the input layer. Two additional Boolean flags determine the weight dependency. In general, each neuron in the layer can have its own set of weights. In this case, the flag neuronDependent is set. In other cases, all the neurons in a layer use the same set of weights. Sometimes a neuron from an output layer can process the inputs from different input layers. In this case, the weights may be independent of the input layer's index or not. For the later, the flag layerDependent is set.

```
struct WEP
{
        int roiWidth,roiHeight; // size of ROI
        int deltaCol,deltaRow; // relative delta of ROI
        int stepCol,stepRow;  // incremental step
        bool neuronDependent;  // weights dependency
        bool layerDependent;

        WEP(...);
        ...
}
```

Fig. 9. Data structure of weight evaluation parameters

The data structure for a layer of neurons is shown in Fig. 10. There is separate memory storage for the data and the weights, the weight evaluation parameters, pointer to neuron activation function, and a set of routines for layer and neuron evaluation.
The layer's evaluation function is straightforward, and is shown in Fig. 11.
The neuron evaluation function is somehow complicated because it is necessary to consider the different index ranges. The ROI of a neuron must be translated into the coordinate space of inputs. The edges of the input array also need to be considered. The indexes may be out of range. The code for a neuron evaluation is presented in Fig. 12.

```
struct Layer
{
public:
        MRegion data;
    MRegion weights;
        WEP wep; // weight evaluation parameters

        float (*actFun)(float x); // activation function

        void evaluateLayer(Layer *inputLayers, int noOfILayers);
        void evaluateNeuron(Layer *inputLayers,
        int noOfILayers,int nrow,int ncol);
        Layer(...) {}
        ...
}
```

Fig. 10. Data structure of layer in ANN

```
void Layer::evaluateLayer(Layer *inputLayers,int noOfILayers)
{
      for(int row=0;row<data.height;row++)
       for(int col=0;col<data.width;col++)
            evaluateNeuron(inputLayers,noOfILayers,row,col);
}
```

Fig. 11. The layer evaluation function.

The main drawback of software implementation may be the inferior speed. Even though the microprocessors are extremely fast, because the common computers can process data only sequentially, the computation of massive ANN may take a long time. In simplified model of a neuron, one multiplication and one addition is required for each input. As well, for each neuron, activation function must be evaluated. There is also an overhead for ROI index calculation. Even longer execution times are required for those models with more complex neuron emulation. This is not a problem if the number of the neurons is small. However, with higher number of neurons and with several layers involved, the total number of computations is very large.

4.2 Parallelization of ANN execution

The problem of slow ANN execution can be somehow mitigated by using the modern microprocessor architectures. Instruction set of the modern microprocessor contains instructions for fused multiply-and-add operations. In this case, multiplication and addition within a synapse are performed at the same time. Sometimes several instructions can be executed simultaneously. In addition, the multi-core architectures of modern microprocessor can be utilized. The processing effort can be divided among different cores. However, the number of cores is relatively small and the speed-up would also be small.

The main idea behind ANN evaluation parallelization is shown in Fig. 13. A group of neurons is used to process a part of data from the input data layer. The neurons are grouped together in such a way that minimizes the amount data that must be read. In this way, the same data can be used with several neurons at the same time. This minimizes the number of memory access cycles where the data can only be read serially. When the neurons have finished their job, another patch of input area is processed. If necessary, another set of weights is used. Several groups of neurons can be executed in parallel by different processing cores. In the hardware implementation as well as with the CUDA solution, each neuron from a group is processed simultaneously with the others.

An attempt was also made to parallelize the training of a multi-layer ANN using back-propagation learning algorithm. It was found that much less parallelization of the code is possible. First of all the training is done in repetitive cycles where different training patterns are exposed to the net. Because of the nature of the back-propagation, the layers had to be evaluated one-by-one; the results from just one of the layers influenced the evaluation of others.

4.3 Hardware implementation

The long execution times for software solutions may be reduced with hardware implementation. It is possible to construct a small digital or analog circuit that represents a single neuron, and then replicate this circuit as many times as possible. Then, instead of evaluating one neuron at a time, several hundred or several thousand neurons may be executed concurrently.

```
void Layer::evaluateNeuron(
       Layer *inputLayers,int noOfILayers,int nrow,int ncol)
{
 // calculate top left data element index
 int row=nrow*wep.stepRow-wep.deltaRow;
 int col=ncol*wep.stepCol-wep.deltaCol;
 float sum=0;
 // sweep ROI of a single neuron
 for(int rrow=0;rrow<wep.roiHeight;rrow++)
 {
  // Calculate the neuron row position in the input data layer
  int irow = row+rrow;
  for(int il=0;il<noOfILayers;il++)
  {
   // Check bounds
   if((irow>=0) && (irow<inputLayers[il].data.height))
   {
    for(int rcol=0;rcol<wep.roiWidth;rcol++)
        {
         int icol=col+rcol;
         // Check bounds
         if((icol>=0) && (icol<inputLayers[il].data.width))
         {
          // Calculate the final coordinates of weight
          int wrow = rrow, wcol = rcol;
          if(wep.neuronDependant)
      {
           wrow+=nrow*weights.height;
              wcol=ncol*weights.width;
              if(wep.layerDependant)
               wrow+=il*inputLayers[il].data.height*wep.roiHeight;
          }
      else if(wep.layerDependant)
              wrow+=il*wep.roiHeight;

          // Get the weight
          float w = weights.getData(wrow,wcol);
          if(w!=0)
           sum += inputLayers[il].data.getData(irow,icol)*w;
         }
    }
   }
  }
 }
 data.setData(nrow,ncol,(activationFun)(sum));
}
```

Fig. 12. Implementation of a neuron evaluation routine

Biological neurons exibits nonlinear behaviour. Therefore it is natural to model them using nonlinear analog circuits. Several such solutions have been proposed in the past (Kong et al., 2008) proposed hardware implementation of the retina. Most of these solutions are only used to implement small to moderate numbers of neurons.
For academic studies and for experimentation, the usage of Field-programmable gate arrays (FPGAs) is more feasible. FPGA is a device that allows for construction of different digital circuits. The construction of these devices is done by writing the schematic circuits or by programing using one of the hardware description languages (HDLs). (Volnei, 2004)
In (Rosado-Muñoz et al., 2011) the authors proposed the FPGA solution for modelling spiking neural networks, which approximates the behaviour of biological neurons. The model of neuron, in this case, requires complex implementation and allows for only small silicon utilization. In our experiments, an attempt was made to utilize the similar FPGA device for a simplified neuron model regarding larger ANNs. This simplified model achieved a much higher density of neurons.

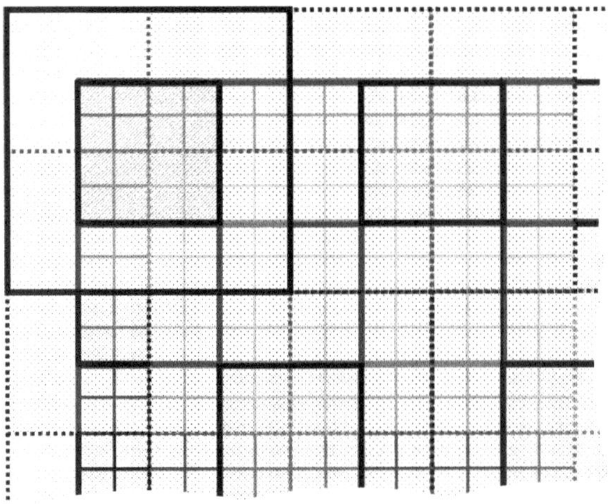

Fig. 13. A concept of ANN evaluation parallelization with 4x4 neurons and 5x5 ROI areas

The schematic of a neuron implementation is shown in Fig. 14.

The neuron consists of 18x18 bit multiplier (MULT18x18), 36-bit adder (Add36), the summation register (Sum), and the activation function parameter's generator (Function). The values are represented as 18 bit fixed floating-point numbers with 13 binary decimal places. A multiplier is implemented with a dedicated device inside FPGA that was used in the experiment. It allows for 18 by 18 bit multiplication over one clock cycle. The adder is also constructed in such a way that was capable of performing a 36-bit addition operation in one clock cycle. The activation function was approximated with a series of linear segments. The activation function generator for a given input produces a series of parameters k, x and n. The actual value of activation function was then calculated by equation y=k*x+n. For addition and multiplication, the same units were used as above. The input data and the weight are read from the dedicated memory blocks inside the FPGA device. Each block can contain up to 1024 values of 18-bit data.

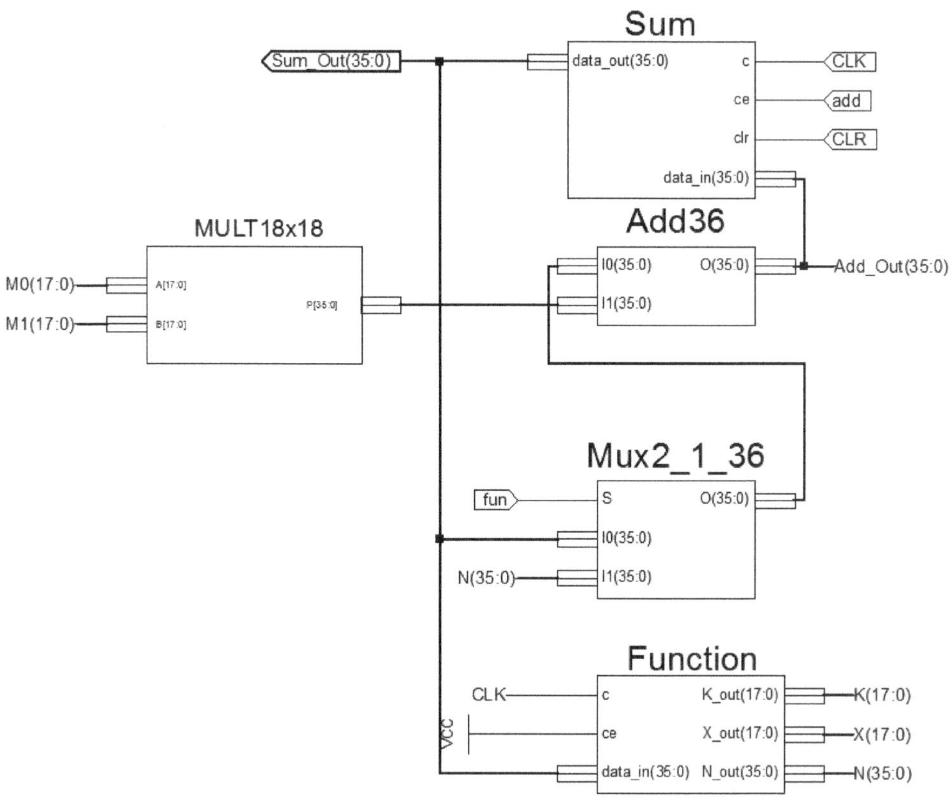

Fig. 14. Schematic of a neuron implemented within hardware

Although, they are extremely fast, there are several obstacles when using FPGA devices for ANN. Firstly, to process data, the inputs have to be provided by the main processor. In our case, the data was loaded into the external memory chip. The same is true as to when the results must be gathered. Secondly, the number of neurons that can be implemented inside FPA device is limited. In our device, 126 multipliers were available. Because of the limited memory available and because some multipliers were used for other purposes, it was only possible to implement 64 neurons. Massive ANN was implemented with the method described in the previous section. Another limiting factor is the amount of available internal memory inside a FPGA device. The weights for each neuron are stored in separate RAM block. The largest ROI available in this case was 32x32 input values. The number of RAM blocks limited. It was necessary to load a new data and weights after each execution cycle of 64 neurons. For this, data was transferred from external to internal memory, by means of software microprocessor implemented inside the device. This process prologues the execution of ANN. The ratio between the times of neuron execution and data transfer was around 1:20 for a typical case.

In earlier experiments, an attempt was made to use the FPGA devices for the training of the ANN. However, the amount of silicon consumption is doubled. In addition, because of the increased memory requirements, only the training for a small ANNs could be implemented.

5. Implementation of artificial neural networks with CUDA

Driven by demands from the mass consumer electronic market, the programmable Graphic Processor Unit or GPU has evolved into device that largely outperforms any general Central Processing Unit or CPU in both computational capabilities and memory bandwidth. GPUs are capable to render high-definition 3D graphics in real-time. In order to do this, the GPU must process millions and millions of pixels in every second. To achieve this, GPUs employ highly-parallel architectures. At first, the process of graphic rendering was predefined and fixed within hardware. Later, it was possible to write simple programs (called shaders) that define how graphic is rendered. Soon after that, programmers tried to harness the great processing power for solving non-graphical problems. This is known as General-Purpose Computing on GPUs or GP-GPU. Instead of rendering a graphic on screen, specially designed shaders were capable of doing the calculations and "rendering" the results into memory. Because of its origin, the GPU was especially well-suited to address problems that could be expressed as data-parallel computations - the same program being executed on many data elements, in parallel. The drawback of this was that the programmers were forced to employ graphical resources of GPUs and use graphical application programming libraries. In order to make GPUs more universal, the manufacturers have changed the architectures of their newest graphical devices, thus enabling them to be used for wider set of applications. Certain silicon areas on GPUs are now devoted to facilitatingthe ease of parallel programming. These parts are never used during picture rendering. The other elements are shared and can be used with a graphical engine and for general parallel programming. GPU based devices now exists with no video output. They are dedicated solely to parallel programming.

The practice of parallel programming is not something new, it has been in use for decades. However, these programs run on large scale and expensive computers (see Herlihy & Shavit, 2008). With recent evolution of GPUs, massive programming has become available for everyone.

The execution of artificial neural networks or ANN is an intrinsically parallel problem; hence, parallel computational architectures lead to a great improvement in speed. Because, the researchers were start to use the GPUs from the very beginning, but what is new is that the newest GPU architectures allow for a more general approach to ANN programming, without taking into consideration the graphical aspects of GPUs.

5.1 CUDA

For the implementation of ANN with parallel programming, a Compute Unified Device Architecture or CUDA was chosen (CUDA, 2011). CUDA is a general-purpose parallel computing architecture developed by Nvidia GPU manufacturer. CUDA comes with a software environment that allows developers to use C as a high-level programming language. In CUDA terminology, the main unit of PC with general CPU is called the *host* and the graphics card is called the *device*. The host and the device communicate with each other through the high-speed bus. CUDA architecture allows several devices to be connected to the same host. However, the programming management of several devices is left to the programmer.

The basic unit of execution in CUDA is the so-called *kernel*. The kernel is a specially marked C routine that may be invoked from the host. The kernel is executed in parallel multiple times by *CUDA threads*. The execution of several threads is grouped into larger collections called *thread blocks*. For even greater parallelism, multiple tread blocks can be run together in a *grid*. Although a kernel code is the same for all threads, each tread has its own programming context (i.e. its own set of registers and local variables). In addition, each tread in a thread block has its unique identification number called *thread ID,* and each block in the grid has its unique *block ID*. The threads inside a block can be organized into one-, two- or three-dimensional structures. This provides a natural way of performing computation across the elements in those domains where data is presented in a vector, matrix, or field form. Similarly, a block within a grid can be organized into one- or two-dimensional structures.

The treads inside a block can communicate with each other by means of shared memory. The shared memory is accessible from all threads within the same thread block. However, the content of the memory is not preserved between executions of different blocks; the content of the shared memory being undefined when the block starts its execution. The access time of the shared memory in CUDA is the same as the access time of the registers.

This device has also a large amount of global memory (up to several GBytes). Its content has the lifetime of an application. In contrast to the shared memory, the access time for the global memory is at least 100 times slower. Therefore, if the same data is accessed several times, it is feasible to copy it first from global memory to the shared one. Alternatively, the data in the global memory can be organized as a texture memory. Textures are a part of typical graphic applications for painting 2D images over 3D objects. In applications, the texture memory provides a faster alternative to general memory because the texture accesses are cached. The drawback is that the textures are read-only and require special access functions. In the newest CUDA devices, access to the global memory is also cached.

There is also a small amount of the so-called constant memory. This read-only memory can be set by the host. This memory is also cached and has a lifespan of the application. Constant memory is ideal when using global application parameters.

CUDA provides synchronization between threads by means of barriers. When a thread within a block reaches the barrier, it waits until all other threads reach the same point of execution. CUDA also provides several atomic operations, which can solve possible race conditions that may happen in parallel programs, where the same global variable is modified by several threads simultaneously.

CUDA programs consist of intermixed code for both the host and for the CUDA device (files with any CUDA code must have a .cu file extension). A special compiler (nvcc) pre-compiles this code and splits it into the host and the device part. Both parts are then compiled separately. Prior to execution, the compiled CUDA code is loaded into the device memory.

On a CUDA device, the threads are executed on a scalable array of multithreaded Streaming Multiprocessors (SMs). Each SM consists of eight Scalar processors (SP) and some other components. All SPs execute the same instruction at a time, and four successive clock cycles are used to execute a single instruction for 32 threads. This is called a warp. Each SM has its own shared memories and access to global, constant and texture memory. All threads in a block are executed by the same SM. The threads are split into warps and executed concurrently, according to the available resources, with hardware-based scheduler. The context switch between warps is performed in zero-time.

5.2 Representation of input and output data

During training and execution, the ANN usually reads the input data from disk or some device (e.g. digital camera). The CUDA device has no direct access to these. Therefore, data must be first loaded into the device. Currently, this can be done only by means of memory-to-memory transfer between the host and the device. The same is true when the results must be stored on the host. Although the transfer bandwidth can be very high, (e.g. with PCI express 16x bus up to 8GB/sec) these transfers introduce a delay which may influence the feasibility of using CUDA. In general, the transfer delay is only justifiable for large ANN. For newer CUDA devices, it is possible to perform memory-to-memory transfer in parallel with kernel execution: part of data space may be processed at the same time when the other part is prepared. It is also possible to execute memory-to-memory transfer between two CUDA devices without CPU involvement.

Very often, the data must be pre- and post-processed. Usually these operations can be parallelized, so it can be done more efficiently on the device. For example, in image processing, the picture is usually enhanced before a computation. This is usually well suited for GPUs and parallel computation.

Natural data types for CUDA are integer and floating point values. Older CUDA devices support only single precision floating-point numbers. This may be the problem if the training algorithm relies on a greater accuracy and may impose some problems with the convergence of some solutions. The newest generation of CUDA devices supports the double-precision FP values. This is the standard for all future devices. However, with double-precision arithmetic, the performance decreases. Because of GPU origins, the data on CUDA devices may be very well organized into one, two or three-dimensional arrays. Algebraic operations, such as vector and matrices multiplications, can be performed very efficiently. There are several programming libraries available for CUDA that implements optimized algebraic operations. The CUDA also supports pointers and, therefore, some sort of dynamic data structures can be used. However, because of the single-instruction-multiple-data nature of program executions, this may be inefficiently.

5.3 Representation of the weights

The weights of the ANN must also be transferred to the device memory. For most ANN topologies, the weights are fixed after the training and remain the same during the execution. For an older device, the best place to put them is in the read-only texture memory. As mentioned before, this memory is cached and provides much greater throughput. The mechanism of the cache is optimized for graphical applications and produces the best results for 2D and 3D access patterns. This is well-suited for ANN topologies where a neuron in some layer is connected to a small-clustered region of neurons on a previous layer. For the new devices, this is not as clear because the main memory is also caches. In most situations, some experimentation is required.

During the training and with ANN topologies where weigh changes continually during the execution, the general global memory of CUDA devices must be used. Because of the large latency, these algorithms that utilize the shared memory should be used. E.g., a multiprocessor takes 4 clock cycles to issue one memory instruction for a warp. When accessing global memory, there are between 400 to 600 clock cycles of memory latency. However, due to the specifics of the CUDA device implementation, the global memory access by all threads of a warp can be coalesced into one or two memory transactions, if it satisfies some conditions (e.g. each thread must access the memory in sequence, the accessed memory must be properly aligned, etc.). The conditions depend on the version of the CUDA devices. Therefore, it is difficult to implement a kernel routine that will be optimal for all boards.

5.4 Summarization of the neuron Inputs

As a part of the ANN execution, the input values of each neuron are multiplied by their appropriate weights and then summarized, before the activation function is applied. This operation is also frequently used in graphical applications and, therefore, GPUs may perform it very efficiently by means of the so-called multiply-and-add (MAD) instruction. The execution time for this instruction is only four clock cycles for single precision floating point numbers.

In parallel computing, this kind of summarization is commonly known as reduction: elements of an array are reduced to a single value by certain operation (see Matson, et al., 2005). This may be achieved differently. It is possible that each kernel performs the multiplication of one input and adds the result to the global sum. However, in this case the workload of the kernel would be very low and insufficient for hiding pipeline and memory latency. In the approach we proposed, the kernel should perform the summation of subset of inputs. Then the intermediate sums are added together. The conceptual diagram of this approach is depicted on Fig. 15. The summation can be done over several stages. At the final stage, when total sum of the weighted inputs is calculated, the activation function is also evaluated. Because several threads must update the same variable (i.e. the sum) simultaneously, the atomic instructions must be used. Intermediate sums should be kept in the shared memory. However, only the newest CUDA devices have implemented the atomic functions that operate with the shared memory.

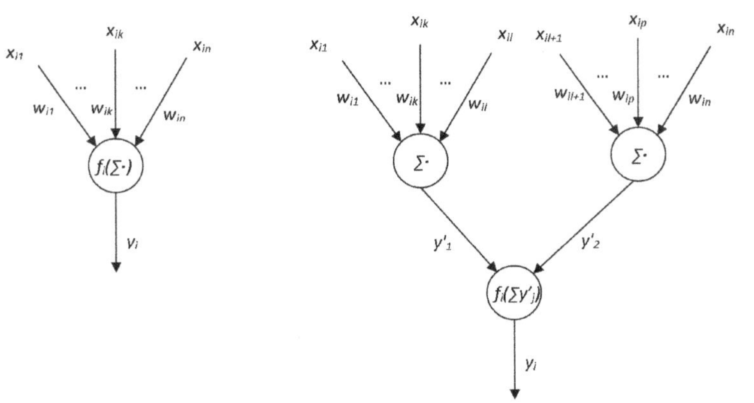

Fig. 15. Conceptual diagram of direct and indirect summation of weighted inputs

5.5 Implementation of the activation function

Some mathematical operations can be executed on the CUDA devices more efficiently than the others. The only intrinsic mathematical instructions implemented in the current version of CUDA devices are: reciprocal, square root, reciprocal of square root, sine, cosine, log, base 2, and exponential, base 2. In addition, some of these functions have limited accuracy. For the implementation of more accurate version of these instructions and for other mathematical functions, run-time routines are needed that consist of a sequence of elementary operations. For example, there is no direct implementation of exponentiation of base e, and must be implemented by other means. There does exists a primitive implementation of natural exponential function (named __expf(x)), where the power of two function is used (__expf(x)=__exp2f(x*LOG2_E)). This function can be implemented with a sequence of simple instructions that take 32 clock cycles. However, this implementation loses significant accuracy for large arguments. There is also a more accurate implementation of this function (named expf(x)), which is much more expensive. Another example is the single-precision floating-point division that takes 36 clock cycles. However, a primitive function __fdividef(x, y) provides a faster version at 20 clock cycles. This function has the same accuracy as the division. However, it produces the zero result for large values of y. The "normal" division reduces the dividend and divisor by one quarter prior the operation if the absolute value of y is too big. Integer division and modulo operation are particularly costly and should be avoided if possible, or replaced by bitwise operations.

The identity and the step activation functions are simple to implement. On the other hand, the sigmoid and the hyperbolic tangent functions are the composite of several basic operations. A big reduction of execution time may be achieved in some situations if a less accurate version of those operations can be used. For example, the activation function is very often implemented by the hyperbolic tangent function and, if accuracy is not a problem, then a code can be used:

```
y = __fdividef(__expf(x)-1/__expf(x),__expf(x)+1/__expf(x));
```

The equation is optimized by the compiler to a single call of the exponentiation function, one call of the reciprocal function, one addition, one subtraction and one division. The reciprocal takes 16 clock cycles, the addition and subtraction take four clock cycles each. Therefore, the exaction time of this activation function can be 76 clock cycles. The more accurate implementation of hyperbolic function would take at least three times more clock cycles.

5.6 Training and execution of ANN

The training and execution of ANN is performed in three steps: preparation of the initial data, transfer of the data to the CUDA device, evocation of the kernel routine, and transfer of the result to the host. The same data may be evaluated using several kernels sequentially and data transfer operations may overlap with the kernel execution. It is also feasible to distribute the workload between several CUDA devices. The upper-end graphical cards incorporate two GPUs, which can be used as two independent CUDA devices. There are dedicated CUDA devices with no graphical output, and with up to 256 processing cores.

6. Experiments

A set of experiments was performed, in order to augment the theoretical research. An attempt was made to implement a massive ANN with software, with hardware and with the

CUDA framework. The first experiment tried to implement a neocognitron. This is an ANN designed to recognize written text (Fukushima et al., 1983). The description of neocognitron presented in the (Fausett, 1994) was used. The idea behind the neocognitron is to mimic the video processing path found in humans and other animals, as described in section 2.4. The data is processed in stages. During the first stage, the basic elements are extracted from input pictures. The basic elements are short lines with different orientations. Several layers of network are trained to respond to these features. At the latter stages, a combination of basic features is processed. Finally, specific characters are recognized. In this specific case, the ANN consisted of 9 layers of neurons. Most layers are further divided into sub-layers, which provide around 160 grouping of neurons. The total number of neurons is around 14 thousand with approximately 160 thousand synapses. The inputs are simple patterns of size 19x19 monochrome pixels. The output is a vector of bits, each corresponding to a specific character (in our case, it was one of the ten decimal digits). The ANN is capable to recognize one character at the time. It cannot be claimed that this is the best ANN for handwritten character recognition. However, it is complex enough to test and compare different implementations. The training of ANN was implemented in software.

During software implementation, the code described in section 4.1 was used. All neurons in a single layer were processed at the same time. The tests were executed on a Xeon processor (W5590 running at 3.33 GHz).

For the hardware implementation, the FPGA solution presented in the section of 4.3 was used. The FPGA device was Xilinx Spartan3A XC3SD3400A, running at 25 MHz (Xilinx, 2011). An extra memory chip with capacity of 4 Mbytes was connected to the device. Within the FPGA, a soft-processor Picoblaze was used to move the data between external an internal memory. The device was connected to the PC via serial communication link.

```
__global__ void
evaluateNeuron(Layer *me, Layer *inputLayers,int noOfILayers)
{
 // calculate actual coordinates of a neuron
 const unsigned int nrow = blockDim.y*blockIdx.y+threadIdx.y;
 const unsigned int ncol = blockDim.x*blockIdx.x+threadIdx.x;
 // access number of threads in this block
 const unsigned int num_threads = blockDim.x;
 ...
 me->data.setData(nrow,ncol,activationFun(sum));
}
```

Fig. 16. The source code excerpt of a kernel for a neuron evaluation

The matrix of 8x8 neurons was used. All necessary data and configuration parameters were put into the external memory, prior the execution. With each execution cycle of ANN, a new set of data and parameters was loaded into the device.

For the implementation with CUDA framework, a slightly modified version of the code was used. In compiler for CUDA, the support for classes is somehow limited. At this time, the CUDA does not support the virtual functions. Similarly, the pointers to the functions are yet unsupported. The kernel routine for the neuron evaluation is shown in Fig. 16.

The class method was converted to the standard C routine. Additional parameter was added that represents the reference to a current layer. All the neurons in a layer are executed at the same time. This eliminates the loop presented in Fig. 11. The indexes of a neuron are determined by block and thread IDs.
For layer-by-layer execution of the ANN as a whole, the kernel was executed several times with different parameters. All the data and parameters were put into the GPU memory prior to the execution.
Firstly, the experiments were implemented with the dedicated CUDA device with compute capability 1.3 (Tesla C1060). Recently, a high-end gamming card has been acquired with the compute capability 2.0 (GeForce GTX595). The level of parallelization in latter device is much higher. The new architecture incorporates global memory cache and it is possible to use two different configuration of shared memory (16KB or 48KB instead of only the 16KB in previous versions). With older devices, each thread block can contain maximum of 512 threads. In our case, this is also the maximum number of neurons per block. Prior the evaluation of the neurons, the inputs could be copied from the global to the shared memory. However, the number of inputs is limited by the maximum amount of shared memory per block. With 16KB of shared memory, 1024 synapses can be implemented. This corresponds to a grid with 32x32 inputs. Nevertheless, the shared memory is also used to store intermediate results during the layer evaluation. At least as much values as there are neurons in a block are needed. At maximum, the number of inputs that may be put into the shared memory is halved. With newer devices and with 48KB of shared memory per block, intermediate values for all 1024 neurons plus another 2048 values for the inputs can be maintained.
In the first experiment, the size of the maximum layer was 19x19 neurons. This is well below the limit of maximum number of threads per block. However, some of the layers needed to process up to 4600 inputs, which is too much for the shared memory. Because of that, direct access to the global memory was used.
The results are shown in Table 1.

Implementation method	**Execution time**
Software	19 ms
Hardware	0.25 s ($5.1 \cdot 10^6$ clock cycles)
CUDA 1.3	47 ms
CUDA 2.0	49 ms

Table 1. The execution times of different neocognitron implementations

The execution time does not include data preparation. However, we include the data transfer time between host and devices in the case of implementation with CUDA. This is the reason why CPU implementation outperforms any CUDA implementation. The duration of data transfer in later cases is around 41ms. The execution time of the newer CUDA device was a little bit slower than with the older one. This is probable due the higher clock frequencies of the older model.
The execution time for the hardware implementation would have been shorter if the faster FPGA device has been exploited. In addition, more capable devices could have been used to

implement larger number of neurons. For example, the Vitrex-7 family of FPGA devices contain up-to 3600 multipliers and as much as that number of neurons can be implemented. Each multiplier is capable of operating at 638 MHz.

There is enough room for additional optimization for both software and CUDA implementation. Instead of the general neuron evaluation routine, a specialized one can be used for each of the different layer configurations. In this way, some loops can be unfolded and some expressions simplified. We have not tried to utilize several processing cores, as yet. Similarly, with several CUDA devices, it would be possible to divide the job into parts and process them simultaneously.

A higher difference was expected between the results of software and CUDA implementation. High-end CPU was used and the number of neurons was too small for significant difference. Because of that, another experiment with much larger number of neurons was conducted.

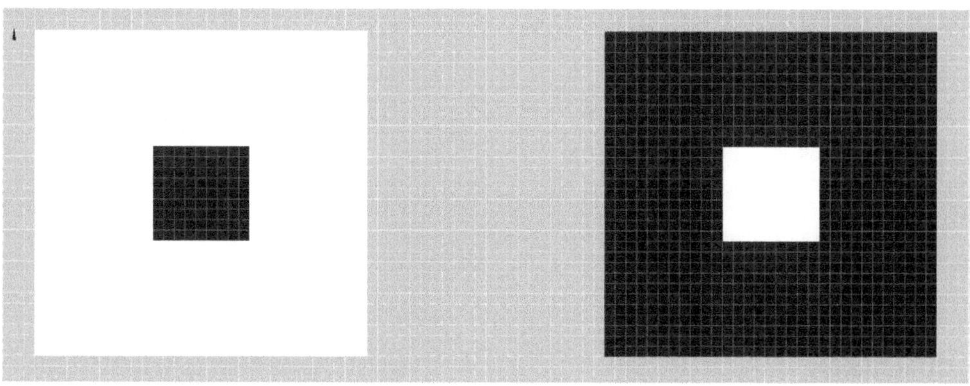

Fig. 17. Weights for off-centre and on-centre ganglion cell simulation

Fig. 18. The original image and the result of the second experiment

In the second experiment, an early stage of visual processing in the retina was implemented. The off-centre/on-centre ganglion cells were realized in the manner described in section 2.4. High-definition photography was taken as an input. The size of the input grid is 5616x3744 pixels. The output grid of neurons is reduced to one third. The ROI of each neuron is 31x31

pixels. The weights were set in advance to mimic the processing of the ganglion cells. A biological ganglion cell has approximately circular ROI. For simplicity, we used the rectangular patterns shown on Fig. 17. The white areas represent the excitatory synapses; the black areas correspond to the inhibitory synapses. The sum of weights is exactly zero. This mimic the ganglion cells response in the diffuse light. The weights pattern is combined with the threshold activation function to get monochromatic (black and white) result. The original and processed picture with off-centre weights is shown on Fig. 18. The result illustrates how edge detection is processed in the human eye.
The results of the second experiment are shown in Table 2.

Implementation method	Execution time
Software	224 s
Hardware	-
CUDA 1.3	670 ms
CUDA 2.0	799 ms

Table 2. The execution times of the second experiment

The amount of data was too high for the hardware implementation. The execution time of the software version was three orders of magnitude slower than the CUDA implementation. The data transfer between the host and the GPU device contributed to about 80% of total execution time.

7. Conclusion

Today, advances in hardware allow the ANN programmers to have efficient implementations of massive neural networks that operate in real-time. For the small ANN topologies, the presented approach is not feasible. It is possible to run several ANN in parallel on a single or several devices. For example, in the case of character recognition, different characters may be evaluated simultaneously.
From all currently available parallel programming environments, CUDA is most suitable for ANN because it is the most mature, and can be used with wide range of products. However, the solution with CUDA is not ideal. The CUDA is bound to a single manufacturer and cannot be used with graphical cards of the others. The CUDA devices are still evolving. Each new generation introduces some new features that make general parallel programming more consistent and easier. At the same time, the optimization approach used with one device's generation may be obsolete or inefficient with another.
Another consideration is the amount of time needed to put the input data into the CUDA device and to get the results. As was shown during the experiments, this can take longer than the processing alone. However, this also means that more complex problems can be solved within approximately the same time.
In the current experiments, we neglected the study of training algorithms for the multilayer neural networks. The main reason is that the training algorithms are very difficult to be

parallelized. The presumption was that the weights would be calculated only once. However, there are ANN topologies where training is executed continuously and must be included in the final implementation.

In the future, we will try to implement other neuronal models for comparison if one model is better than another or if one model allows for solving larger classes of problems than the others.

8. References

BBP (2011). Home page of the Blue Brain Project. Available at http://bluebrain.epfl.ch/

CUDA(2011).Cuda Developers Page, Available from http://developer.nvidia.com/category/zone/cuda-zone

Dayan, P. & Abbot L.F. (2001) *Theoretical Neuroscience: Computational and Mathematical Modeling of Neural Systems*. MIT Press, Cambridge. ISBN 0-262-04199-5

DMOZ (2011). Portal to different ANN resources. Available at http://www.dmoz.org/Computers/Artificial_Intelligence/Neural_Networks/

Fausett, L. (1994). *Fundamentals of Neural Networks: architectures, algorithms, and application.* Prentice Hall, New Jersey.

Fiete, I. R. (2003) *Learning and coding in biological neural networks.* PhD thesis. Harvard University, Cambridge, Massachusetts.

Fukushima K., Miyake S. & Ito T. (1983) Neocognitron: a neural network model for a mechanism of visual pattern recognition. *IEEE Transactions on Systems, Man, and Cybernetics*, Vol. 13, 826–834.

Gerstner, W. & Werner, M.K. (2005). *Spiking Neuron Models, Single Neurons, Population, Plasticity*, Cambridge University Press, Cambridge, UK

Hebb, D. O. (1946), *Organization of Behavior: A Neuropsychological Theory* (Wiley, Inc., ADDRESS, 1949).

Herlihy, M. & Shavit, N. (2008) *The Art of Multiprocessor Programming.* Elsevier, Burlington.

Kong, J., Sung, D., Hyun, H. & Shin, J. (2008) A 160×120 Edge Detection Vision Chip for Neuromorphic Systems Using Logarithmic Active Pixel Sensor with Low Power Dissipation. *In: Lecture Notes in Computer Science.* Springer, Berlin, ISBN: 978-3-540-69159-4, pp: 97-106.

Matlab (2011). Matlab ANN Package, Available at http://www.mathworks.com/matlabcentral/fileexchange/976-ann

Matson, T.G., Sanders, B.A. & Massingill, B.L. (2005) *Patterns for Parallel Programming.* Pearson Education, Boston

Nicholls, J. G., Martin A. R., Wallace B. G. & Fuchs, P. A. (2001) *From Neuron to Brain: A Cellular and Molecular Approach to the Function of the Nervous System.* Sinauer Associates, 4th ed. ISBN: 0878934391

Rosado-Muñoz, A., Fijałkowski, A.B., Bataller-Mompeán, M. & Guerrero-Martínez J. (2011). FPGA implementation of Spiking Neural Networks supported by a Software Design Environment. *Proceedings of 18th IFAC World Congress*, Milano, Italy.

Volnei, A. P. (2004) *Circuit Design with VHDL.* MIT Press. Cambridge. ISBN: 0-262-16224-5

Xilinx (2011). Home page of Sparatn 3a FPGA devices. Available at http://www.xilinx.com/products/spartan3a/index.htm

Part 6

Transportation

14

Modelling of System for Transport and Traffic Information Management in Republic of Croatia

Dragan Perakovic, Vladimir Remenar and Ivan Jovovic
University of Zagreb, Faculty of Transport and Traffic Sciences
Croatia

1. Introduction

The traffic system is expected to provide faster, safer, more reliable, more comfortable and less expensive movement with enabling of maximal personal mobility. One of the necessary preconditions to realize these expectations is the real-time information of all those participating in the traffic processes with all the relevant data. Consequently, it is necessary to design and construct a traffic system that will, with the application of advanced information communication technologies (ICT), insure the backbone that will integrate the users' services based on the principles of intelligent transport systems (ITS), in order to be able to provide the system users with reliable, precise and timely information necessary for a more efficient realization of the transport process.

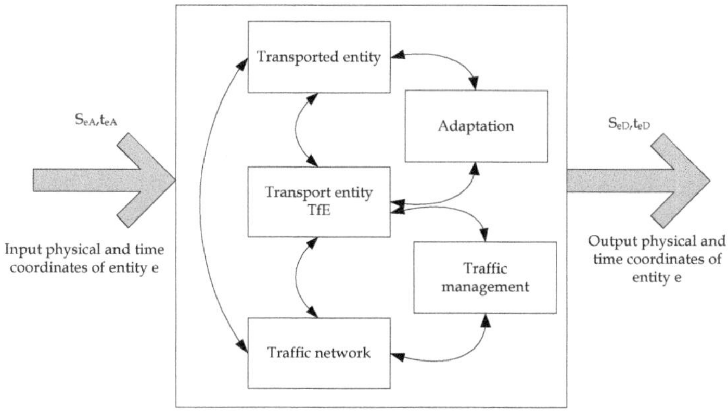

Fig. 1. Generalized model of the traffic system

With the application of modern ICT, the system for traffic information management collects, performs the adaptation process, fusion and processing (verification and addition of weights to each collected information) and carries out the distribution to all the interested users. Thus, new services may be offered that directly contribute to the improvement of traffic system quality by improving the change of space - time (s, t) coordinates (Figure 1) of the transported entity which is transported or transferred in adequate transport entity and moves with it along the transport network, with better efficiency of the consumed energy and less polluted environment.

The application of terminal user equipment and the developed program package adapted to single users and terminal equipment, allows input of information on the events by the information source or their request for certain information. The system implies the implementation of services of advanced fix and mobile communication systems, and naturally the Internet. By implementing separate modules it will be possible to process different input information received by WEB provider, WAP protocol, E-mail and SMS/MMS messages and the so-called CALL centre. With the previously mentioned communication technologies, RDS-TMC technology for user information by means of radio receiver can be applied in the distribution of data.

The information service providers can use output data from the system for traffic data management and thus expand their activity or be exclusive suppliers of traffic information with or without added value. Based on this, with intermodal pre-travel and on-travel informing of the traffic system users, significant improvements can be achieved in dynamic vehicle fleet routing, forecasting of travel time, etc.

Modern mobile communication systems offer services which allow realization of real-time informing of all the traffic system users, with the possibility of realizing financial profit of the system for traffic and transport information management (STTIM). Different business models of operation STTIM operation will attract a large number of those interested in performing this, until now in the Republic of Croatia (RH) unrecognized activity.

2. Assumptions for the application of ICT and ITS technology in the traffic system of the Republic of Croatia

Modern traffic system needs to be planned and expanded according to the principles of intelligent transport systems in order to take advantage that such a system can insure.

Intelligent transport system means the implementation of new ICT and sensor technologies in traffic and transport in order to improve the quality of traffic, transportation and transport for all participants in the process. ITS is a system that provides services to the users by means of the distributed information system using user-friendly interfaces, either in private or public sector. ITS is adaptable and open, on the one hand it offers the implementation of different technologies of interactive and multimedia characteristics, and on the other hand it guarantees full action on the entire area, from the micro location (streets, city), to regions, nations and the world as a whole.

The efficiency of the traffic system is an extremely important element of the strategic planning in which two complexes of indicators are distinguished: quality and productivity on the one hand and allocation and environmental dimension on the other. The mobility that does not meet the additional requirements of allocation and environmental efficacy i.e. which does not cover entirely the external (social) costs, is considered as unefficient.

ITS implementation in the Croatian traffic systems results in higher satisfaction both of the national population as well as the visitors, business people or tourists, i.e. in overall prosperity of the environment. The ITS role in the development of the transport of goods and services, especially tourism as one of the most perspective export products, has strategic importance of the highest level.

Obviously, modern approach to the development of the traffic systems places the emphasis on the raising the level of safety and security of the traffic system users and on the more efficient transport network. The above mentioned has to be supplemented by the solutions that will contribute to national security, the more so since at present the malicious activities are frequent worldwide, directed to disturbing the national security, by disturbing the security of the traffic system users.

The national ITS architecture should provide a general orientation in order to provide compatibility / interoperability of the system, products and services, without restricting the provided options. It gives us a common structure of developing intelligent transport systems. This is the frame around which multiple approaches to the development can be developed, out of which each one is specifically adapted in order to satisfy the individual needs of the users, at the same time keeping the advantages of common architecture. ITS is expected to be capable of action, as well as of the growth, regardless of the change in operation, organisation or technical conditions. The basic precondition that has to be insured is the system standardization, from the layer of architecture concept (i.e. referent model) to physical implementation.

The purpose of defining ITS architecture is to establish an integral architecture of the system so that individual components represent the subsystems which are used to realize the set objectives of ITS, at the same time supporting the necessary range of services, with compatibility and interoperability at all state levels. The possibility will also be provided to expand and modernize the system at affordable costs.

In defining ITS architecture it is necessary to take into consideration the following features:

- structure modularity in such a way that the functions of user services can be distributed to subsystems;
- transparency of data in relation to subsystems and services;
- standardization of mobile users interfaces, in order to realize the integrity of the services as part of an integral national and international system;
- the possibility of implementing the system in the existing traffic and telecommunication infrastructure, with the possibility of upgrading by new technologies;
- structural characteristic of the architecture enabling the implementation of a wide range of communication and information systems and protocols;
- the flexibility of the system architecture by being able to adapt to centralized or decentralized activity in order to meet different functions, different preferences and different strategies of supervision and control.

Regardless of the methodology which is used to construct the ITS architecture, the following models are used for the start:

- logic, and
- physical architecture.

Logic architecture represents the functional aspect of ITS user services. This aspect is separated from the possible implementations and requirements of the physical interface. It defines the functions or specifies the processes that are necessary to perform ITS user services and flows of information or data that need to be exchanged between these functions. The logic model is the basis for the definition of the physical architecture which can be used as the basis for the construction of the system. It is independent of the implementation and physical requirements of software and hardware. The logic model defines first of all the functions that need to be supported in order to realize the ITS user services, as well as the model of information and data exchange between these functions.

The physical architecture presents how the system will perform the defined functions. It can be divided into transport and communication subsystem that connects the transport layer elements.

3. Model of information management for the traffic system users

Over time it became obvious that some of the originally defined ITS user services were too wide in scope to be suitable for the planning of actual implementation. Therefore, a finer classification of ITS services known as "market packages" has been defined. They are harmonized so as to correspond individually or in combination to problems and needs of the transportation and transport in the real world.

Since even this finer classification of services is rather inflexible and rigid, the work presents a new model of traffic information management for the traffic system users that is based on the analysis of international experiences in the implementation of the current users services, the so-called market packages in ITS and the users requirements recognising the specific features of the Republic of Croatia.

3.1 The basic information requirements

The model predicts the meeting of all the basic requirements that refer to ITS (Federal Highway Administration, 2004), and were studied in detail in previous scientific and research works[1]. They define everything that is necessary for the configuration and realization of a reliable system that will provide the users with updated, precise and timely information.

The requirements set according to the model of information distribution to the traffic system users, i.e. the system for the management of information on the condition in the traffic system, which is based on the implementation of the developed model, refer to:

a. availability and quality of service;

Model of information distribution to the traffic system users should provide the backbone for the configuration and integration of the users services in order to enable:

- precise,
- reliable,
- safe,
- fast
    - delivery, and
    - exchange of information among the users,

timely (in real time) and in the economically justified manner.

b. interoperability of services;

The model has to enable interoperability of services regardless of the service provider involved.

c. continuity of service;

The backbone necessary for the continuity of the service between different service providers who offer the same or similar services needs to be insured.

d. growth, flexibility, and expansion of service;

---

[1] Refers to independent works or works co-authored by the author of this chapter, listed among literature

The system which applies the proposed model should support the growth, flexibility and expansion of the user services with interaction with the external service elements that do not belong to ITS. The system has to have the possibility of adaptation and implementation of the existing technologies and infrastructure to the currently existing and future ITS users services, as well as enabling the interaction with services that do not belong to the ITS services. It should also insure acceptable level of integration and continuity of operation.

e. unbiased services and support;

The model should insure unbiased provision of services and the pricing possibilities and information charge. A wide demographic segment of the society (senior people, the disabled, etc.) need to be included, as well as different geographic areas (urban, rural environments). The economic justified work and maintenance need to be supported in order to adapt it to the means of the service providers and the users' needs.

f. evolution and service;

The system should support the evolutionary nature of the users' services in order to be able to adapt to the improvement of the technology of transport and communication infrastructure, as well as the development of the means, partner arrangements of public and private sector.

g. variations in the configuration of services;

It is necessary to allow for variations in relation to the configuration of services, as well as the variations in the operation and the technologies. Also, mutual exchange of service components should be promoted (e.g. GIS maps, etc.) which has been provided by another producer.

With the abovementioned, the system for the management of information about the condition in the traffic system has to be modular, highly flexible for all the possible changes in it. Also, the principle of the availability of the equipment that is planned to be used has to be recognized, i.e. the price of the basic equipment necessary to use the new services has to be affordable in order to make the system accessible to a maximum number of users.

3.2 Presentation of negative characteristics of applying current methods of increasing the information level of traffic system users

This chapter focuses on the observed drawbacks of the existing information systems for the traffic system users with concrete examples in the city of Velika Gorica and the city of Zagreb.

The example in Figures 2 and 3 shows the information system intended for the citizens about the communal works in the city of Velika Gorica. Not only does the sticking of notices on the trees fail to comply with the minimal visual and ecological standards, this notice does not even show the complete text so that it is not clear whom the information is intended to nor is it clear what it is about.

Regarding the traffic flow within the traffic system, from this information it remains unclear whether the works will have impact on the undisturbed traffic flow at locations at which these are performed.

In order to contribute to the development of the traffic system it is necessary to develop a model whose application provides management of information about the condition in the traffic system (e.g. information on road conditions, traffic accidents, traffic congestions, bans, road works, etc.) to the interested and those participating in traffic flows. This is especially important for individual users since the application of the model contributes also to the overall satisfaction of the users.

Fig. 2. Example of informing the citizens about the works and their influence on the traffic flow (city of Velika Gorica)

Fig. 3. Example of informing the citizens about the works and their influence on the traffic flow (city of Velika Gorica)

Often, during big traffic jams, dissatisfied users can be encountered who make attempts to solve the traffic problem by themselves, which almost always results in an even greater problem. Figures 4 and 5 show the behaviour of motorists dissatisfied with the condition in the traffic system of the city of Zagreb.

Fig. 4. Example of the behaviour of citizens in traffic (city of Zagreb)

Fig. 5. Example of the behaviour of citizens in traffic (city of Zagreb)

3.3 Existing information systems

Not one of the currently available systems for information on the condition in traffic in the Republic of Croatia supports the users' requests for high quality of information, updatedness (high updating frequency of the actual information), simplicity of access (e.g. by implementation of the developed applications for mobile terminal devices), simplicity of searching real-time information, etc.

There are several providers of information on the traffic conditions, but all of them act within isolated areas of interests or they take over the information from the Croatian Automobile Club (Hrvatski autoklub - HAK) that provides information via Internet server, with textual information on the traffic condition and the possibility of displaying video-recordings of certain critical points at certain sections. This causes high redundancy and availability of old, no more actual information on the traffic condition.

Other subjects that have some information on the events in traffic (e.g. floods, traffic accidents, roadworks) publish these isolated and distribute them by radio broadcasts, web portals, etc. The numerousness of such portals confuses the potential users and questions the quality and updatedness of the published information.At the moment there is no unique portal in the Republic of Croatia that would serve for the purpose of managing information about maximum number of events that may affect the traffic flow.

The model resulting from this work provides the basis for the implementation of such systems for traffic information management (STTIM) that as the product of this work would generate and distribute the information on the traffic system included in the traffic flow, in order to improve their operation and travelling along the network.

For instance, the service of automated informing of the car motorists that they are approaching a section of the road covered by ice is not a service for the users of mobile communication systems and LBS-based but rather the so-called VMS (*Variable Message Signs*) traffic signs along the road. Although the section manager has such information, there is no developed system of distribution by means of SMS, TMC/RDS or the developed applications for terminal devices.

Currently there is a series of services that are not available in the Republic of Croatia and that the foreign users have learned to use in the countries they come from, e.g. in the area of tourist information, and the application possibilities studied in the past research by the author of this work. The additional functionality, namely, with the aim of better and enhanced information of users regarding tourist resources, is the information of the users about other facilities and activities, and full personalization of the services on the basis of users' habits. The user, i.e. the tourist staying in Croatia can obtain information about the requested tourist resources in their vicinity shortening thus the path through the traffic network and the time required otherwise to arrive to the concrete object (Peraković, Jovović, & Forenbacher, 2010).

Mobile information and communication technology determines the society and methods of behaviour since it represents a component of the personality and method of communication and operation. The potential provided by the hardware can be fully used only be adjusting to the users' requirements, by improving the software part. Modern applications have to provide the user with what they want, anywhere they want it and in the best possible way, and one of the examples is also the reminder based on the location of the user (LBS - location-based service) such as the GpsALARM application (Peraković, Remenar, & Husnjak, 2011).

It should be emphasised that the realization of a wide spectrum of services in ITS means the implementation of modern computers, sensors and communication systems. Therefore, especially important is the telecommunication infrastructure which is responsible for the ensurance of maximum distributed backbone for interconnection of terminal equipment, signaling and sensor equipment in buildings, control and information spots, as well as mobile terminals in vechicles and users in movement. Since the transport system of today cannot be imagined without information and managing systems that represent the backbone of traffic security and transport, risk management plays an important role in the development and management of all transportation systems. Identification of security risks is a process that allows quality and more cost-effective decision-making regarding the promotion and improvement of (Peraković, Kuljanić, & Šipek, 2011).

3.4 Types of information on the condition in traffic system included in the model

In order to contribute to the improvement of the traffic system, the information have to be provided for the traffic system users based on which they will be able to change from the classical sequential model of selecting the mode, route, and time of transport to a dynamic demand model.

The research in this paper is directed to the study of the source of possible impacts of, e.g.:

- incidents in traffic (traffic accidents, special transport, etc.),
- preventive and corrective maintenance of the network infrastructure (e.g. repair and asphalting of certain sections, cleaning of traffic lights, etc.),
- preventive and intervention/corrective maintenance that can have influence on the traffic flow (e.g. mowing the grass area along the road, maintenance of electrical, gas or telecommunication networks, etc.),
- planned events (e.g. soccer games, open-air concerts, transport and stay of protected persons on certain sections, etc.),
- weather conditions and weather forecasts,

on the traffic process, i.e. impact on the traffic flow along the network infrastructure.

The improvement of the existing routing systems of traffic entities along the network is possible if during the phase of interactive generation of new route plans by using heuristic methods the real-time traffic information about the potential routes are taken into consideration as well. Currently, directing of the traffic entities in the network is based either on the knowledge of the driver or the driver is left to rely on the navigation algorithms in determining the travel route. The travel routes are calculated by means of algorithms that do not take into consideration the current conditions on the roads but rather determine exclusively the shortest, simplest or fastest route. The real-time collection of data from traffic and the collection of data about the movement of each individual traffic entity (as sensor in the traffic network about the system condition) make it possible to obtain good information on the traffic condition. The mentioned information would be distributed by STTIM and they would be included in the advanced determination of routes in order to shorten the travel time, reduce road congestion, reduce the overall fuel consumption and reduce the pollution generated by fuel combustion.

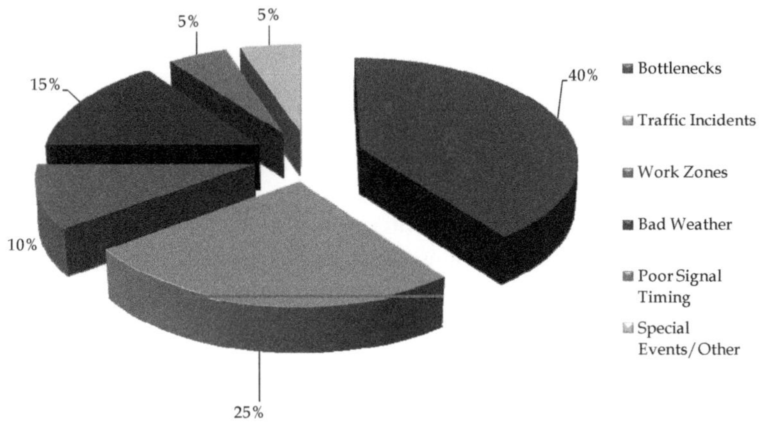

Fig. 6. Sources of traffic flow congestion

Numerous studies that have been carried out until now, have shown the impact of certain sources on the traffic flow congestion in the road (and urban) traffic flow (Cambridge Systematic, 2004). Figure 6 shows that the research scope of this work covers a significant part of potential sources of problems in road traffic. The scope of research that had not yet been covered by the actual version of the model forecasts further improvement precisely in this direction. The "bottleneck" problematic can be significantly reduced by better real-time information of the motorists based on the data collected by means of , for instance, sensors, cameras, etc. about the traffic volumes, traffic flow density, etc.
The up-to-datedness or time of information update about a certain event and the speed of receiving it by STTIM for further processing within the system is the basis for further information handling.
Based on the results of the research project Actual and Dynamic MAP for Transport Telematic Applications (ActMAP) Figure 7 shows the recommended frequency of information updates within the time frame of the observation (Dr. Bernd Thomas, 2007).

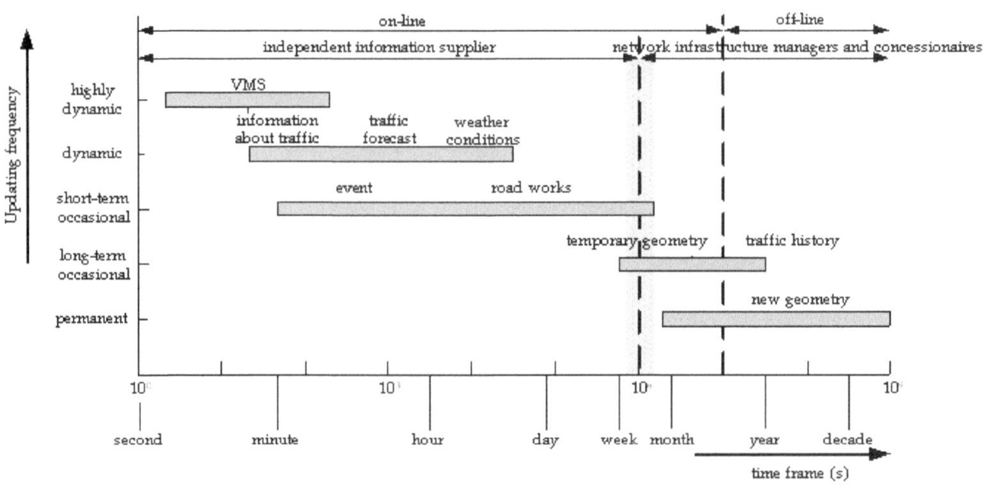

Fig. 7. Frequency of updating single types of information in relation to the time frames of observation

It may be observed that the information can be collected by independent suppliers, using online ICT technology in real time. The network infrastructure managers and concessionaires can use also the offline technologies in publishing individual, predefined and known classes of data, e.g. new road routes and archive data on the network traffic. For the temporary changes in geometry, online methods need to be implemented by all means. These methods mean the so-called internet business model as the most valuable and modern way of company operation.
Certainly, the time delay in the process of creating information within STTIM needs to be taken into consideration (collection, fusion, processing and distribution of information) that implements the developed application package IS STUP (Peraković, 2006). The time delay has been explained by Figure 8.

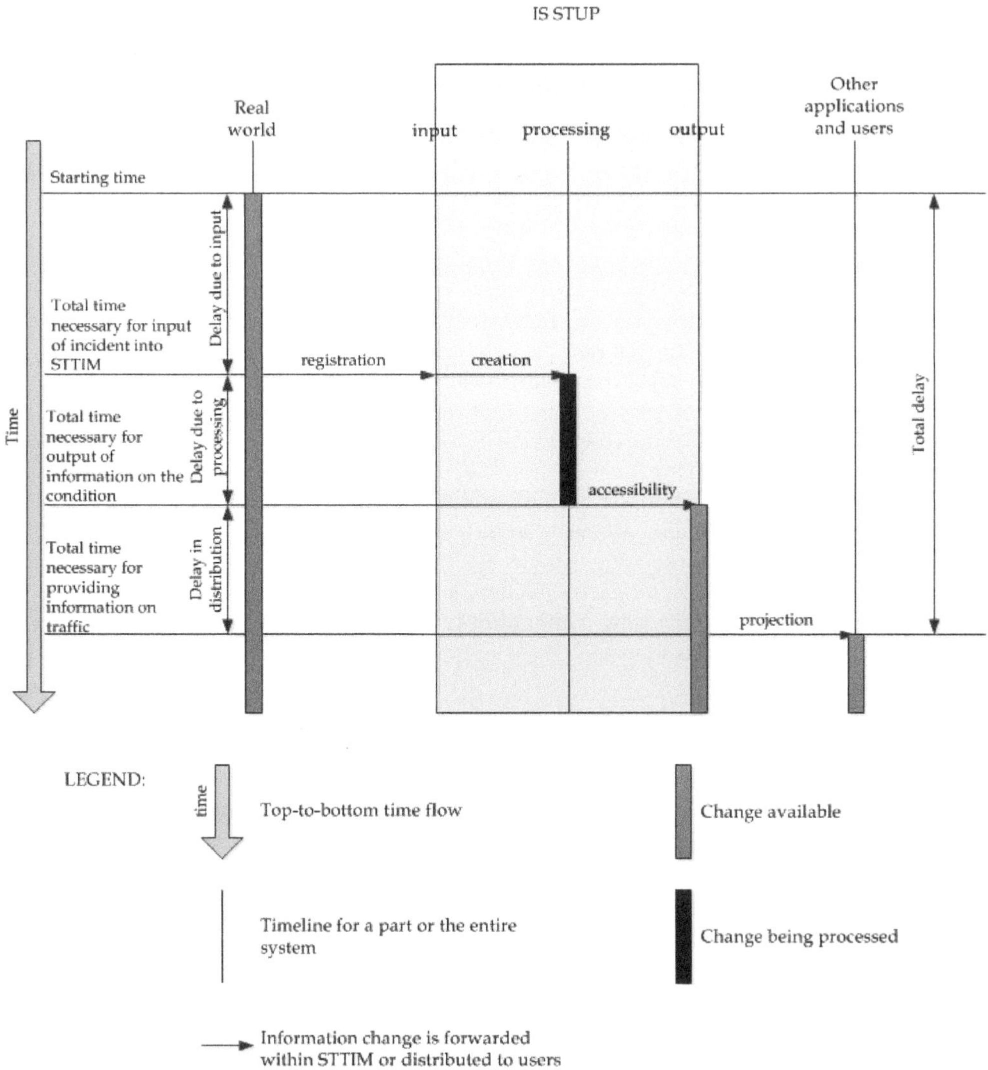

Fig. 8. Time delay in collection, processing and distribution of information

Further improvements of the proposed model need to be directed to research of the possibility of implementing advanced sensor technologies in collecting the traffic information, with emphasis on the information about the traffic flow. With certain adjustment for the adaptation of the input information and the implementation of these modules, it is possible to use the information about the traffic flow in the model of this work. The traffic parameters that the sensors need to collect are classified into two groups: traffic parameters for motorways and traffic parameters for intersections (Jelušić, Protega, & Carić, 2002).

The traffic flow parameters for the highways can be:

- intensity or volume of the traffic flow is the number of vehicles that pass the detection point within a defined time interval; usually expressed in the number of vehicles per hour (veh./h),
- average speed at the detection point is the average speed of all the vehicles within the defined period (km/h),
- occupancy is defined on a certain point, and it says how much time within the defined time period the vehicle physically occupies (covers) this point; expressed in percentages,
- the density of vehicles is the number of vehicles on a certain section of motorway and it is expressed in the number of vehicles per kilometre for every traffic lane (veh./km),
- presence (stopped vehicle) on the ramp (entry to motorway) has the value 1 if the vehicle is present and value 0 if not,
- the queue on the ramp says how many vehicles are waiting to merge the motorway traffic (vehicles) etc.

The traffic flow parameters for the intersection are:

- traffic parameters: intensity, average approach speed, vehicle density, vehicle presence, occupancy and classification are defined in the same manner, but for each approaching direction to the intersection and every lane,
- the length of the approaching queue is the number of vehicles waiting at the intersection for each cycle, for each direction, and each lane,
- the profile of the approach flow is the number of vehicles approaching the intersection in a group in every traffic lane, etc.

3.5 Potential information suppliers

The identification of all the potential users represents a precondition during the development of the model for data management on a modern traffic system. The users are selected in several classes. The problematic of identification of the potential ITS users can be found in several projects in Europe and in the world.

According to guidelines resulting from the CONVERGE (*Telematics Sector Consensus and Support Project*), the potential users of ITS services are classified into four main categories (Jesty, 1998).

These are the users who:

- want ITS systems, that will solve (or reduce) the traffic problems, or supply with traffic information the information management system about the state in the traffic system STTIM;
- build ITS systems, such as: system integrators, transport means manufacturers, telecom operators, information service providers, etc.
- use ITS systems, such as:
    a. primary users, who will benefit from the useful information generated by ITS by its operation (commuters, business users, users out of entertainment and fun, travelling salespersons, passengers with special needs, etc.), and
    b. secondary users (traffic controllers, emergency services, etc.) control the ITS system and provide a part of input data in ITS.
- define ITS and manage it (municipal, local and national authorities that are responsible for regulating the regime according to which the system will be implemented and used).

The research on the possibilities of implementing mobile Internet in intelligent transport systems (Peraković, 2003) and on the basis of KAREN's (*Keystone Architecture Required for European Networks*) (Bossan, 1999) project and ITSWAP (*Intelligent Transport Systems Over Wireless Application Protocol*) project, have resulted for the needs of the operation of the model from this work in the classification of the potential STTIM users in Croatia into eight initial classes.

By experimental implementation of the allocation IS STUP package, and additional analyses and continuous research of the users' requests and the satisfaction in experimental work, taking into consideration the specific characteristics of the Republic of Croatia as a transition country, and based on the recommendations issued by US DOT (*US Department of Transportation)* expressed in the development of national ITS architecture(Architecture Development Team, 2007), a new and expanded identification of potential STTIM users divided into nine categories has been made. Detailed identification of the users is a necessity in order to be able to form the weight value (ponder) in the development and implementation of the model, which is used to estimate a certain source of information and the quality of the information itself.

Figure 9 shows the classification of users in institutional, transportation and communication ITS layers.

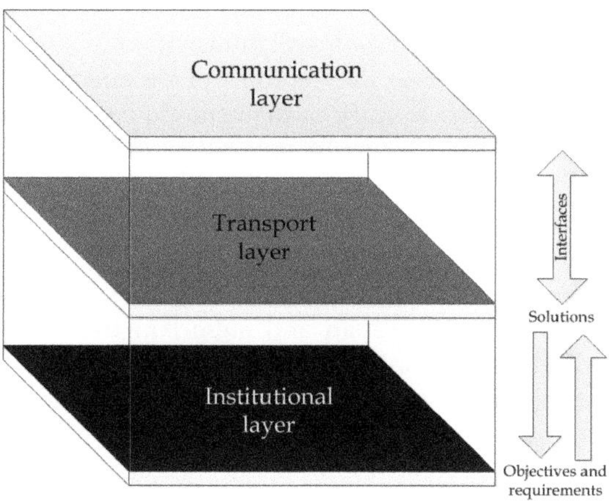

Fig. 9. ITS layered architecture

Institutional element reflects the borders of authorities (city, county, state and government offices, public institutions) and organization borders within the authorities, and includes private companies, public-private enterprises and public enterprises. This element insures financing for the performed users' services. Every authority (jurisdiction) and organization needs to determine which services should be offered and at what cost. Private companies can offer services that can exceed the jurisdiction borders regarding the market influence.

Consequently, the users can be classified into:

- Government administration and public institutions

In compliance with the Constitution of the Republic of Croatia and the regulations on the organization of work of the government bodies the users can be identified such as government administration, county administration, district and municipal authority as well as institutions in city areas.

They plan, manage and regulate traffic transport needs in their limited area, and for the performance of these activities they use, and during work generate information on the condition in traffic within a limited area.

- Providers of information services and traffic-related contents

In the Republic of Croatia there is an increasing problem in the numerousness of the information about the condition in the traffic system (Peraković, Protega, & Jelušić, 2004). For the moment, there is not one economic subject whose core activity is the provision of high-quality, updated and precise information in the field of traffic.
The information contents providers appear also as users of another provider, i.e. they are users of some information services which results in the question of the quality of the input data which they further use and publish. The problem increases with the taking over of information from unverified source and through further distribution to the interested users at the expense of quality and up-to-datedness of taking over and distributing. This very often results in the publication of information that are modified or fail to be of value any more.

- Personal users

Personal users of the traffic system services can be drivers of passenger cars, passengers in public transport means and persons with special needs (the disabled, handicapped and senior persons) who will use STTIM in order to increase their mobility in realizing their private, business or tourist interests. In order to raise the satisfaction level of the users, it is necessary to create services of pre- and on-travel information.
Taking into consideration the fact that the Republic of Croatia is at the moment one of the most desirable tourist destinations, a traffic system should be built that will be able to serve daily the passengers and transport of goods, as well as to increasingly interested tourists especially since they have already learned in their home countries or countries they had visited how to use certain classes of services.

- Public transport operators

Public transport operators can appear as users in all traffic system modes as air operators, rail operators, companies of public urban and suburban traffic, companies for intercity international road line and free transport of passengers and taxi services, VIP taxi at hotels i.e. companies that operate public transport in passenger traffic and that will use STTIM to improve the efficiency of operation and for information of their users (passengers). Moreover, they can serve as source of information (e.g. about the incidents which involve their transport means or as witnesses of a certain incident) to systems for information management regarding the traffic condition.

- Operators of commercial transport means

Operators of commercial transport means (so-called cargo transport agencies) in any transport mode are air operators, rail traffic operators, companies for intercity and international road transport of goods and logistics, companies for road transport of special cargo, postal services and companies that perform courier and package deliveries on a wide area and companies that perform package deliveries within the city area, i.e. companies that transport commercial goods, packages, business documentation, etc. and can use single ITS services to improve the efficiency of their operation in the city, intercity and international traffic. As in Item 4 (Public transport operators), all of them can serve as source of information, e.g. about the incidents which involve their transport means or they can act as witnesses of an incident.

- Public security activities

The public security activities mean those activities that would benefit from ITS by providing better services as for instance the army, police, firefighting units, emergency medical services, towing services, property insurance companies and companies providing safe transport of persons, money and securities. These are then organizations and companies that manage the fleet of emergency vehicles and that will use ITS to improve their operation. By analogy from the previous items and by recognizing the fundamental activities e.g. of police or Services for emergency medical interventions, it is obvious that they are also a good source of information on the traffic condition.

- Companies managing a fleet of vehicles for their core activities

There are numerous companies in the Republic of Croatia that have the need to use their own, big fleets of vehicles to perform their core activities, such as the companies for the delivery of their products, or ambulatory sales, companies for communal activities and companies that are engaged in maintenance and construction of road infrastructure, electrical and gas networks. They can expect higher efficiency with ITS technology application in their operation, and each of them can of its own accord join the dissertation model and become the source of information.

- Telecom operators

These are the companies that provide access to the telecommunication networks for all forms of transmitting sound and data in order to enable communication among ITS elements and the service users.

Telecom operators can participate also in the distribution of information during programs by broadcasting the news or by means of RDS system thus contributing to the level of information of the drivers by means of the so-called RDS-TMC channel (Radio Data System - Traffic Message Channel)[2] via FM radio-diffusion, for the users who have the user's device (radio receiver) that supports RDS reception.

- Independent companies

This group of users includes the companies that manufacture or maintain and improve the transport means, follow the ITS development, deal with installations, sales and maintenance of ITS equipment.

The model presented in the work assumes that all the potential ITS service users in the Republic of Croatia can also be the source of information for STTIM operation.

The generalized model of traffic data manipulation on the basis of which the STTIM operation model has been developed is presented in Figure 10.

The model understands assigning of different weight values to the information from single sources. For instance, any driver of the transport means, traffic system participant or even a accidental passer-by can report an incident situation. However, such information certainly has to be verified. If a known driver is involved (or a user of this service), the information may be assigned a higher weight value. If for instance an ambulance (for emergency medical assistance) or taxi vehicle are involved in the event there is small probability that the received information needs to be verified, i.e. that it fails to correspond to the actual situation in the traffic system. If the police wants to inform other users about the incident event on the roads, we can state with certainty that it is precise, updated and verified information of the highest weight value.

[2] For more detailed cf. official Internet server of TPEG project at the address http://www.tisa.org/

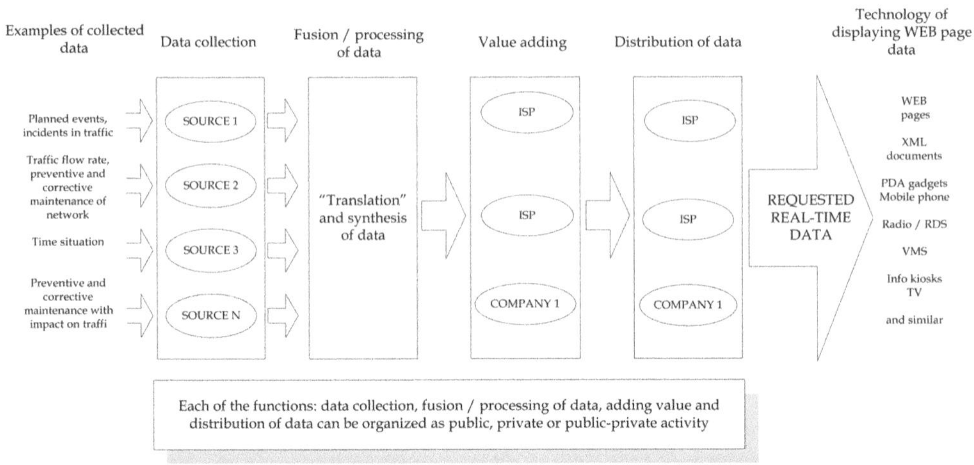

Fig. 10. Generalized model of collecting and processing of traffic data

The same analogy can be used also for the event in the category of preventive and corrective maintenance of network infrastructure and other similar events that may generate influence on the traffic flow. For the sake of example, if in the city of Zagreb the Zrinjevac Ltd. Company, whose core activity is to take care of the green areas, publishes the information in the city about the mowing of the grass areas along the road, the authenticity of this information need not be checked. If the information is not published, every conscientious citizen can report the event in which the intensity of the impact of events on the traffic flow is described, but then such information should be assessed and it should be assigned a certain weight value.

Figure 11 shows a generalized presentation of information and communication connection of the traffic system users included in the processes of *production* of real-time information on the traffic system condition.

The area of research encompasses the problematic of the fusion and processing of data on a generalized level, and the emphasis is oriented to the identification of the source, technology of collecting and distributing information to the interested parties. It should also be mentioned that it is precisely the problems of fusion and processing that are extremely important in this process. Numerous further studies will have to be oriented to this area, so that the users would receive high-quality timely information.

It should be emphasised that each function in the work of the generalized model can be organized as public, private or public-private activity.

Recommendations for the development of legal standards and definition of business models in which also the private sector may participate in the development of telematic-based systems of real-time information of all those interested in traffic condition (*Traffic and Travel Information* – TTI) were given by the European Commission on 4 July 2001, No. 201/551/EC (European Commission, 2001).

The business model for traffic information management has been analyzed in several research works and can appear in various forms (US Department of Transportation, 2004). The model which stands out with its advantages, and is applicable in the Republic of Croatia is the public-oriented business model presented in Figure 12.

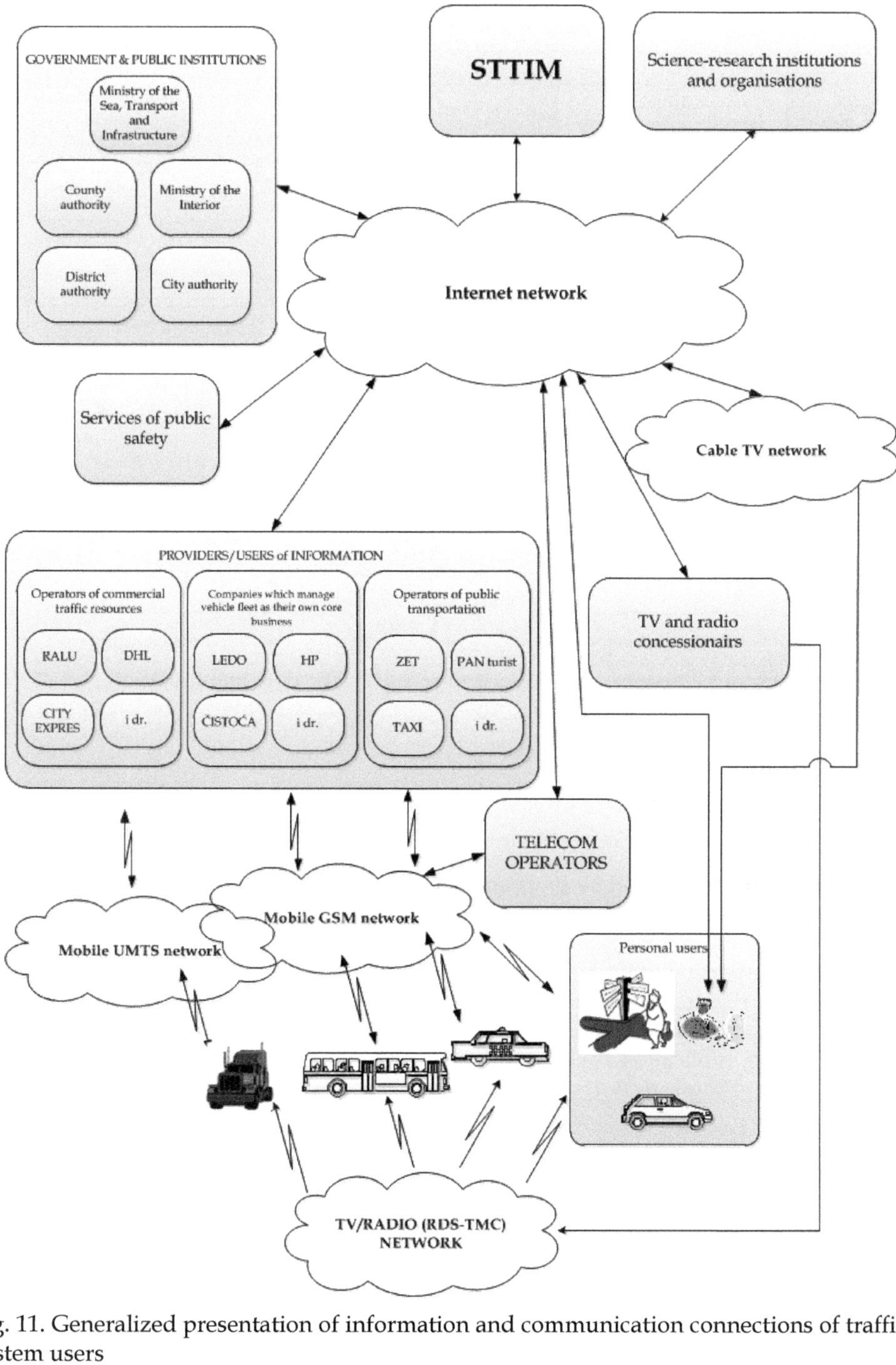

Fig. 11. Generalized presentation of information and communication connections of traffic system users

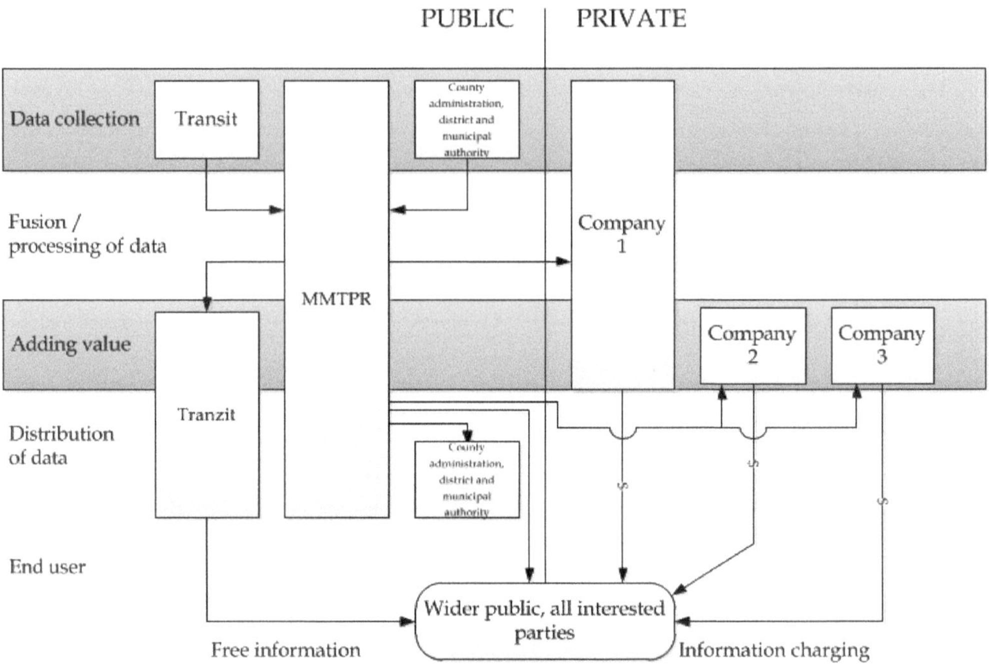

Fig. 12. Public-oriented business generalized model of traffic data manipulation

4. UML Model with notation

In this section the model, which is the result of the research using UML notation, is presented.

4.1 Presentation of the model by diagram of usage case

In the generalized presentation Figure 13 shows different groups of users who use the services of the traffic information management system (STTIM). The basic categories of services include reporting of individual events depending on the authorities and requests on the condition in the traffic system which are distributed with or without charge. The synchronization and harmonization of work of all the programme package modules are preformed by the application called STUP (Cro. Stanje u prometu - Condition in traffic) (Peraković, 2006). The protection of data during input, processing and distribution for the services of special purposes such as police, army, etc. has also been planned.

4.2 Presentation of the modular structure of STUP application

Figure 14 gives a presentation and the logic model of the operation of the developed software package called STUP AP (Peraković, 2006). Eight synchronized applications follow the input, processing and distribution of information of the known and unknown users and users with special requests regarding data security.
The basis of the STUP operation is the database into which every input of a new event and every new request for a service is entered.

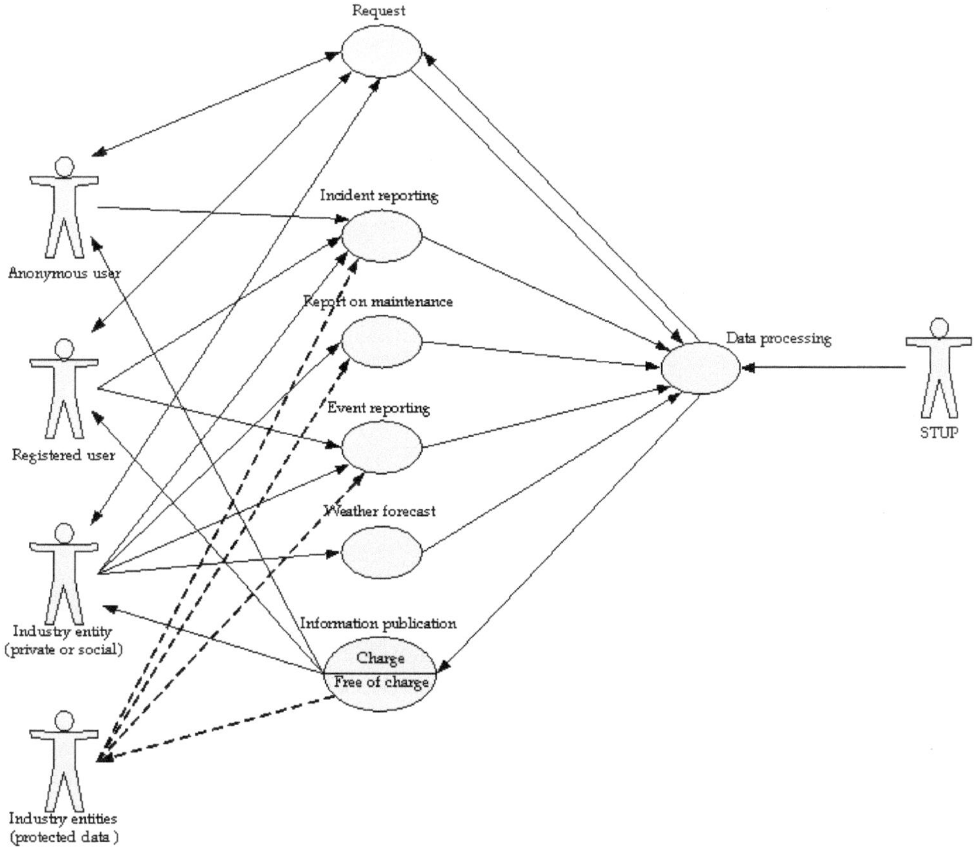

Fig. 13. Model presentation by UML diagram application of case of usage

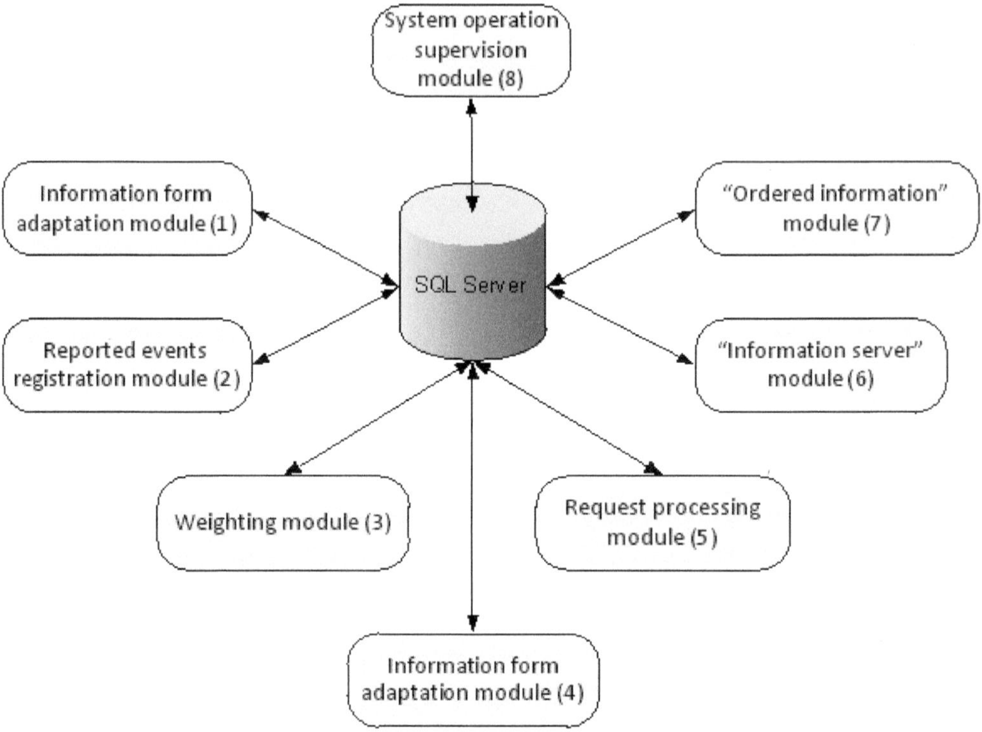

Fig. 14. Modular structure of STUP application

5. Conclusion

The work presents the results of several years of research of the very actual problematics regarding information of traffic system participants with emphasis on the possibility of applying advanced information and communication systems and services with the aim of providing reliable, precise and timely information necessary for harmonized operation and sustainable development of the transport system.

The traffic condition information is the base for creating the ITS service classes such as pre-travel information for travel planning and on travel information, information about public transport, traffic information for motorists during driving about road conditions, services of mobile charging, tourist information, etc.

A presentation is given regarding identified and classified potential sources of information, as well as of the traffic system users and the potential new users of ITS information services.

As the biggest network of computer networks, the Internet is the most suitable infrastructure which provides the possibility to realize access to ITS information subsystems. The factors of success and spreading of the Internet lie in its exploitation properties, flexibility and ease of implementation and application, and simple and relatively financially inexpensive access with the application of the today easily accessible terminal equipment.

With the development of mobile communication systems and the provision of new services, the Internet users can be both the motorists and the passengers, i.e. users on the move, and not just stationary persons. The advanced mobile communication systems provide services that make it possible to realize the real-time information of mobile users of the traffic system, with the possibility of realizing the financial benefits in the STTIM operation.

The work proposes and describes one of the possible business models of forming and operation of STTIM (public-oriented business model). Further research need to be oriented to research of the possibility of implementing others as well, such as: business model based on the franchise contract and market-oriented business model of traffic data manipulation.

As the transport system of today is unimaginable without information and managing systems that represent the backbone of traffic security and transport, IT risk management plays an important role in the development and management of all transportation systems. Identification of security risks is a process that allows quality and more cost-effective decision-making regarding the promotion and improvement of security.

Global business process automation, and mass introduction of information technologies in business companies from the transport environment, sets major challenge to quality management of IT systems which are dependent on the business processes.

By the advent of new services in the communication systems, based on the XaaS principles, the management problematic and end distribution and (optional) charging traffic information gain a new dimension and require continuation of research in this direction.

XaaS is a very suitable platform, both for private companies and corporations, and for the government. The advantages such as reliability, conditional security, high availability, low maintenance level (depending on the model) and scalability, are extremely interesting characteristics of XaaS platforms. However, attention should be paid to the problems of security and privacy of data and to the SLA agreement conditions. Regarding the delicate nature of data handled by ITS, additional efforts should be invested in defining the Service Layer Agreement (SLA), End User Licence Agreement (EULA) and Terms of Service (TOS) with special reference to data security and privacy. This method of negotiating services (by defining SLA, EULA and TOS) can only be observed by telecom operators so that in this case they are ideal partner for the provision of STTIM XaaS platform

6. References

Architecture Development Team . (2007). National ITS Architecture Mission Definition. Washington: Federal Highway Administration, US Department of Transportation.

Bossan, R. (1999). KAREN Project (TR 4108). Telematics Application Programme.

Cambridge Systematic, I. (2004). Traffic Congestion and Reliability: Linking Solutions to Problems - Final Raport. Washington D.C.: Cambridge Systematics for Federal Highway Administration, US Department of Transportation.

Dr. Bernd Thomas, N. A. (2007). D3.2 ActMAP White Paper and Interfaces to the FeedMAP framework. ERTICO, NAVTEQ, TELEATLAS, BMW Group Forschung und Technik, DaimlerChrysler, Centro Ricerche FIAT, Volvo Technology, Magneti Marelli Sistemi Elettronici, NAVIGON, PTV, ICCS, Swedish Road Administration, Oberste Baubehörde im Bayerischen Staatsministerium.

European Commission. (2001). WHITE PAPER – European transport policy for 2010: time to decide. Office For Official Publications Of The European Communities, European Communities.

Jelušić, N., Protega, V., & Carić, T. (2002). Značaj senzora u ITS sustavima cestovnog prometa. Proceedings of 10th International Symposium on Electronics in Traffic ISEP 2002. Ljubljana.

Jesty, P. H. (1998). Guidelines for the Development and Assetment of Intelligent System Architectures. Framework IV Transport Telematics Projects CONVERGE.

Peraković, D. (2006). Model of distribution of information to users of transport systems, Ph.D. thesis. Zagreb: Universitiy of Zagreb, Faculty of Transport and Traffic Sciences.

Peraković, D. (2003). Mogućnost primjene mobilnog interneta u inteligentnim transportnim sustavima, Master's thesis. Zagreb: Universitiy of Zagreb, Faculty of Transport and Traffic Sciences.

Peraković, D., Jovović, I., & Forenbacher, I. (2010). Improving Croatian Tourist Information via Location-based Public Service Through Mobile Phones. MeTTeG10 - 4th International Conference on Methodologies, Technologies and Tools enabling e-Government. Olten, Switzerland.

Peraković, D., Kuljanić, M., & Šipek, K. (2011). Research issues of governance, risk management and compliance in information and communication infrastructure of ITS. Proceedings of the 19th International Symposium on Electronics in Traffic ISEP 2011. Ljubljana.

Peraković, D., Kušan, I., & Špoljarić, M. (2004). Application of New Communication Ethnologies in Improving the Tourist Offer of the Sisak-Moslavina County. PROMET-TRAFFIC-TRAFFICO. Proceedings of International Scientific and Professional Conference New Technologies in Improving Traffic Safety. Vol. 16, (str. 29-35).

Peraković, D., Protega, V., & Jelušić, N. (2004). Influence Of The Number Of Traffic Information Providers In Croatia On The Level Of Information Among The Traffic Participants. Proceedings of 12th International Symposium on Electronics in Traffic ISEP 2004, (str. U12). Ljubljana.

Peraković, D., Remenar, V., & Husnjak, S. (2011). Reminder based on the users's location. ICTS 2011 - Maritime, Transport and Logistics Science, (str. 1-9). Potrorož.

US Department of Transportation. (2004). Washington State Transportation Center: Choosing The Route To Traveler Information Systems Deployment, Decision Factors For Rating Public/Private Business Plans. Washington D.C.: US Department of Transportation.

Part 7

Water Plant Technology

15

Modelling of Critical Water Quality Indicators for Water Treatment Plant

Adam Rak
Opole University of Technology,
Opole,
Poland

1. Introduction

A technology system water treatment requires pre-determined information on critical quality indicators for the water. A neural networks can be designed to assess this.
Use of an artificial neural networks [ANN] to analyse unit and technology processes within water management, treatment and distribution led to consideration of the possibilities of using an ANN for forecasting water quality and creating a model which would permit this. Siwoń et al. (2008) examined the results of experiments carried out using an ANN and indicated that the first use of an ANN in modelling and forecasting water system operation occurred in the 1990s. A lot of work was done to assess whether the ANN might be useful in modelling and forecasting the distribution as well as the sale and production of water. A similar analysis was carried out by Bardossy and his colleagues (2009). Additionally, a Variable Input Spread Inference Training [VISIT] programme was created, which allowed for the automatic examination of the ANN model being suggested for implementation.
Artificial neural networks were also used to establish the essential doses of chloride needed within large water systems and to forecast the contamination amount of left over chloride. Koo et al. (2008) examined the possibility of using an ANN to forecast the amount of chloride remaining in water pipes. A model was then created. It analysed five scenarios regarding the use of chloride subject to a dose of chloride being put into the water system. An established model indicated the amount of chloride to put into a water system, subject to the water temperature and the amount of water being put through the system, so that the amount of chloride remaining does not exceed the allowed dose. The Pearson correlation coefficient, R, between the examined and observed data was calculated; its value for the model being analysed was $R = 0.959$.
An ANN has recently been introduced for the operation of a water system. It has been widely discussed not only in foreign literature (Camarinha-Matos & Martinelli, 1998; Zhou et al., 2000, 2002), but local works as well. For example, Sroczan and Urbaniak (2004) as well as Zimoch and Kłos (2003) suggest using an ANN to monitor, steer and operate water supply systems, and for water protection. Dawidowicz (2005) carried out a number of numerical tests which allowed verification of the idea of using an ANN in assessing the 'Diametre Nominal' (DN) of the water pipes and for taking any hydraulic measurements. Licznar and Łomotowski (2004) obtained very good results in forecasting the daily amount of water distributed in a large scale system of water pipes using various ANN technologies.

Deveughèle and Do-Quang (2005) supplied the results of using an ANN to forecast the use of coagulant in a surface water treatment process. A model was created which enabled the optimal dose of coagulant to be established subject to the parameters of the raw water. The model's prototype was fully implemented in a French water treatment plant. It enabled the amount of coagulant being used to be reduced by approximately 10%. Cougnaud et al. (2005) used an ANN to forecast the absorption capacity of active carbon subject to the concentration of pesticides in the water. It was noted that the operating parameters - contamination, concentration and pace of filtration - had an effect on the efficacy of reducing contamination. The ANN model established the relationship between the features of active carbon and the adsorption of pesticides. There was a mutual relationship noted between the concentration of pesticides, pace and filtration time. It was a strong linear relationship; the coefficient of determination, R^2, being 0.985. An ANN was also used to assess the maximum level of water contamination. Brion and Lingireddy (2003) indicated that an ANN could be useful when examining the micro-biological changes of water. The ANN provided the ability to examine the mutual relationships between large numbers of parameters. It allowed this information on the predicted micro-biological features of the water to be supplied to the system operator. Using an ANN for modelling allowed the highest level of micro-biological water contamination to be assessed with 90% accuracy. Gavin et al. (2003) used an ANN to model water supply systems. The networks were used to predict the salinity of rivers in South Australia. The salinity was forecast 14 days in advance using a linear model. Strugholtz et al. (2009) used an ANN to examine a water treatment process as a subject for both raw water quality and technology process. It established the optimal technology process and predicted its course. That was one stage of the examination. The other stage developed a model to optimise operational costs. The ANN was used to set the parameters of the filtration and to optimise the doses of the reactants; the result was a lowering of the costs by 15%.

These various examinations and analyses have led to the conclusion that ANN models are as good as statistical ones.

2. Experimental

2.1 The site

The Sosnówka reservoir was constructed in the 1990s to supply water to the municipality of Jelenia Góra. It supplies raw water to a local water treatment plant. Sosnówka reservoir retains water from the catchment areas of the Czerwonka and Sośniak streams and the Sosnówka tributary. The total catchment area is 15.3 km^2.

The flow of the Czerwonka stream within the reservoir's section, Q_{SN}, is 0.038 m^3·s^{-1}, with the average flow, Q_{aver}, being 0.192 m^3·s^{-1}. The Czerwonka stream catchment area is 5.5 km^2 and it is separated from the Sośniak stream by the reservoir catchment area. It indirectly supplies water to the object under study (Figure 1).

The total capacity of the reservoir is 15.4 million m^3 and it has a maximum surface area of 178 ha. The depth at the lower dam is 13.5 m and the total average depth is 8.15 m. The amount of water from the reservoir used for consumption is 11 million m^3 (Photo 1). It is assumed that at least 70% of the total amount of water required can be supplied from the Podgórna river catchment area to the reservoir in the even that the Sosnówka reservoir water is completely used (Photo 2). So the Podgórna River is the main source of water for the reservoir (Rak, 2008).

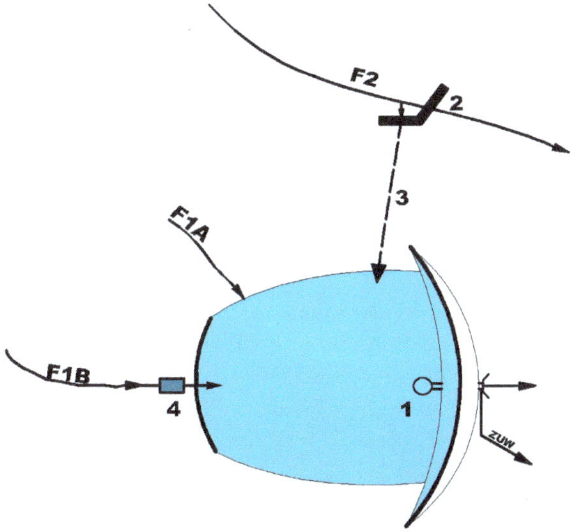

F1A -Czerwonka stream catchment area; F1B -Sośniak stream catchment area
F2 - Podgórna River catchment area.

1. Extraction construction of the reservoir including the water catchment area.
2. Construction of water distribution system at Podgórna River.
3. Transfer canal.
4. Initial reservoir including a pumping station.

Fig. 1. Schematic of the water distribution of the 'Sosnówka' reservoir:

Photo 1. Retention reservoir "Sosnówka"

Photo 2. Podgórna River

In the reservoir area there were few anthropogenic contamination sources. The area is a mix of woods and meadows, the soil comprises mountain peat deposits (Institute of Environmental Protection Wrocław [IMGW], 1986). Factors which have a positive influence on changes in reservoir water quality include the wooded character of the terrain, a low annual water exchange rate and an absence of indirect contamination sources. The Schindler index value of 1.39 ranks the reservoir as being in the first category regarding susceptibility to degradation. However, given the increasing amount of water being taken for consumption purposes, more water is being supplied from the Podgórna River. The water factors have change and the Schindler index value has risen to 4.62. This has resulted in the reservoir being downgraded regarding susceptibility to degradation to the second category.

2.2 Methodology

Some parameters of the raw and treated waters at the treatment plant are automatically monitored. These include the temperature, pH, muddiness, colour and conductivity. Between November 2007 and October 2008, the first operational phase of the water treatment plant, certain physical and chemical examinations were conducted every few days. These tests included measuring the water temperature, turbidity, colour, pH, general hardness, alkalinity, iron, manganese, chlorides, ammonium and nitrates levels, oxidisability, dissolved oxygen content, conductivity and phosphates. The technology system allowed examination of such unit processes as sieving, pre-ozonation, coagulation, correction of pH, flocculation, accelerating anthracite and sand deposit filtration, derivative ozonation, sorption on active carbon, final correction of pH and hardness as well as disinfection of the treated water (see Figure 2).

The technology system is designed to allow different combinations of the various unit processes depending on the quality of the water.

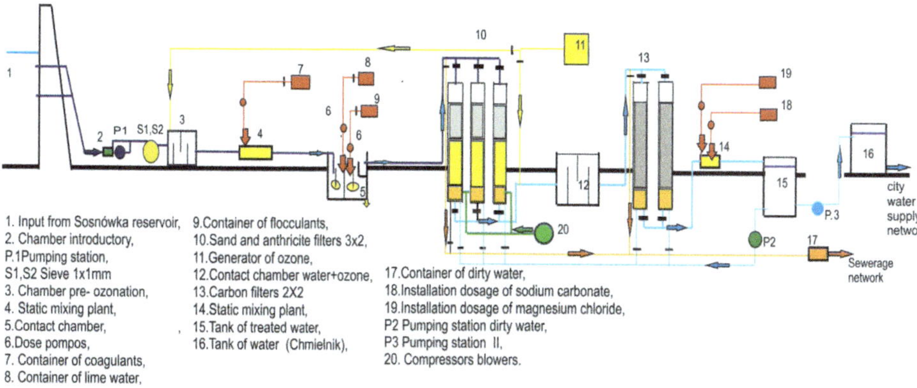

Fig. 2. The technology of the water treatment process

The unit processes available within the technology system are presented in Table 1.

Unit processes	Technology system					
	W1	W1A	W1B	W2	W2A	W3
Sieving with a 1 mm by 1 mm sieve	+	+	+	+	+	+
Pre-ozonation	+	+	+	+	+	+
Coagulation with aluminium sulphate	+	+	+			
Flocculation	+	+	+			
Correction of pH	+	+				+
Anthracite and sand deposit filtration	+	+	+	+	+	+
Derivative ozonation	+		+	+	+	+
Active carbon deposit filtration	+	+	+	+	+	+
Final correction of water pH	+	+	+	+		+
Disinfection	+	+	+	+	+	+

W1, W1A, W1B, W2, W2A and W3 – different possible sequences or systems of unit processes used for treatment of the water

Table 1. The separate unit processes of the technology system

The results of the technology examinations conducted during the first operating period of the water treatment plant resulted in the decision to opt for the W2 sequence. This included such unit processes as sieving, pre-ozonation, accelerating filtration on anthracite and sand deposits, derivative ozonation, sorption on active carbon, final correction of water quality and disinfection. When the pH value was low, technology system W3 was implemented as this included correction of the pH of the water.

3. Results

During the first period of operation of the treatment plant, the water taken from the reservoir was analysed to establish the characteristics of the raw water. Table 2 displays the typical minimum and maximum values of the contamination factors of this water.

Water quality indicators	Unit	Sosnówka reservoir			
		S_{min}	S_{max}	S_{aver}	Standard deviation(σ)
Water temperature	°C	3	15	9.2	1.620
Turbidity	$mgSiO_2 \cdot dm^{-3}$	0.57	3.81	1.56	1.052
Colour	$mgPt \cdot dm^{-3}$	5	14.9	8.21	2.791
pH	pH	7.2	8.4	7.63	0.366
General hardness	$mval \cdot dm^{-3}$	0.62	1.03	0.85	0.072
Nitrate nitrogen	$mgN\text{-}NO_3 \cdot dm^{-3}$	2	5	2.44	1.300
Chlorides	$mgCl \cdot dm^{-3}$	4	6.8	4.95	1.321
Oxidability	$mgO_2 \cdot dm^{-3}$	1.0	5.22	3.458	1.532
Conductivity	$\mu s \cdot cm^{-1}$	98.28	122.50	110.88	6.692
Dissolved oxygen	$mgO_2 \cdot dm^{-3}$	9.4	12.4	11.18	1.172
Fe	$mgFe \cdot dm^{-3}$	0.05	0.136	0.066	0.027
Mn	$mgMn \cdot dm^{-3}$	0.012	0.036	0.021	0.007

S_{min} – minimum value of the parameter for the Sosnówka reservoir,
S_{max} – maximum value of the parameter for the Sosnówka reservoir,
S_{aver} – average value of 2005-2006 samples of the parameter for the Sosnówka reservoir.

Table 2. Typical values of chosen water quality indicators during initial period of operation

During the first period of operation of the plant, given the good quality of the water being extracted, the technology tests that were carried out within the technology system included sieving, pre-ozonation, accelerating filtration on anthracite and sand filters, derivative ozonation and sorption on active carbon (W2A - excluding final correction of the water pH). During the second period, final correction of water quality was included in the technology system; the W2 system was applied. The flow rate of the water during treatment was between 172 $m^3 \cdot h^{-1}$ and 202 $m^3 \cdot h^{-1}$ and the speed of filtration on the anthracite and sand filters and with the active carbon was between 4.9 $m^3 \cdot h^{-1}$and 5.7 $m^3 \cdot h^{-1}$. Certain amounts of ozone and other substances were applied within the technology system. These included:

- in the pre-ozonation process, between 1 and 2 mg $O_3 \cdot dm^{-3}$ were added,
- in the derivative ozonation process, up to 1 mg $O_3 \cdot dm^{-3}$ was added,
- to correct the pH of the treated water, 1.0 mg dm^{-3} of sodium carbonate was added,
- for stabilization of the treated water: 1.0 mg dm^{-3} of magnesium chloride was added,
- in the disinfection process, between 0.8 and 0.9 mg $Cl_2 \cdot dm^{-3}$ (in the form of sodium hypochlorite) was added.

By examining both the raw and the treated water it was possible to establish the levels of reduction of the indicators being monitored for a particular water temperature and the dose of ozone given relative to the raw water colour. Following the W2 technology system, the water colour was reduced from 65% to 98% at water temperatures between 4°C and 15°C. The average colour reduction level was 80%. The relationship between water colour and the temperature of the treated water was noted. When the temperature was lower, there was a 60% reduction in the water colour. However, when the temperature was 6°C the reduction in water colour increased to 80%. Further reduction in water colour was noticed at temperatures higher than 14°C. The possibility of constant monitoring allowed both the water temperature and pH to be analysed in week long cycles. In spring time, when the level of water in the reservoir was high, the pH value of the water was quite high. In April the pH value of the water was greater than 8.85 while the water temperature was quite low (between 6°C and 7°C). The high pH value required a technology process without a pH correction stage in it. In May when the temperature increased to 15°C, weather conditions stabilised, and the pH level of the raw water was considerably reduced. However, it was still sufficiently high that a technology system without a pH correction stage could be used. After each of the unit processes within the water treatment process it was noted that there were changes in the pH of the water. With every test carried out the pH changed. For example, after the pre-ozonation process the pH was increased by about 10%. The filtration processes, in contrast, decreased the pH to between 6.55 and 6.90. Disinfection with sodium hypochlorite increased the pH slightly - to 7.05 (see Figure 3).

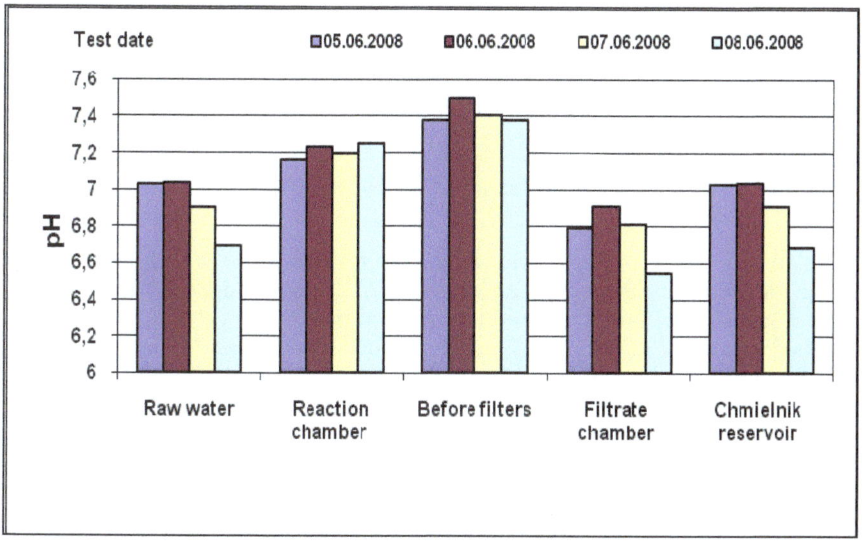

Fig. 3. Changes in the pH of the water at various water treatment processes

Changes in pH, the low general hardness of the water and its alkalinity during the treatment process were analysed to assess the stability and aggressive character of the treated water. The pH at saturation (pH_S) and the Langelier (I_L) and Rezner (I_R) indices were determined. The values of the water temperature, pH, pH_S, I_L and I_R, are shown in Figure 4. It is clear that both the raw and the treated water are aggressive in character in the W2A technology system. These indicators changed their values, particularly in May when the temperature increased from 8°C to 15°C. However, until May when the water temperature was between 4°C and 8°C, there was little increase in the aggressiveness of the character of the treated water compared with that of the raw water. Moreover, with higher temperatures and increasing pH values (up to pH = 8.4), the treated water became unstable and more aggressive in character. At that time the I_L index was negative (- 1.6), while the I_R index value was 10.5. The results of the examinations were the basis for implementing a water stabilization process by dosing the treated water with magnesium chloride and sodium carbonate before it was disinfected. The tests in the W2 technology system were carried out using fixed doses of $MgCl_2$ and Na_2CO_3 - 0.5, 0.75 and 1.0 $mg \cdot dm^{-3}$.

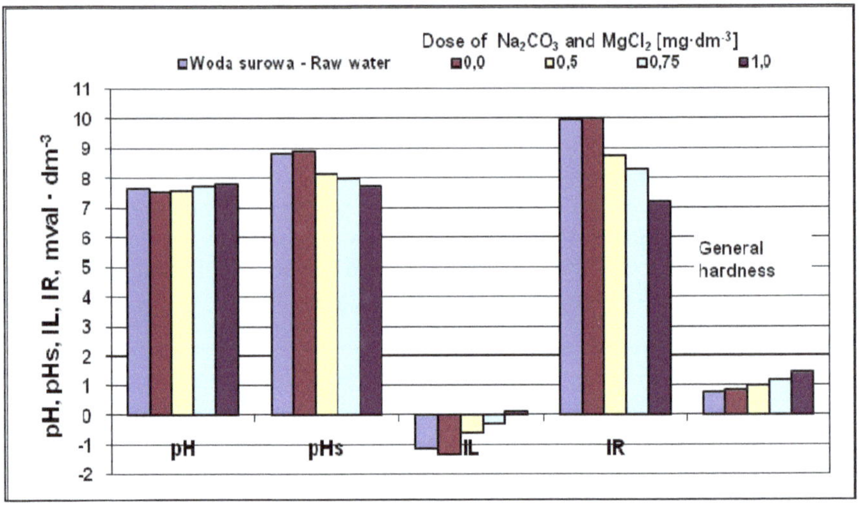

Fig. 4. Changes in pH, general hardness, and the pH_S, I_L, and I_R indices of the water for various doses of $MgCl_2$ and Na_2CO

The results obtained (see Figure 4) show that the treated water is stabilized by increased doses of magnesium chloride and sodium carbonate. Using 0.75 $mg{\cdot}dm^{-3}$ of both magnesium chloride and sodium carbonate resulted in a general hardness level of 1.2 $mval{\cdot}dm^{-3}$. The I_L index reached a value of -0.3 and the I_R index reached a value of 8.3. However, both indices changed their values when the amounts of magnesium chloride and sodium carbonate added reached 1.0 $mg{\cdot}dm^{-3}$ becoming 0.1 for I_L and 7.2 for I_R.

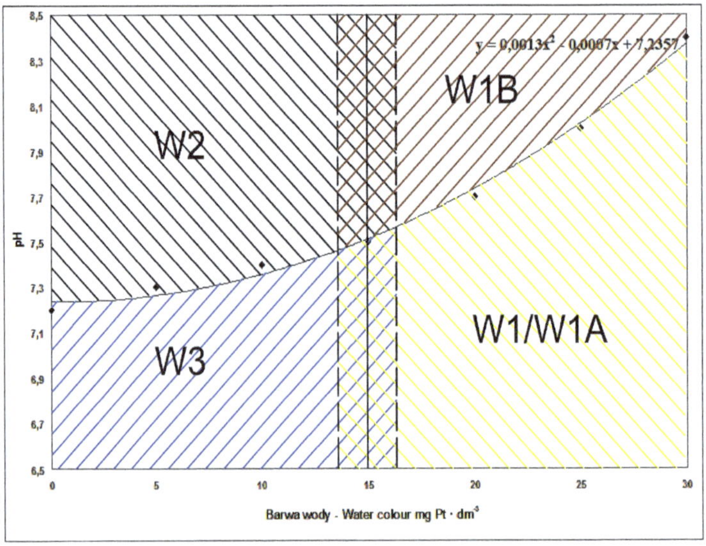

Fig. 5. Nomogram for selecting the technology system subject to raw water colour and pH values

The research identified limit values for the colour and pH of the raw water where there is a possible choice of treatment process. Those areas are shown in Figure 5. The nomogram governs the selection of the appropriate technology system subject to the colour and pH value of the reservoir water. Each of technology systems, W1, W2 and W3, can be modified with the inclusion of the derivative ozonation process, subject to water temperature. In winter time when the temperature drops to between 9°C and 12°C, derivative ozonation does not have to be implemented.

4. Discussion

The analysis led to the conclusion that the critical indicators that have an effect on the selection of the technology system to apply are the water colour, pH and temperature. Other contamination indicators are not relevant to the choice of the technology system of water treatment processes. The study indicated that both the raw and treated water were aggressive and unstable in character. Therefore, each of the technology systems includes processes which make the water stable and harder.

From a water management and distribution points of view, the quality of the water supplied to recipients is as important as the operational costs of the system. The costs depend on the technology system being used within the water treatment process. The technology system selected will generate operational costs subject to certain quality parameters of the raw water and the length of time that they are prevalent. These costs depend on the duration of the water treatment process within the appointed technology system. Therefore, it is important that the water treatment plant's operator be supplied with a prognosis of the contaminants of the raw water, as these determine the choice of technology system to implement. This forecast is critical for the water treatment plant under consideration. Pawełek and Bergel (2008) suggested a methodology to establish the duration time of a critical indicator. This led to the development of a readiness indicator, K_g, for critical water quality indicators. This readiness indictor is defined as:

$$K_g = \frac{T - \sum t_i}{T} \tag{1}$$

where T is the examination time. For operational cost purposes this is taken to be 365, $\sum t_i$ is the total time for any limit values of the critical indicators designated for a certain type of a technology system.

Based on equation (1), readiness indicators were calculated for the critical indicators of water colour, pH and water temperature. These are shown in Figures 6, 7 and 8. The total duration time of the limit values was calculated for each of the critical indicators as a curve function of time per examination year.

The values for the readiness indicator for pH indicate that technology systems W2 and W3 (Figure 6) can be used for 302 days a year. For the rest of the year, the W1 technology system should be applied as coagulation needs to be implemented. Both W2 and W3 technology systems can be modified with the use of the pH correction. Figure 7 shows that the water pH correction needs to be implemented for 232 days per annum (63%). All technology systems (W1, W2 and W3) can be modified subject to the temperature of the water being treated. Figure 8 shows that the longest that the water treatment process W1A (excluding derivative ozonation) can be applied is 151 days a year.

The readiness indicators for water colour, pH and water temperature allow factors to be set to adjust the operational costs of the technology system appropriate to the critical values of indicators as given in the chart of Figure 5.

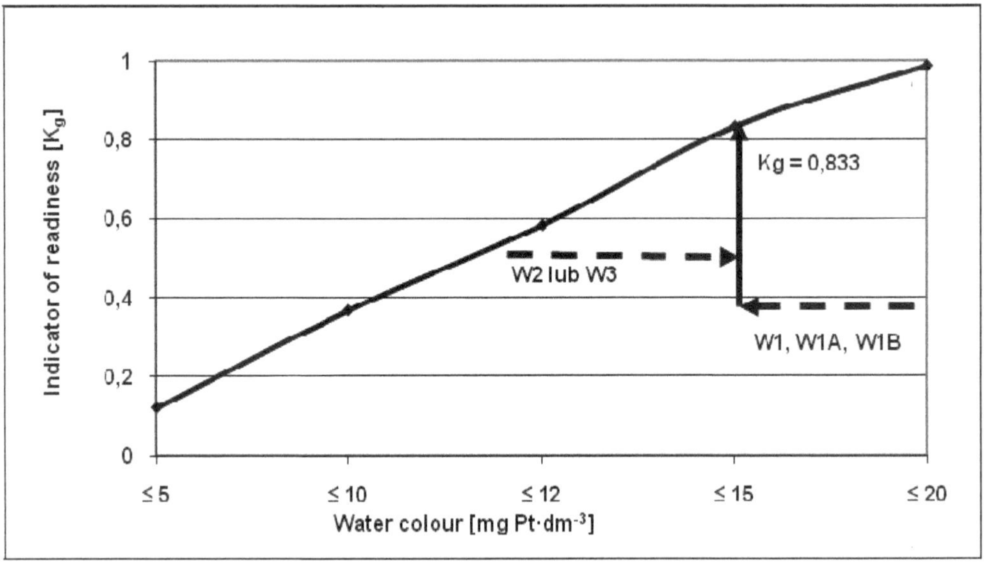

Fig. 6. Readiness indicator for water colour

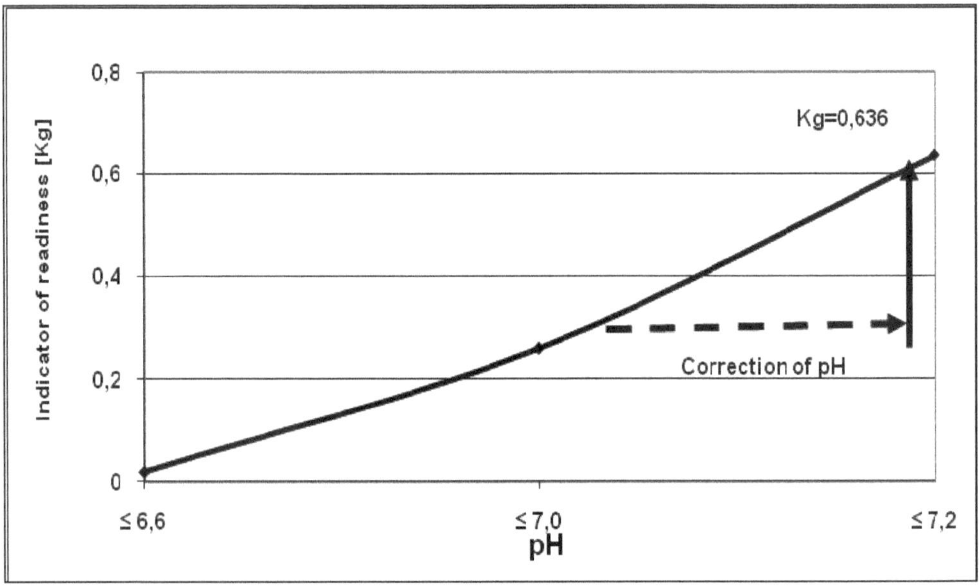

Fig. 7. Readiness indicator for water pH

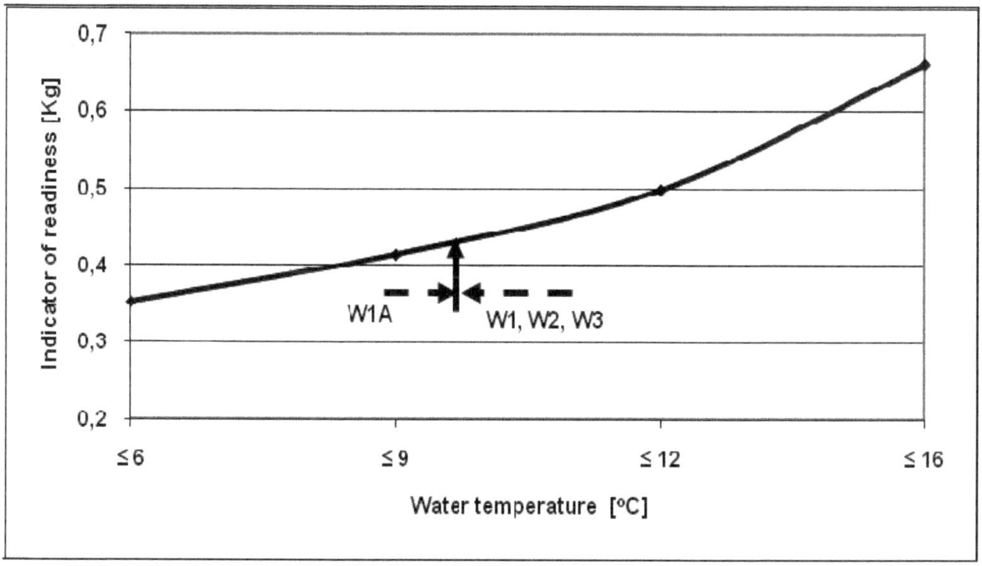

Fig. 8. Readiness indicator for water temperature

It is essential that the operator of the water treatment plant has a prediction of the water quality parameters which govern which type of technology system is set in motion. Short term prognoses of water quality can be used in the modelling of water management in various situations depending on water conditions, the ecological status of the inflow and the amount of water taken out for consumption.

Flexible Bayesian models (FBM) are the neural networks which were used to forecast the reservoir water quality indicators (Neal, 2000). A numerical analysis applied to a regression model in which the target variables, such as colour (S_1), turbidity (S_2), pH (S_3) and water hardness (S_4), are subject to input variables such as time (Z_1), reservoir water level (Z_2), daily precipitation amounts in the catchment area (Z_3) and water temperature (Z_4). Adjustments to the network were based on historic data gathered from assessments carried out for 365 days from November 2007 till 31 October 2008. Verification of the regression model was carried out using the same data. The parameters of the network were established at such levels as to accommodate the lowest values of the predicted errors. The parameters of the network architecture were set at values which allowed acquisition of the smallest values of the prediction errors by control of the rejection rates (value 0.5) of the chosen hyper-parameters that make the network adjustment optimal. A numerical simulation was carried out with 250 iterations after the first 20% of the burn-in steps were rejected.

There are two criteria used to set the final parameters of the neural network. The first is the root mean square error (RMSE). The other is the correlation (R) between the predicted indices and the observed ones. The correlation coefficient shows the strength of the linear relation between two variables. The coefficient of determination, R^2, shows how well future outcomes are likely to be predicted by the model.

Figures 9, 10, 11, 12, 13 and 14 show the results of examinations carried out using of ANN FMB.

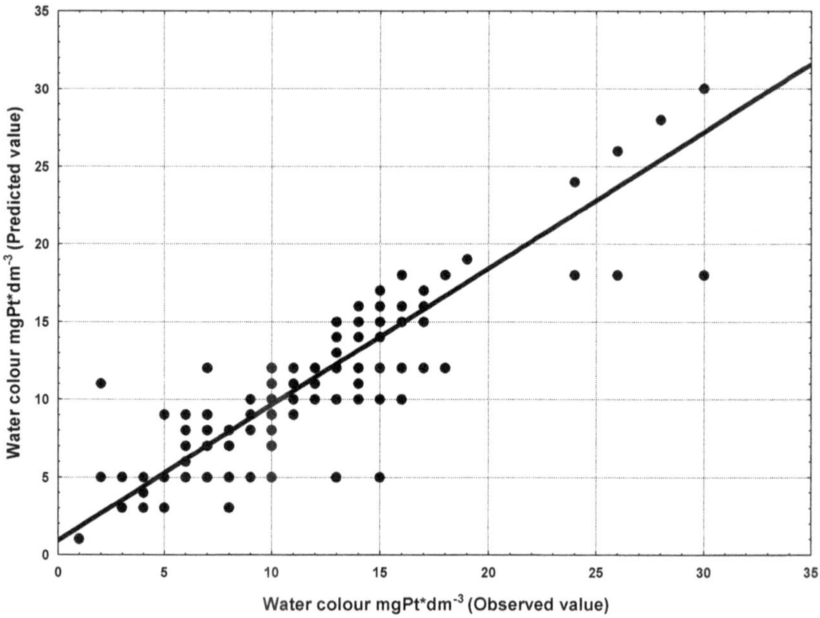

Fig. 9. Correlation between forecasted and observed values for water colour

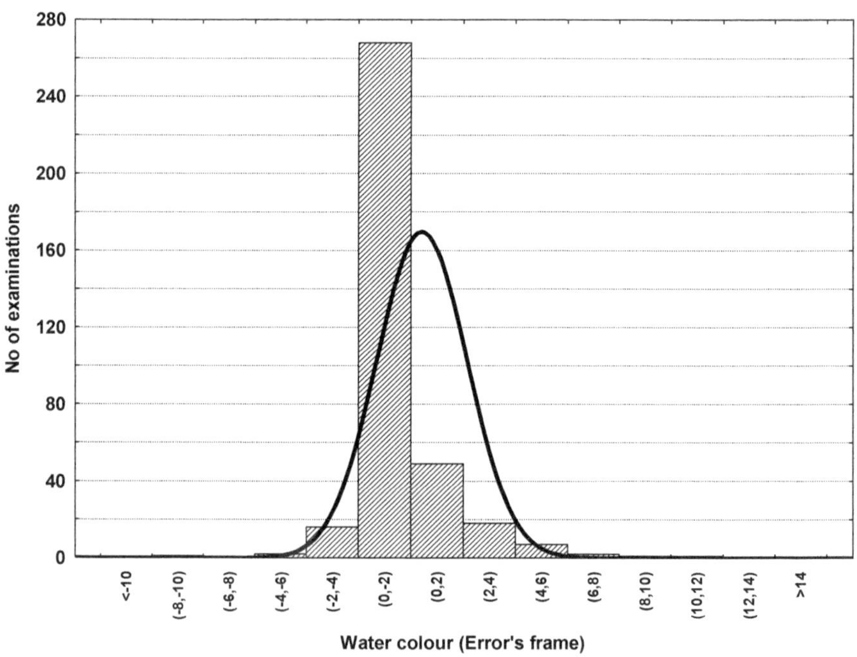

Fig. 10. Histogram of errors for forecasted water colour

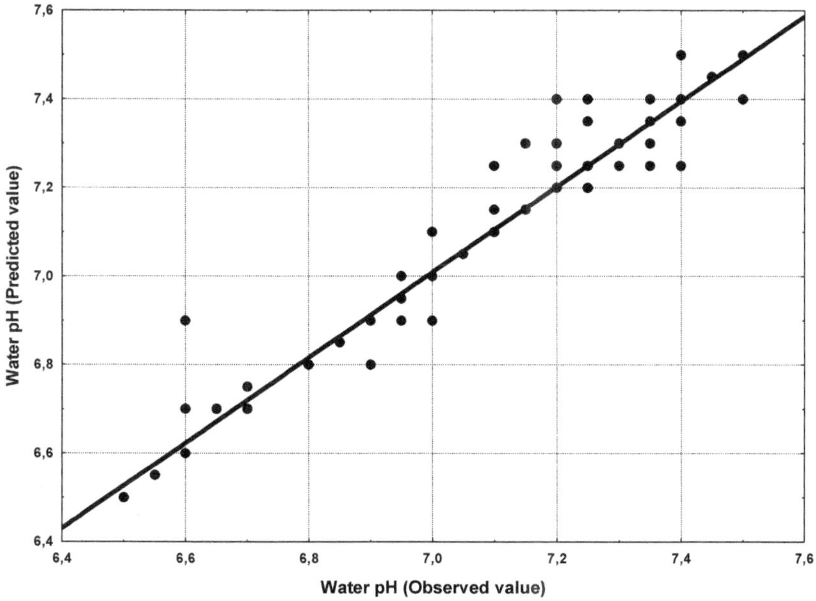

Fig. 11. Correlation between forecasted and observed values of water pH

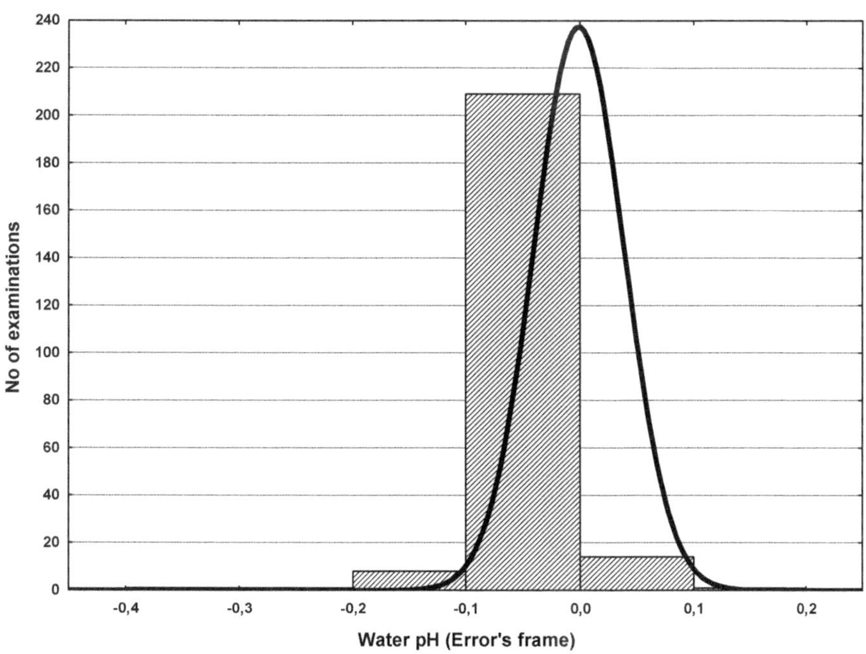

Fig. 12. Histogram of errors for forecasted water pH

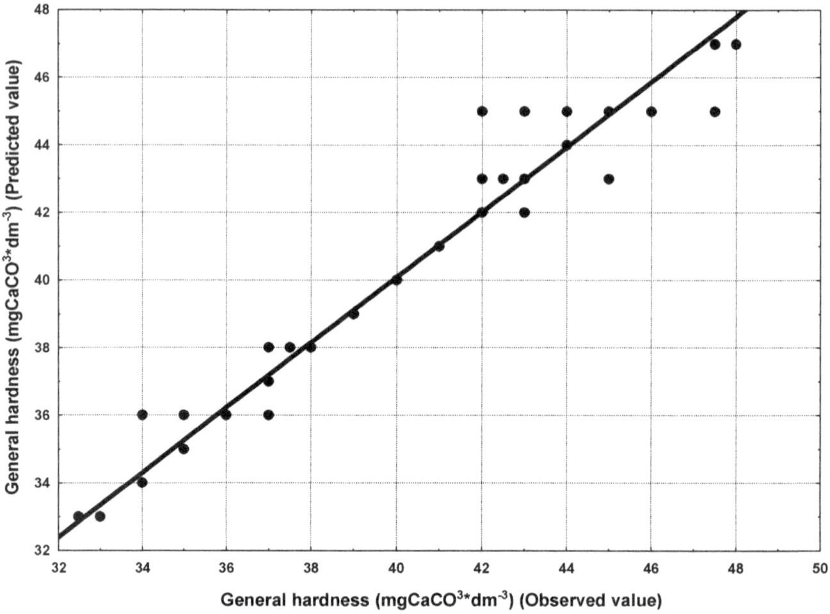

Fig. 13. Correlation between forecasted and observed values of water general hardness

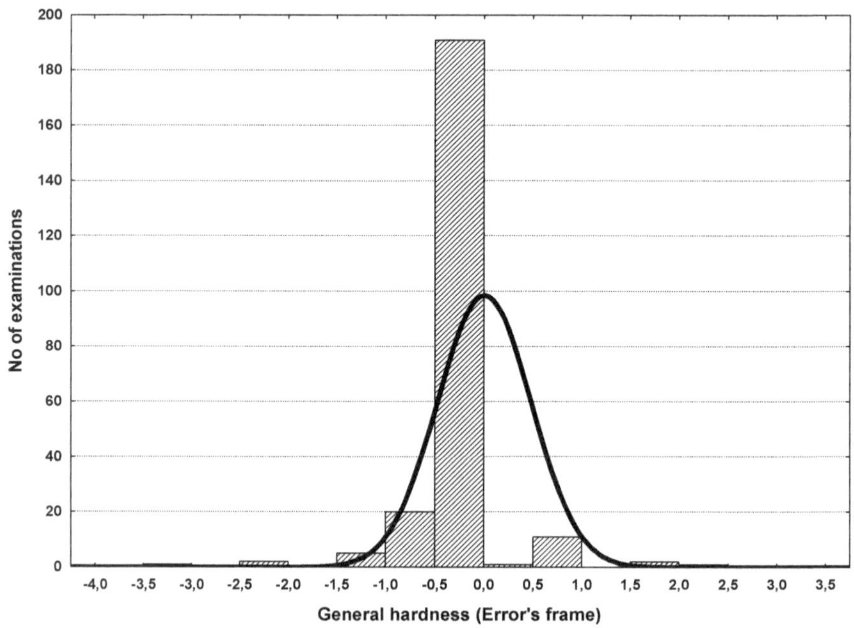

Fig. 14. Histogram of errors for forecasted water general hardness

The calculated RMSE values were 1.69 mg Pt·dm^{-3} for water colour, 0.83 mg SiO_3·dm^{-3} for water turbidity, 0.062 for pH and 1.38 mg $CaCO_3$·dm^{-3} for the general hardness of the water. The correlation coefficients between the assessed and observed indices were 0.9318 for water colour, 0.9448 for water turbidity, 0.9475 for water pH and 0.9177 for the general hardness of the water. Indicators of the chosen water quality parameters were assessed based on the chosen ANN model. The results obtained were grouped into three subsets based on the water temperature. The results are shown in Table 3.

Set	Conditions	Mark	Colour	Turbidity	pH	Hardness	Alkalinity
			mg Pt·dm^{-3}	mg SiO_2·dm^{-3}	pH	mg $CaCO_3$·dm^{-3}	mval·dm^{-3}
Subset I	Temp 17-22°C, V_Z 6.4-9.2 mln m^3 rainfall P_{max} = 54 mm·d^{-1}	S_{max} observed	30	12	7.2	48	0.75
		S_{max} forecast	30	12	7.2	45	0.7
		S_{min} observed	10	3	6.9	35	0.5
		S_{min} forecast	5	3	6.9	35	0.5
Subset II	Temp. 9-16,9°C, V_Z 6.5-8.9 mln m^3, rain fall P_{max} = 16.6 mm·d^{-1}	S_{max} observed	14	12	7.5	44	0.65
		S_{max} forecast	14	12	7.4	44	0.65
		S_{min} observed	3	3	6.5	33	0.55
		S_{min} forecast	3	3	6.6	33	0.55
Subset III	Temp. 3-8,9°C, V_Z 7.8-8.4 mln m^3, rainfall P_{max} = 25.7 mm·d^{-1}	S_{max} observed	16	12	7.5	4.8	0.95
		S_{max} forecast	15	12	7.4	48	0.9
		S_{min} observed	1	1	7	38	0.5
		S_{min} forecast	1	1	7	39	0.5
Total set	Temp. 3-22°C, aver temp. 11.5°C V_Z 6.6-9.2 mln m^3, V_{averr} 8.2 mln m^3, rainfall P_{max} = 54 mm·d^{-1}	S_{max} observed	30	12	7.5	48	0.95
		S_{max} forecast	30	12	7.4	48	0.9
		S_{min} observed	1	1	6.6	33	0.5
		S_{min} forecast	1	1	6.6	33	0.5
		S_{aver} observed	10.2	5.2	7.09	41	0.64
		S_{aver} forecast	9.9	5.1	7.1	41.5	0.64

Table 3. Comparison of results forecasted by an ANN model and the indicators of water quality observed at the retention reservoir

5. Conclusions

The effectiveness of an ANN FBM in establishing the critical indicators of water contamination at the reservoir was relatively good, providing high quality predictions with each of the models. Analysis of the predicted values of the quality indicator variables of the reservoir water and the observed ones was undertaken using Microsoft Access 2007.

The observed and forecasted values of the analysed indicators of water quality were aggregated into a data base for each ANN FBM model. Using the 'filter' function allowed the predicted values of the relevant indicator of water quality to be found once the observed variables were defined.

A suggested methodology for designing an ANN model resulted in an optimal configuration of the particular ANN type in order to establish the predicted values of the quality indicators of the water. The chosen ANN model could replace the current algorithms used in the design of modern systems which are responsible for water management at retention reservoirs. It could also be used to control and direct the processes controlling the abstraction and treatment of water used for consumption and industrial and agricultural purposes.

Designing a model to forecast indicators of water quality requires extensive examination of each water management establishment. The results of the analysis should determine the critical indicators of water quality. These would determine the character of the water being analysed and the characteristics required given the current and planned uses of the water.

A review and analysis of the literature has led to the following conclusions:

- Management of surface water treatment requires constant optimisation given the large number of indicators which influence water quality. It is essential to establish a model which predicts the critical values of the water quality indicators which, in turn, determine the water treatment technology system to apply.
- Having a large set of critical water quality indicators means that it is not possible to operate with just one technology system of water treatment processes. For the retention reservoir considered in this paper, the critical indicators were water temperature, water colour, pH and the general hardness of the water.
- The suggested methodology for creating and verifying ANN models permits the development of an optimised configuration for a particular ANN type to best forecast the quality indicators of the water held in the reservoir.

6. References

Bardossy, G.; Halasz, G. and Winter, J. (2009). Prognosis of urban water consumption using hybrid fuzzy algorithms. *Journal of Water Supply: Research and Technology-AQUA* 58(3), pp. 203–211.

Brion, G.M. and Lingireddy, S. (2003). Artificial neural network modelling: a summary of successful applications relative to microbial water quality. *Water Science & Technology* 47(3), pp. 235–240.

Camarinha-Matos, L.M. and Martinelli, F.J. (1998). Application of machine learning in water distribution networks. *Intelligent Data Analysis* 2, pp. 311-332.

Cougnaud, A.; Faur-Brasquet, C. and Le Cloirec, P. (2005). Neural networks modelling of pesticides removal by activated carbon for water treatment. *Water Supply* 4(5-6), pp. 9–19.

Dawidowicz, J. (2005). Methods of assessing the DN of the water pipes with use of artificial neural networks. In: Proceedings of the Waste and Water Economy Issues at Industrial and Agricultural Regions Conference. Białowieża, 5-7 June 2005. Environment Engineering Committee Monographs of the Committee for Environmental Engineering PAN. 30, pp. 345-360.

Deveughèle, S. and Do-Quang, Z. (2005). Neural networks: an efficient approach to predict on-line the optimal coagulant dose. *Water Supply* 45-6, pp. 87–94.

Gavin J., Graeme C., Dandy, Holger R. (2003). Data transformation for neural network models in water resources applications. *Journal of Hydroinformatics* 5, pp. 245-258.

Institute of Environmental Protection Wrocław branch [IMGW]. (1986). Examination within quality and technology and forecasting of the water quality with regard to planned Sosnowka reservoir to be located at Czerwnoka Potok rivers for Jelenia Gora district water supply purposes.

Koo, J.; Inakazu, T.; Koizumi, A.; Arai, Y.; Kim, K. and Ahn, J. (2008). Application of Artificial Neural Network for Reducing of Chlorine Residual Concentration in Water Distribution Network. *Water Practice & Technology* 3(2), 2008.032.

Licznar, P. and Lomotowski, J. (2004). Forecasting daily water demands with use of artificial neural networks. In: Proceedings of the Water Supply and Water Quality Conference, 6-8 September 2004, Poznań, pp. 175-183.

Neal, R. (2000). Flexible Bayesian Models on Neural Networks, Gaussian Processes, and Mixtures v 2000-08-13 University of Toronto, Toronto.

Pawełek, J. and Bergel, T. (2008). Characteristics of higher muddiness of water in small mountain rivers. *Gas, Water and Sanitary Technology* 9, pp. 5-29.

Rak, A. (2008). Changes in the quality of the water retained at a mountains reservoir for consumption purposes. *Gospodarka Wodna* 3 (711), pp. 115-121.

Siwoń, Z.; Łomotowski, J.; Cieżak, W.; Licznar, P. and Cieżak, J. (2008). Analysis and prognosis of the water demand in water systems. PAN Committee for Inland and Water Engineering. Instytut Podstawowych Problemów Techniki, No 61, pp. 32-38.

Sroczan, E.M. and Urbaniak, A. (2004). Use of artificial intelligence methods in monitoring, steering and operating within the water supply and water protection systems. In: Proceedings of the Water Supply and Water Quality Conference, 6-8 September, 2004, Poznań, 695-704.

Strugholtz, S.; Panglisch, S.; Gimbel, R. and Gebhardt, J. (2009). Neural networks and genetic algorithms in membrane technology modelling. *Journal of Water Supply: Research and Technology – AQUA* 57(1), pp. 23–34.

Zhou, S.L.; Macmahon, T.A.; Walton, A. and Lewis, J. (2002). Forecasting operational demand for an urban water supply zone. *Journal of Hydrology* 259, pp. 189-202.

Zhou, S.L.; Macmahont, T.A.; Walton, A. and Lewis, J. (2000). Forecasting daily urban water demand: a case study of Melbourne. *Journal of Hydrology* 235, pp. 153-164.

Zimoch, I. and Kłos, M. (2003). Use of IT to forecast eutrophication of surface water, "Dobczyce" reservoir as an example. *Environment Protection* 25(3), 73-76.

Permissions

The contributors of this book come from diverse backgrounds, making this book a truly international effort. This book will bring forth new frontiers with its revolutionizing research information and detailed analysis of the nascent developments around the world.

We would like to thank Constantin Volosencu, for lending his expertise to make the book truly unique. He has played a crucial role in the development of this book. Without his invaluable contribution this book wouldn't have been possible. He has made vital efforts to compile up to date information on the varied aspects of this subject to make this book a valuable addition to the collection of many professionals and students.

This book was conceptualized with the vision of imparting up-to-date information and advanced data in this field. To ensure the same, a matchless editorial board was set up. Every individual on the board went through rigorous rounds of assessment to prove their worth. After which they invested a large part of their time researching and compiling the most relevant data for our readers. Conferences and sessions were held from time to time between the editorial board and the contributing authors to present the data in the most comprehensible form. The editorial team has worked tirelessly to provide valuable and valid information to help people across the globe.

Every chapter published in this book has been scrutinized by our experts. Their significance has been extensively debated. The topics covered herein carry significant findings which will fuel the growth of the discipline. They may even be implemented as practical applications or may be referred to as a beginning point for another development. Chapters in this book were first published by InTech; hereby published with permission under the Creative Commons Attribution License or equivalent.

The editorial board has been involved in producing this book since its inception. They have spent rigorous hours researching and exploring the diverse topics which have resulted in the successful publishing of this book. They have passed on their knowledge of decades through this book. To expedite this challenging task, the publisher supported the team at every step. A small team of assistant editors was also appointed to further simplify the editing procedure and attain best results for the readers.

Our editorial team has been hand-picked from every corner of the world. Their multi-ethnicity adds dynamic inputs to the discussions which result in innovative outcomes. These outcomes are then further discussed with the researchers and contributors who give their valuable feedback and opinion regarding the same. The feedback is then collaborated with the researches and they are edited in a comprehensive manner to aid the understanding of the subject.

Apart from the editorial board, the designing team has also invested a significant amount of their time in understanding the subject and creating the most relevant covers. They scrutinized every image to scout for the most suitable representation of the subject and create an appropriate cover for the book.

The publishing team has been involved in this book since its early stages. They were actively engaged in every process, be it collecting the data, connecting with the contributors or procuring relevant information. The team has been an ardent support to the editorial, designing and production team. Their endless efforts to recruit the best for this project, has resulted in the accomplishment of this book. They are a veteran in the field of academics and their pool of knowledge is as vast as their experience in printing. Their expertise and guidance has proved useful at every step. Their uncompromising quality standards have made this book an exceptional effort. Their encouragement from time to time has been an inspiration for everyone.

The publisher and the editorial board hope that this book will prove to be a valuable piece of knowledge for researchers, students, practitioners and scholars across the globe.

List of Contributors

Branko L. Dokic
University of Banja Luka, Faculty of Electrical Engineering, Bosnia and Herzegovina

Richard Zemann, Philipp Walter Reiss, Paul Schörghofer and Friedrich Bleicher
Vienna University of Technology, Institute for Production Engineering and Laser Technology, Austria

Ivana Gudlin Schwarz and Stana Kovačević
University of Zagreb, Faculty of Textile Technology, Department of Textile Design and Management, Croatia

Constantin Volosencu
"Politehnica" University of Timisoara, Romania

Catarina B. Santiago, Lobinho Gomes and Armando Sousa
Faculty of Engineering, University of Porto and INESC-Porto, Porto, Portugal

Luis Paulo Reis
School of Engineering, University of Minho and LIACC, University of Porto, Porto, Portugal

Maria Luisa Estriga
Faculty of Sports, University of Porto and CIFI2D, Porto, Portugal

Bojan Grˇcar, Peter Cafuta and Gorazd Štumberger
University of Maribor, Faculty of Electrical Engineering and Computer Science, Slovenia

José F. Díez-Higuera, Francisco J. Díaz-Pernas, Miriam Antón-Rodríguez, David González-Ortega and Mario Martínez-Zarzuela
Department of Signal Theory, Communications and Telematics Engineering, Telecommunications Engineering School, Valladolid University, Valladolid, Spain

Radek Matušů and Roman Prokop
Department of Automation and Control Engineering, Faculty of Applied Informatics, Tomas Bata University in Zlín, Czech Republic

Karel Perutka
Tomas Bata University in Zlin, Faculty of Applied Informatics, Czech Republic

Grzegorz Pastuszak
Warsaw University of Technology, Poland

Anastasios Fragopoulos, John Gialelis and Dimitrios Serpanos
Industrial Systems Institute (I.S.I.), Greece

Nora Ali, Hany ElSayed and Magdy El-Soudani
Cairo University, Egypt

Hassanein Amer
American University in Cairo, Egypt

Ramez Daoud
KAMA Trading, Egypt

Domen Verber
University of Maribor, Slovenia

Dragan Perakovic, Vladimir Remenar and Ivan Jovovic
University of Zagreb, Faculty of Transport and Traffic Sciences, Croatia

Adam Rak
Opole University of Technology, Opole, Poland